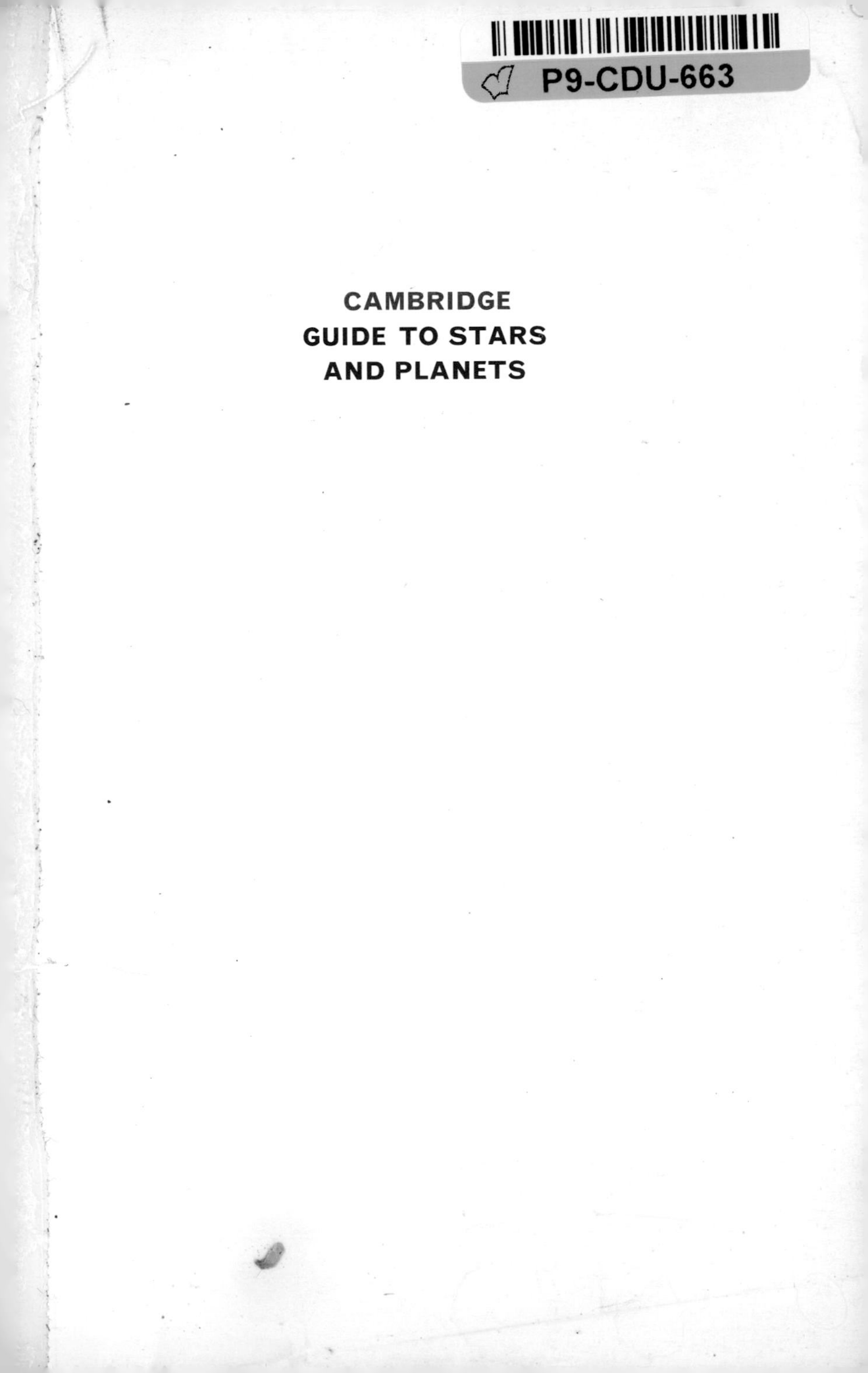

CAMBRIDGE GUIDE TO STARS AND PLANETS

CAMBRIDGE
GUIDE TO STARS AND PLANETS

PATRICK MOORE AND WIL TIRION

Title-page illustration
The Eagle Nebula, M16, in the constellation Serpens.

PUBLISHED BY THE PRESS SYNDICATE OF THE UNIVERSITY OF CAMBRIDGE
The Pitt Building, Trumpington Street, Cambridge CB2 1RP, United Kingdom

CAMBRIDGE UNIVERSITY PRESS
The Edinburgh Building, Cambridge CB2 2RU, United Kingdom
40 West 20th Street, New York, NY 10011–4211, USA
10 Stamford Road, Oakleigh, Melbourne 3166, Australia

First published in Great Britain in 1980 by Mitchell Beazley under the title
Patrick Moore's Pocket Guide to the Stars and Planets

This revised and expanded edition first published in 1993 by George Philip Limited;
Second edition published 1997

First published by Cambridge University Press 1997

Reprinted 1997, 1998 (twice), 1999

Printed in Hong Kong

This edition only for sale in the United States of America and Canada

Library of Congress Cataloguing in Publication data available

ISBN 0 521 58582 1 paperback

CONTENTS

PREFACE

This pocket guide is a working manual for serious amateur observers all over the world. It is divided into 10 sections: the opening one, Introduction to Astronomy, covers astronomical terminology and observational techniques. Eight sections on the planets and the stars provide a detailed guide to the heavens. They incorporate not only the latest information from the space probes, in color photographs and data, but also all the background essential to understanding the stars and planets. They include many tables giving, for the near future, important phenomena of the observable planets, the dates of lunar and solar eclipses, and the times of return of comets and meteor showers.

The Constellations section opens with whole-sky and seasonal star maps, which enable a particular constellation to be located. All 88 constellations are then mapped in detail, with information about their leading stars and interesting doubles, variables and non-stellar objects. The descriptive comments accompanying the maps and charts are based on perfect viewing conditions.

Tables at the end of the book outline the history of astronomy and space research, comprehensively list the constellation names and their meanings, and provide a glossary. The book concludes with an extensive index.

Throughout this book, distances are given in kilometers. To convert kilometers to miles, simply *divide* by 1.61. To convert miles to kilometers, *multiply* by the same figure. Focal lengths of binoculars are generally given in millimeters. Apertures of telescopes are quoted in centimeters, though dealers often give them in inches; to convert inches to centimeters, *multiply* by 2.54; to make the reverse conversion, *divide* by that figure. There are, of course, 10 millimeters to a centimeter.

I

INTRODUCTION TO ASTRONOMY

Understanding the skies

Astronomy is the study of the sky and everything we see there. It is the oldest science in the world, and fascinates all kinds of people – young and old, amateurs and professionals.

The Earth is a planet, 12,756 kilometers in diameter, which moves round the Sun once every 365.86 days at a mean distance of 150 million kilometers. It is a member of the Sun's family or Solar System and there are eight other planets in the system, some of which have satellites orbiting them. The Earth has one satellite, the familiar Moon. The stars themselves are suns, and these are so remote that even their light, moving at 300,000 kilometers per second, takes years to reach the Earth.

Because the stars are so far away, their individual or "proper" motions are very slight, and the patterns or constellations now known to astronomers are the same as those that must have been seen by prehistoric man. It is convenient to picture the Earth as being surrounded by a sphere, the celestial sphere, which has a center the same as the center of the Earth. The various objects in the sky may then be positioned on this imaginary sphere.

In the sky the equivalent of latitude on Earth is known as declination: that is, the angular distance north or south of the celestial equator. The celestial equivalent of longitude is right ascension, and

Though the Earth seems immense to us, it is a mere speck in the Solar System and the wider Universe beyond. But it is a unique planet: nowhere else we know of possesses abundant quantities of water and an oxygen-rich atmosphere.

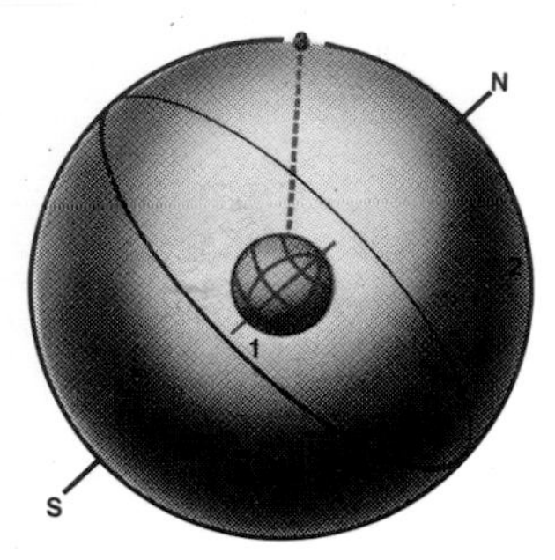

The celestial poles (N and S) are the points on the celestial sphere indicated by the direction of the Earth's axis; the north celestial pole is marked to within 1° by the bright star Polaris; there is no equivalent star to mark the south celestial pole. The celestial equator (1) is the projection of the Earth's equator on to the celestial sphere. The meridian (2) is the circle on the celestial sphere that passes through the poles and the zenith, or overhead point (3).

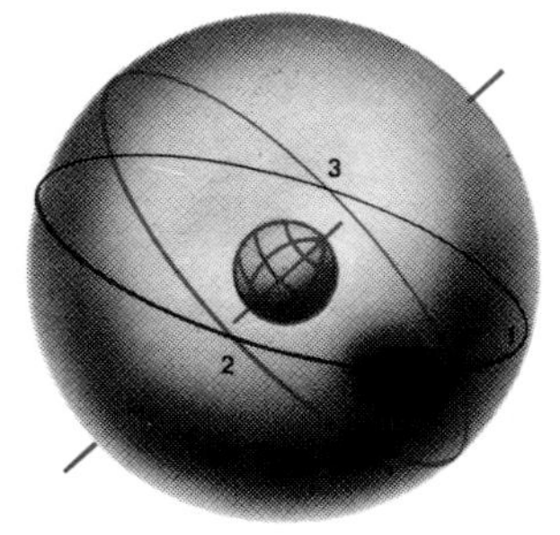

The ecliptic is the apparent yearly path (1) of the Sun among the stars. Its inclination to the celestial equator is the same as that of the Earth's axis to the orbital plane: thus the Sun stays six months in the Northern Hemisphere and six months in the Southern. The points where the Sun crosses the equator are the equinoxes: the vernal equinox (about 21 March) is also called the First Point of Aries (2); the autumnal equinox (about 22 September) is called the First Point of Libra (3).

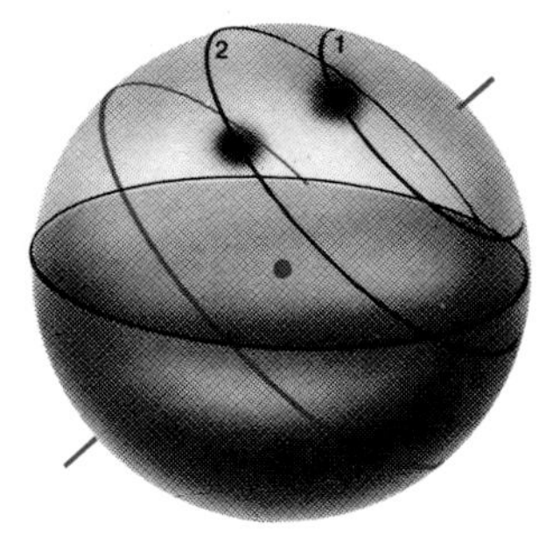

The altitude of the celestial pole above the horizon is equal to the observer's latitude. From the Earth's pole the celestial pole will be at the zenith – altitude 90°. For an observer at latitude 40°N the north celestial pole will be at altitude 40°. Stars close to the pole (1), such as those in Ursa Major seen from central Europe and the northern United States, are circumpolar and never set. Stars further from the pole (2) go below the horizon.

this is the angular distance of a body from the First Point of Aries (see diagram above), measured eastward; it is usually measured in units of time rather than in degrees. As the Earth rotates, a star will seem to rise, reach its highest point or "culmination" and then set. Any other fixed point on the celestial sphere also culminates once a day. The right ascension of an object is defined as the time interval between the culmination of the First Point of Aries and the culmination of the object concerned.

Declination for celestial bodies corresponds to latitude on the Earth. It is defined as the angular distance, as viewed by an observer on the Earth, north or south of the celestial equator (1). Thus the north celestial pole has declination 90° N; the declination of the celestial equator is zero. The declinations of all celestial bodies take values between 0° and 90°, described as positive for an object north of the celestial equator and negative for an object south of it.

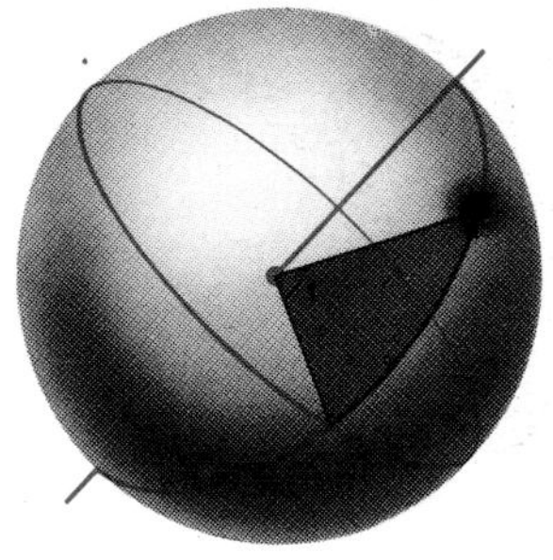

The prime meridian (1) is the great circle running north–south on the celestial sphere through the First Point of Aries. The angle between the prime meridian and a celestial body is known as right ascension. The right ascension of a star (2) is the angle (3) between the First Point of Aries (4) and the point on the celestial equator vertically below the star (5). Declination and right ascension jointly define the position of any point on the celestial sphere.

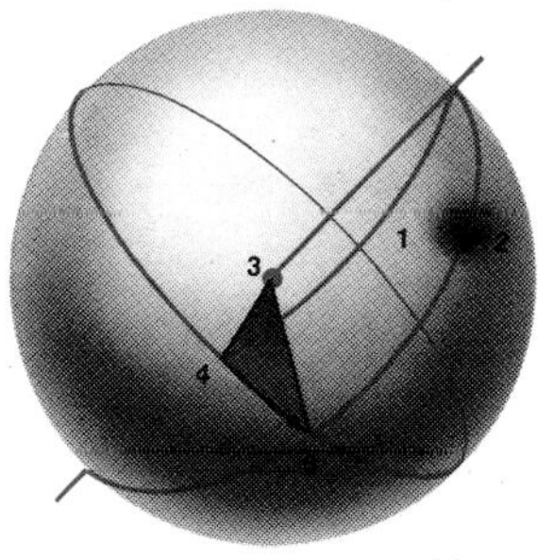

Right ascension is measured in time units, where 24 hours equals 360°. When the First Point of Aries (1) culminates the time is 00h 00m. The star Sirius (2), which is then below the horizon, culminates 6h 44m later, having RA 6h 44m. This apparent 360° rotation of the celestial sphere is called the sidereal day. It is, however, about 4 minutes shorter than the solar day because the Sun, which completes the 360° in 365 days, moves eastward by about 4 minutes a day.

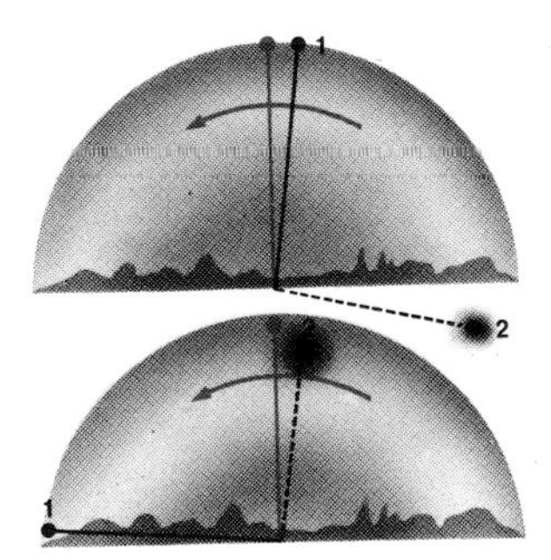

Apart from the effects of "precession," a minor shift in position of the celestial poles and equator due to a slight "wobbling" of the Earth's axis, the right ascensions and declinations of the stars do not change appreciably. Thus a star's celestial coordinates can be sufficiently well established using the celestial equator as the reference plane.

Over a sufficiently long period of time, however, the effects of precession become appreciable. At the moment astronomical catalogs give stellar positions for "epoch 2000," but older catalogs give them for "epoch 1950" – so when really accurate setting of a telescope is needed, make sure that you have an up-to-date catalog.

Astronomers are used to vast distances and immense spans of time. The Moon, at a mean distance of 384,400 kilometers from the Earth, is its closest neighbor; even the Sun (150 million kilometers away) is relatively very near the Earth. The average Sun–Earth distance is called the astronomical unit (a.u.) and is used in describing distances on the scale of the Solar System; but in measuring the distances of the stars, even this unit becomes hopelessly inconvenient.

Instead, astronomers use the light-year, which is the distance traveled by light in a year. Light moves at 300,000 kilometers a second, so one light-year is equal to 9.46 million million kilometers. Even the nearest star beyond the Sun is over four light-years away; this is why the "proper" motions of the stars – their apparent motions across the celestial sphere – are so slight. The star with the greatest known proper motion (Barnard's Star, a faint red dwarf) takes 180 years to crawl a distance equal to the apparent diameter of the full Moon.

Another unit used by astronomers is the parsec. One parsec is equal to 3.26 light-years. It is defined in such a way that a star at a distance of one parsec would appear to move through one second of arc during the year, as a result of the Earth's motion around the Sun (see below).

A "constellation" has no real meaning, as the stars are not at equal distances from the Earth; it is simply a pattern of stars that happen to lie in much the same direction, as seen from Earth. For example, the stars of the Big Dipper are part of the constellation of Ursa Major. Though they all look equally far away, their distances actually differ enormously, ranging from 59 light-years (Mizar) to 108 light-years for Alkaid, so Alkaid is much further from Mizar than might be thought. From a different vantage point in the universe, the constellation patterns would look quite different.

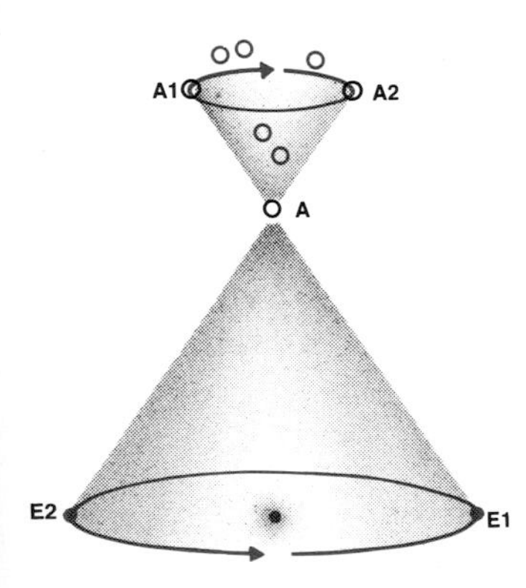

The distances of the nearer stars (left) may be measured by the method known as trigonometrical parallax. A relatively close star (A) is observed from opposite sides of the Earth's orbit – in other words, at intervals of six months. From the Earth's first position (E1), the star will be seen in position A1 against the background of more remote stars. From the second position (E2), the star will appear at A2. Knowing the diameter of the Earth's orbit (E1–E2), and knowing the apparent angular shift, the observer can calculate the star's distance from the Earth.

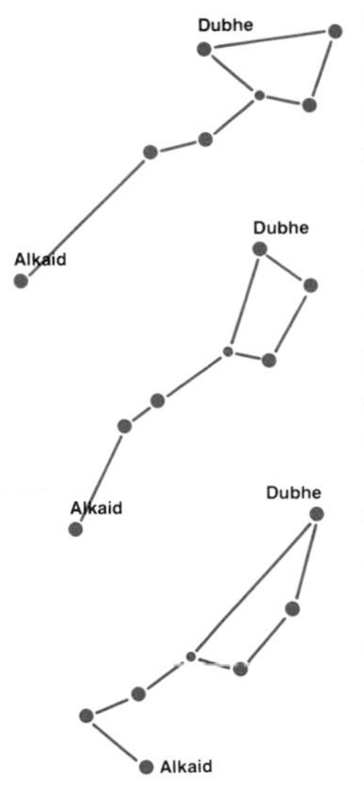

The motions of the stars (left) over a period of millennia alter the constellation patterns as seen from the Earth. Of the seven stars that today form the Big Dipper (center), two stars (Alkaid and Dubhe) are more remote than the rest and are moving in different directions. Thousands of years ago this group looked very different (top). In the future it will look different again (bottom).

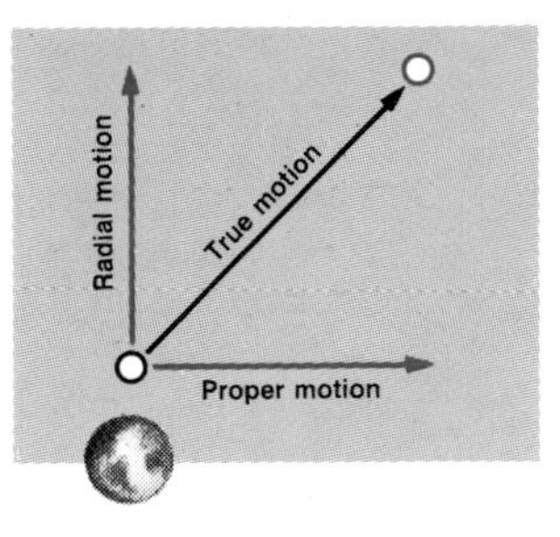

The motion of a star (above) is a combination of its proper ("sideways") motion and its radial motion (toward or away from us).

The stars are suns, but there is a tremendous range in both size and luminosity. The apparent magnitude of a star is a measure of its brightness as seen from Earth. The scale works like a golfer's handicap, with the more brilliant performers having the lower values. Thus bright stars are of magnitude 1 or even less (the four brightest stars in the sky have negative magnitudes), while stars of magnitude 6 represent the limit of naked-eye visibility under good conditions, and large telescopes can reach well below magnitude +20.

Everything, however, depends upon the star's distance: for example, Vega (magnitude 0.04) looks much brighter than Deneb (magnitude 1.3), yet Vega is a mere 55 times as luminous as the Sun, while Deneb equals perhaps 60,000 Suns. Deneb is more remote, and its absolute magnitude is far brighter. (Absolute magnitude is defined as the apparent magnitude that a star would have at a distance of 10 parsecs.)

Telescopes and observing

There are many worthwhile astronomical observations that can be made with the naked eye, but obviously the real enthusiast will soon want to graduate to using an optical aid. For the astronomer, a telescope fulfils two main functions. First, by collecting more light than the unaided eye, it makes it possible to detect fainter objects; and, secondly, it reveals fine details that cannot be seen with the unaided eye. This latter function is known as a telescope's resolving power or resolution.

There are two main types of telescope: first, the refractor, which collects its light by means of a special lens known as an object glass or objective. The lens bends or "refracts" the rays of light and brings them to a focus. An image of the object is formed, which can then be enlarged by an eyepiece – this latter being a magnifying glass of special design. Initially a telescope produces an inverted image, so for terrestrial use an extra lens system or prism is used to correct the picture. (This feature is also present on all binoculars.) However, each time light passes through a lens it is slightly weakened. In an astronomical telescope, therefore, the correcting lenses are omitted and images are seen inverted. (This is why the Moon maps, on pages 26 to 43, are arranged with south at the top to suit northern-hemisphere observers using telescopes.)

The second kind of telescope is the reflector. In the Newtonian pattern (so called after its inventor, Sir Isaac Newton), the rays of light

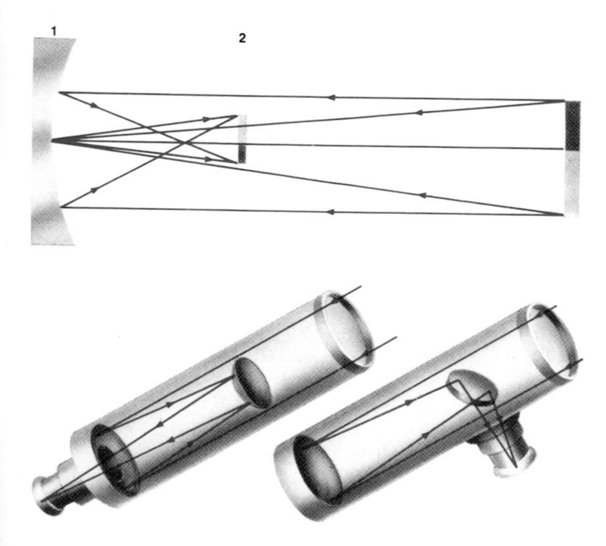

In a reflector the light is collected by a concave primary mirror (1) and brought to a focus, forming an image (2). The size of the image depends on its distance from the mirror.

In a Cassegrain reflector (far left) light from the primary mirror is reflected by a convex secondary back through a hole in the primary. In the Newtonian reflector (left) there is a flat secondary, set at 45°.

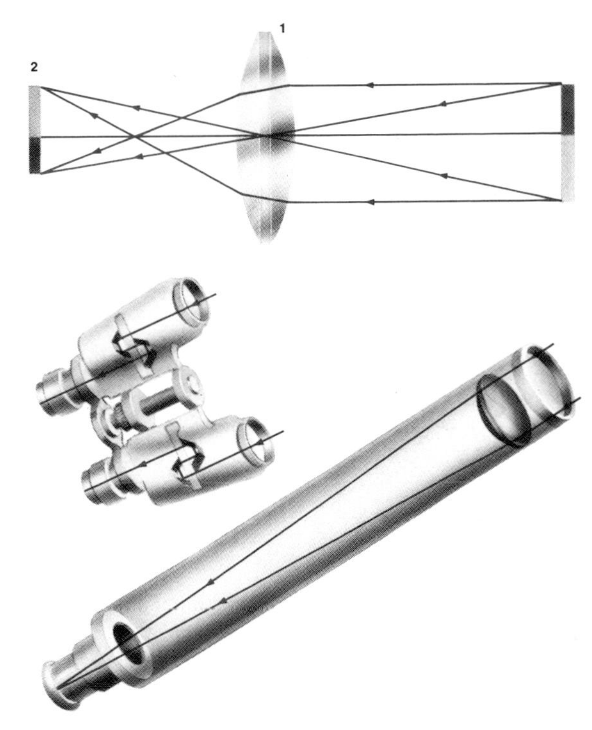

In a refractor (top) the light is collected by a lens (1) and brought to a focus, forming an image (2). An eyepiece is placed beyond the image to produce a magnified image (bottom). Unless an extra lens or prism is used, as in binoculars, the image is inverted. Aperture for aperture, a refractor is more effective than a reflector; it is also easier to maintain, and will need little attention, whereas the mirrors of a reflector have to be regularly recoated. Against this, a refractor is much more expensive than a reflector of equivalent performance.

strike a curved mirror and are reflected back up the open tube and on to a smaller flat mirror, which directs the light rays to the eyepiece set in the side of the tube. Therefore the observer looks into the tube and not up it. In an alternative system, which is known as the Cassegrain, the main mirror has a hole in its center so that all light rays are reflected back through it to the eyepiece.

It is unwise to spend a great deal of money on a refractor with an aperture less than 7.5 centimeters in diameter or on a reflector with a main mirror smaller than 15 centimeters. A very small telescope, which may be relatively inexpensive, will have a small field of view and will be awkward to handle, as well as being low-powered. Thus it is probably better to invest in a good pair of binoculars, possibly of the 7 × 50 variety (that is, with a magnification of 7, with each lens 50 millimeters in diameter). Binoculars will not be capable of showing, for example, the rings of Saturn, but will certainly be useful.

Refractors are more efficient than reflectors, and are less liable to damage, but they are more expensive and tend to give a certain amount of false color around brilliant objects such as stars.

TELESCOPE MOUNTINGS: Using a good telescope on a poor mounting is rather like trying to use a good record player with a faulty stylus. If the telescope mount is unsteady, the object under study will seem to dance a wild waltz in the heavens, and any useful observation will be impossible. Therefore the mounting must be completely rigid and the telescope must be able to move smoothly and regularly.

The simplest mounting is the altazimuth, with which the telescope can be moved either in *altitude* (vertically) or in *azimuth* (horizontally).

With an equatorial mount, the telescope is set upon an axis that is parallel to the axis of the Earth. This means that when the telescope is moved from east to west, the "up and down" movement looks after itself; and with a clock drive, the telescope will move to follow a target.

Equatorial mountings are of various patterns, each of which has its advantages and drawbacks. If you are buying or making a mounting, the first essential is to ensure that it is really firm. With a larger telescope, on a massive mounting, the installation must be permanent. Generally speaking a Newtonian reflector with an aperture of 20 centimeters (that is, with a mirror 20 centimeters across) is on the limit of portability, even if the tube is a skeleton construction.

It would be wrong to claim that all telescopes powerful enough to show significant detail must be equatorially mounted, but certainly an equatorial mounting is much superior to the simple altazimuth.

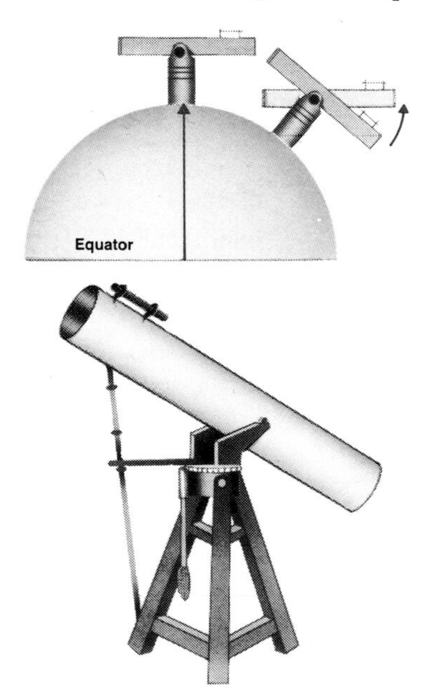

A telescope mounted on an altazimuth head, positioned at the north or south pole, only needs to turn in azimuth to follow a star. But at any other latitude on the globe two movements are required, in altitude and azimuth, to keep the stars in view. An altazimuth mount is usually a tripod, whch allows free movement in any direction. It cannot be guided very easily because of the two separate motions involved. This means that, by comparison with the equatorial mount, the altazimuth is mainly suited to the smallest astronomical telescopes (bottom). It is more commonly used for terrestrial observation, with a refractor.

A telescope mounted on an equatorial head always has its polar axis pointing at the celestial pole, wherever it is positioned on the globe. When the telescope is moved from east to west, it automatically compensates for the changing altitude.

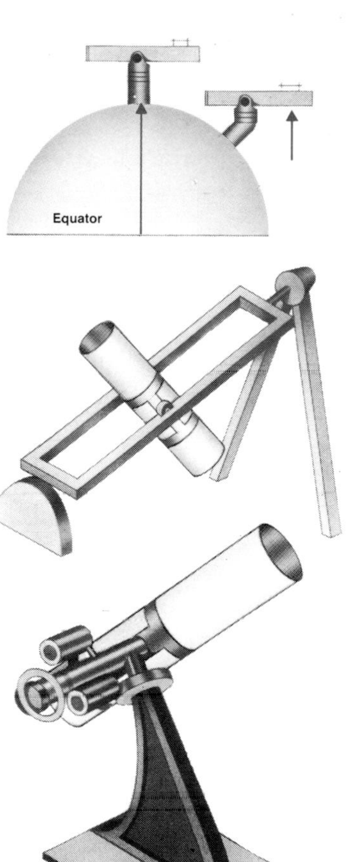

The yoke design has two pillars carrying an axis on which the telescope is mounted. Its main advantage is that it does not need a counterweight. However, the polar region is inaccessible and so, to a certain extent, is the area below it at any time. The mounting is unsuitable for refractors.

The German mount is particularly suited to refractors. The telescope is mounted on one end of the axis with a counterweight at the other. It can be used in all latitudes, it is relatively simple to construct, accessories are easily added and the whole sky can be observed. The main disadvantage is the necessity for a counterbalance, which increases the weight.

The fork mounting is the most suitable for reflectors. There is no cumbersome counterweight, and the design is particularly stable. If the polar axis and fork are made sufficiently rigid, the mounting is probably the best that can be used in conjunction with the Newtonian reflector, one of the more popular amateur telescopes; it is used in many of the world's largest observatories.

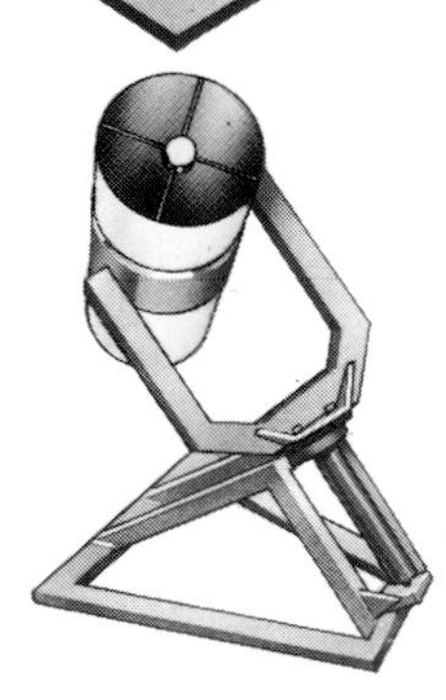

HOME OBSERVATORIES: The owner of a large or moderately sized telescope (say, above 24 centimeters in aperture) will require some sort of observatory. Erecting and dismantling a telescope every time observations are to be made is obviously a tedious business and, sooner or later, something is likely to be dropped. Reflectors, in particular, can only too easily go out of adjustment if they are moved around frequently. Leaving a telescope in the open is even worse. The mirror of a reflector will become tarnished, and will have to be recoated with a thin layer of silver or aluminum (this has to be done periodically in any case). The mechanical parts of the mounting will rust and may cause uneven movement of the telescope. Covering everything with a tarpaulin is at best an unsatisfactory compromise, so that in every respect an observatory is necessary.

There are difficulties, however. More often than not high trees or other obstructions may lie in the direction in which the observer is most interested. The amateur interested only in the Moon and planets will not care too much about the northern sky (assuming, of course, that he lives north of the Earth's equator) and so the part of the sky most needed will be the south, although it is clearly desirable to cover as much of the sky as possible. The variable star enthusiast, however, will ideally require as comprehensive a view of the sky as possible and will have to work out which site will give him the most reasonable scope.

Another nuisance is artificial lighting. Viewing will be hindered if the sky is illuminated by street lamps in the direction of your observations. There is not much that can be done about this apart from erecting screens.

The type of telescope used will affect the design of an observatory. If the telescope is a refractor, the run-off roof arrangement is excellent,

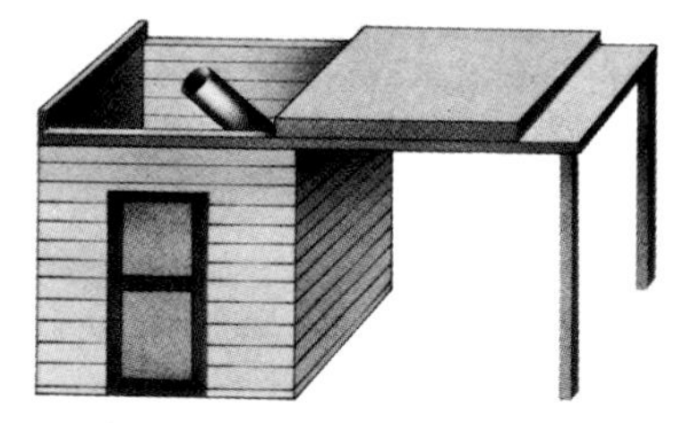

The run-off roof is designed to slide back out of the way, leaving a shelter for the observer. The design is best suited for observers of the Sun: with a reflector, the view of the sky is restricted.

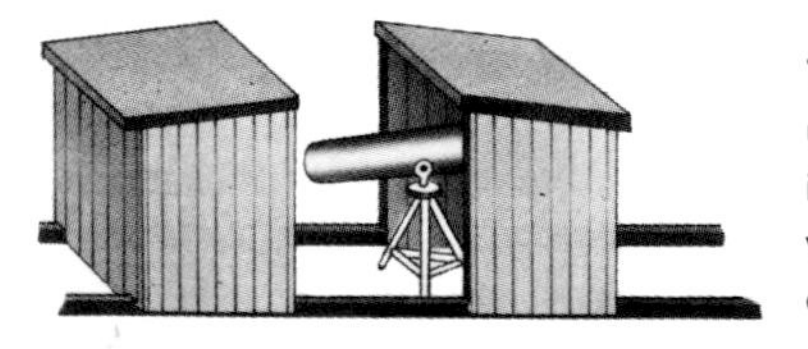

The run-off shed is easy to make and use; however, the viewer has to work in the open and endure both cold and wind as well as scattered light. The best design consists of two parts.

The dome-shaped observatory is difficult to make. The upper part, the hemispherical dome itself, moves round so that the slit can be directed to any point in the sky. Part of the problem of construction is making the rail, on which the dome rotates, absolutely circular.

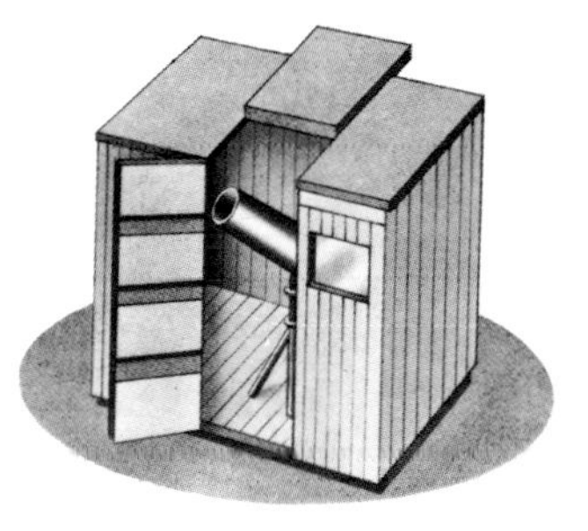

In the total rotator design (right) the whole observatory moves round a circular rail on wheels. This means that there can be a conveniently placed door, but it poses the minor problem that the weight of the entire observatory makes it rather more difficult to rotate. Also, as with the dome observatory, there is the requirement that the rail must be shaped accurately. However, for a reasonably large telescope the design can be very practical.

but not for a reflector, which is mounted lower down. Observers of the Sun will have to ensure that everything is light-tight, and for this a dome structure is certainly the best method. So far as the materials are concerned the observatory may be of wood, hardboard or plastic with a firm frame. The roof may be rolled back by a pulley arrangement; observers with real practical skill may prefer to motorize it. If well made, the whole construction should be fully weatherproof.

It is not necessary to site an observatory in a high part of the garden. Going up 3 or 4 meters is generally of little advantage, apart from the occasional possibility of avoiding obstructions such as tall trees or adjacent buildings. An observatory perched precariously on a roof will be subject to vibration and will receive the full force of any prevailing wind. Worse still is the problem of rising hot air. If the house is inhabited, it will be warm inside; hot air will swirl up over the roof with disastrous effects on the observing conditions.

MAKING OBSERVATIONS: It is often thought that no useful or even satisfactory observations can be made without the help of a powerful and very expensive telescope. This is not so. Of course, the larger the telescope the greater the range, but things are not nearly so daunting as the beginner sometimes believes. For instance, binoculars will give splendid low-power views of the Moon's surface as well as rich star fields, clusters and nebulae. The phases of Venus are easily discernible, and the four large satellites of Jupiter can also be seen. If the binoculars have a magnification higher than 12, however, the field is so small, and the binoculars themselves so heavy, that it becomes difficult to hand-hold them without the aid of some kind of rest. Second-hand telescopes of good quality are rare, for they may have defects that are not obvious at first glance.

Small refractors with object glasses from 2.5 to 6.5 centimeters across are of limited value. The minimum useful aperture for a refractor is probably 7.5 centimeters; but although refractors will show more detail than binoculars, they are difficult to keep steady. With a good telescope of this size, even if mounted on an altazimuth stand, the Moon's surface is revealed in considerable detail. Stars below the 10th magnitude can be seen so that observations of real value may be made of variable stars. Refractors of aperture above 7.5 centimeters are very costly, and lens-making is beyond the capability of most people.

Most amateurs choose a reflector, and again there is little point in spending money on a very small one – that is, one with less than a 15-centimeter aperture. Many amateurs make valuable contributions using a 15-centimeter reflector, particularly if the telescope can be equatorially mounted and clock-driven. For example, the features on the surface of Jupiter are striking and, because of the planet's rapid rotation, they may be followed and timed as they drift across the disk. Saturn's rings are glorious, and when the system is tilted at a suitable angle the ring separation is prominent. Mars is more of a problem, and to see it well a relatively high magnification must be used – up to 250 or 300 on a 15-centimeter reflector, for example.

It is important to remember that each time an image is enlarged, it becomes fainter and definition suffers. It is far better to study a smaller, sharper picture than a larger diffuse one. Eyepieces are interchangeable from one telescope to another. The ideal is to have a whole series, ranging from the low-power wide-field type (for star clusters and similar views) up to high powers with small fields (for detailed observations of the Moon and planets). As soon as the image starts to become blurred, a change should be made to another eyepiece with a lower magnification.

Especially when observing faint objects, "dark-adaptation" is very important, as the eye becomes much more sensitive to very faint

The difference in resolution between three different telescopes can be seen in these drawings of Jupiter made by the author. The view with a 7.5 cm refractor (left) was made with a magnification of 120. It shows the main belts as well as the Great Red Spot. The view with a 15 cm reflector (× 230) reveals finer details (center). With a 38 cm reflector (right) the details of the belt system are clearly seen.

impressions the longer it remains in the dark. Once exposed to bright light the adaptation will be lost until a further period has been spent in darkness. Very faint objects can often best be detected by directing the eye slightly to one side, while the observer's attention is concentrated on the spot where the object is believed to be.

Another problem concerns "extinction." When a star (or any other object) is low over the horizon, its light is received after passing through a relatively thick layer of the Earth's atmosphere. Since the atmosphere is dirty and turbulent, definition falls away and a star will twinkle strongly rather than remain constant. A planet, which appears as a tiny disk, twinkles less. The effect of twinkling is thus due entirely to the atmosphere and has nothing to do with the stars themselves. If conditions are poor, it may well happen that a smaller telescope will give better results than a larger one.

Larger telescopes in amateur hands are very often of the reflecting type. A 30-centimeter reflector can reach stars below the 14th magnitude, so that hundreds of variables come within range. Lunar craters less than 2 kilometers across can be detected, the disk of Jupiter is well delineated and many of Saturn's satellites and the ring system can be seen.

Any astronomer – amateur or professional – will inevitably tend to specialize, and this will determine the choice of telescope. An observer of the Sun will certainly prefer a refractor to a reflector, while the planetary enthusiast will need as much magnification as possible. On the other hand a profitable, but laborious, field of research is to hunt for new comets and new exploding stars (novae). This means using a relatively low power but a wide field and some specialists use powerful mounted binoculars rather than telescopes.

37
1971
[April 18] 5in OG x 260.

	ω^1	ω^2	S.
27. 2347 est [Apr 17]. Centre of Red Spot	...	009.5	–
28. 2357 est. F. of Red Spot.	...	015.5	–
29. 0001 [Apr 18]. Projection from NEBs.	317.2	...	4
30. 0009. Second part of proj. 29 from NEBs.	322.1	...	
31. 0011. Projection from SEBs.	...	024.0	
32. 0015. Small white spot in NTrZ.	...	026.4	
33. 0020. F. of long projection 29/30 from NEBs.	328.8	...	
34. 0037. C. of long dip in SEBs.	...	039.7	
35. 0047. C. of slanting condensation in SEBs.	...	045.7	
36. 0052. F. of condensation 35 in SEBs.	...	048.7	
37. 0103. Small dip in NEBs. (Doubtful!)	355.0	...	
38. 0105. F. of long dip 34 in SEBs.	...	056.6	
39. 0109. Small white spot in NTrZ.	...	059.6	
40. 0111. Small projection from NEBs.	359.9	...	
41. 0113. F. of small white spot 39 in NTrZ.	...	061.4	5
42. 0147. Small projection from NEBs. (Dubious!)	021.8	...	4-5
43. 0216 est. White spot in STeZ.	...	099.5	
44. 0222 est. White spot in EqZN.	043.1	...	
45. 0226 (12½ in. refl. x 200). F. of white spot 43 in STeZ.	...	105.5	5-

Apr. 18, 0052. 5in OG x 330.
ω^1 = 348.3 ω^2 = 048.7 S=4.

There was an extra belt in the STrZ, apparently in the latitude of the middle of the Red Spot.

The dark feature 35-6 was interesting. It formed part of the long, rather ill-defined dip in the S. border of the SEB, and was definitely angled at a slant. I could not see a white spot in the dip itself but this may have been due to the consistently bad conditions.

Observations of Jupiter are recorded in this notebook kept by the author. With experience, the observer can learn to discern very subtle variations in the giant planet's features.

OBSERVER'S NOTEBOOK: Stargazing will prove infinitely more satisfactory if an accurate and systematic record is made after each observation. A spiral notebook is best for this, or even a series of notebooks, one for each object. The notebook should always include the name of the observer, the type and aperture of the telescope, the magnification and the time of observation, as well as the seeing conditions, based on the Antoniadi scale. This scale is in roman numerals and ranges from I (perfect seeing) to V (very bad seeing). It is important to complete any drawings made at the telescope as soon as possible after observation. All temptation to "leave it until tomorrow" should be resisted, as errors of fact or interpretation are easily made and a faulty

95

RU PEGASI

k = 9.8

e = 9.0 h = 9.9 n = 11.2 r = 12.7 u = 13.7

f = 9.3 l = 10.5 p = 12.0 s = 13.1 w = 14.0

g = 9.5 m = 11.2 q = 12.5 t = 13.5 x = 14.2

1971						
Jly 17	2240	1	12½ in ×96	0.1<m, 0.1<n	11.3	
Jly 20	0020	1	"	=p, 0.5>q	12.0	
Jly 21	2330	5	" ×116	=q	12.5	Clouding.
Jly 26	0040	1	"	0.1<q	12.6	
Jly 29	2300	1	"	=q	12.5	
Aug. 15	2340	4	"	=q	12.5	
Aug. 16	2350	1	"	0.1<q	12.6	
Aug. 24	2300	1	"	=q	12.5	
Aug. 26	2300	1	"	0.1>q	12.4	Also with 5in OG.
Aug. 31	2010	1	" ×96	0.1>l, 0.7>m.n	10.4	
Set 19	2200	4	" ×116	=q	12.5	
Set 20	2020	4	"	=q	12.5	
Set 21	1940	2	"	=q	12.5	
Set 22	1930	5	" ×96	=q	12.5	Misty.
Set 25	2220	3	" ×116	=q	12.5	
Set 26	2200	1	5in OG ×116	=q	12.5	
Set 28	2100	1	12½ in ×116	=q	12.5	
Oct. 5	2025	5	"	=q	=12.5	Moon.
Oct. 6	2010	4	"	=q	12.5	"
Oct 7	1900	3	"	=q	12.5	"
Oct 10	1910	1	"	=q	12.5	
Oct. 12	2110	3	"	=q	12.5	Mist.
Oct. 17	2030	3	"	=q	12.5	
Oct. 19	2100	1	"	=q	12.5	
Oct 20	2000	3	"	=q	12.5	
Oct. 22	2200	2	"	0.1<q	12.6	
Oct. 24	1940	1	"	=q	12.5	
Oct. 26	2000	1	"	>l, 0.7>m	10.5	

Variable stars are worthwhile objects of study for the amateur. In this page from one of the author's notebooks, two sudden increases in the brightness of RU Pegasi are recorded.

observation is worse than useless. Times are given in Greenwich Mean Time (GMT), which is known as Universal Time (UT) by astronomers. The extract from the author's Jupiter notebook (*above left*) shows a sketch of that planet and six columns filled with notation; there is also a comments column.

The second extract (*above right*) is from the author's notebook on variable stars. RU Pegasi is a dwarf nova, or SS Cygni star, which is normally of magnitude 12.5 but, as has been noted, every now and again brightens to magnitude 10. A list of ordinary stars has also been included (column 5) so that a comparison can readily be made with the estimated magnitudes of RU Pegasi (column 6).

2

THE SUN'S FAMILY

The Solar System

The Solar System is our home in space. It is almost certainly not unique; other suns may well have planetary systems of the same kind – but as yet we have no proof of their existence.

The Solar System is the Sun's family. The Sun is an ordinary star; it appears so brilliant and hot only because it is relatively close to Earth (150 million kilometers). The principal members controlled by the Sun are the planets. Mercury, Venus, the Earth and Mars form an inner group; then comes a wide gap, in which move thousands of dwarf worlds known as asteroids or minor planets; beyond, there are the giant planets Jupiter, Saturn, Uranus and Neptune. Finally there is Pluto, a body which is much smaller than the Moon and which may not be a true planet.

The inner planets are solid and rocky, but only the Earth has an atmosphere suitable to support life. The giant planets are quite different: their surfaces are gaseous and are always changing. The Solar System also contains bodies of lesser importance: the satellites of the planets, comets, meteoroids and a great deal of interplanetary "dust."

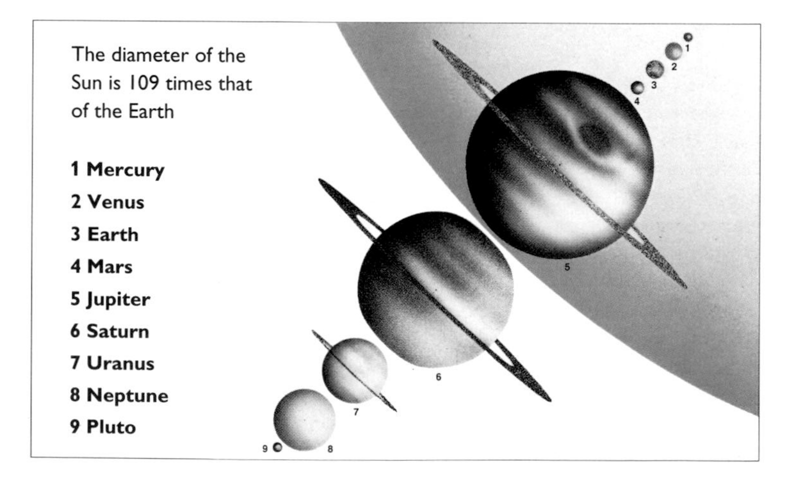

	Mercury	*Venus*	*Earth*	*Mars*	*Jupiter*	*Saturn*	*Uranus*	*Neptune*	*Pluto*
Distance from Sun (million km)									
max.	69.7	109.0	152	249	815.7	1,507	3,004	4,537	7,375
mean	57.9	108.2	149.6	227.9	778	1,427	2,870	4,497	5,900
min.	45.9	107.4	147	206.7	740.9	1,347	2,735	4,456	4,425
Orbital period	89.97d	224.7d	365.3d	687.0d	11.86y	29.46y	84.01y	164.8y	247.7y
Synodic period	115.9d	583.9d	–	779.9d	398.9d	378.1d	369.7d	367.5d	366.7d
Rotation period (eq.)	58.646d	243.16d	23h56m4s	24h37m23s	9h50m30s	10h13m59s	17h14m0s	16h7m0s	6.375d
Orbital eccentricity	0.206	0.007	0.017	0.093	0.048	0.056	0.047	0.009	0.248
Orbital inclin. (°)	7.0	3.4	0	1.8	1.3	2.5	0.8	1.8	17.15
Axial inclin. (°)	2	178	23.44	24.0	3.1	26.7	98	28.8	122.5
Escape vel. (km/s)	4.25	10.36	11.18	5.03	60.22	32.26	22.5	23.9	1.18
Mass (Earth=1)	0.055	0.815	1	0.11	317.9	95.2	14.6	17.2	0.002
Vol. (Earth=1)	0.056	0.86	1	0.15	1319	744	67	57	0.01
Density (water=1)	5.44	5.25	5.52	3.94	1.33	0.71	1.27	2.06	2.03
Surface grav. (Earth=1)	0.38	0.90	1	0.38	2.64	1.16	1.17	1.2	0.06
Surface temp. (°C)	+427	+480	+22	−23	−150	−180	−214	−220	−230
Albedo	0.06	0.76	0.36	0.16	0.43	0.61	0.35	0.35	0.4
Equatorial diam. (km)	4,878	12,104	12,756	6,794	143,884	120,536	51,118	50,538	2,324
Maximum magnitude	−1.9	−4.4	–	−2.8	−2.6	−0.3	+5.6	+7.7	+14

The planets were formed from a "solar nebula," a cloud of material associated with the youthful Sun. They all move around the Sun in the same direction, but some comets are retrograde, as are some satellites of the planets. Moreover, the orbits of most planets are in roughly the same plane, with inclinations of less than 4 degrees. The two exceptions are Pluto, with an inclination of 17 degrees, and Mercury, with 7 degrees.

The revolution periods – the time that it takes for a planet to complete one orbit around the Sun – range from 88 days for Mercury to almost 248 years for Pluto. Most of the orbits take the form of near-circular ellipses. Again Pluto is exceptional: its orbit is much more eccentric. At perihelion, or point of closest approach to the Sun, it comes within the orbit of Neptune. At present Neptune, not Pluto, ranks as the outermost planet in the Solar System.

3
THE MOON

Earth's companion world

The Moon is our companion in space, and stays together with us as we travel round the Sun. It is much nearer than any other natural body in the sky, which is why it appears so brilliant.

The Moon's origin is still a matter for debate, but few astronomers now believe that it broke away from the Earth, as used to be thought. They hold rather that the Moon has always been an independent body.

The Moon is officially ranked as the Earth's satellite, but since it is relatively large and massive, with a diameter of 3,476 kilometers and a mass of 0.012 that of the Earth, it may better be regarded as a companion planet. The mean distance from the Earth is 384,400 kilometers. The Moon's low escape velocity (2.4 kilometers per second) means that it has been unable to retain any appreciable atmosphere. Indeed, analysis of the lunar material brought back by the Apollo astronauts and the Russian automatic probes has confirmed that no life has ever existed there.

The Moon takes 27.3 days to complete one revolution in its orbit. It takes exactly the same time to spin once on its axis (this is known as captured or synchronous rotation). The rotation of the Moon has been slowed down by the action of the Earth until by now it keeps the same face toward Earth. However, the Moon does not keep the same face toward the Sun, so that day and night conditions are the same in each hemisphere. To an observer on the Moon's near side, the Earth would seem to stay almost motionless in the lunar sky.

The Moon's orbit is not circular; the Moon moves fastest when closest to the Earth, so that the position in orbit and the amount of axial spin become "out of step." The Moon thus seems to sway slightly, allowing observers to see beyond alternate limbs (libration in longitude). This means that from Earth 59 per cent of the total lunar surface may be seen at various times. The remaining 41 per cent remained unknown until the circumlunar flight of the Soviet probe Lunik 3 in 1959. Since then there have been the Apollo missions; the first manned landing was that of Apollo 11 in 1969. Apollo 15, 16 and 17, in particular, included detailed geological surveys that provided knowledge of the formation of the crust, mantle and core.

An eclipse of the Moon occurs when the Moon passes into the cone of shadow cast by the Earth. If the Moon partially enters the cone there is a partial eclipse; if it wholly enters the cone the eclipse is total. Eclipses do not happen at every full Moon because the lunar orbit is appreciably inclined.

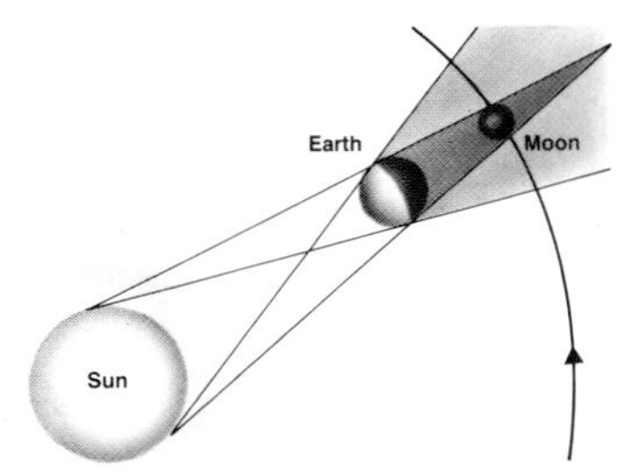

Lunar eclipse	*Mid-eclipse, UT*	*Type and duration*
1999 Jul 28	11h 34m	Partial: 40%, 2h 22m
2000 Jan 21	04h 45m	Total: 1h 16m
2000 Jul 16	13h 57m	Total: 1h 0m
2001 Jan 9	20h 22m	Total: 0h 30m
2001 Jul 5	14h 57m	Partial: 49%, 1h 19m

The Moon has one-quarter the diameter and 0.012 the mass of the Earth. Its core is considerably smaller, relatively, than the Earth's. Because its center of gravity is eccentric, its crust is thinner on one side than the other, and it is also thinner under the maria.

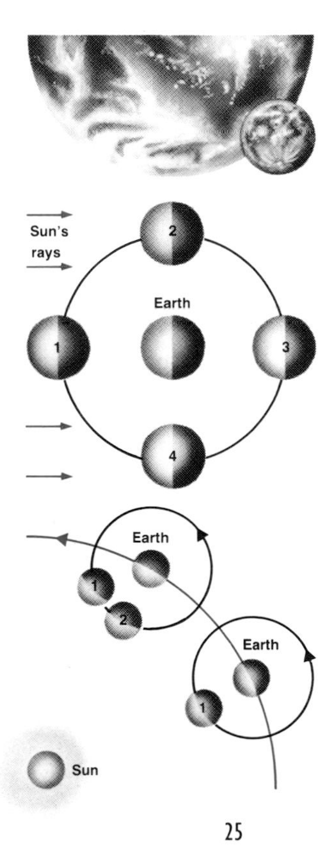

The familiar lunar phases occur because the Moon does not always turn its daylight side toward the Earth. When it is almost between the Earth and the Sun, the Moon is new (1) and its dark side is turned toward the Earth. It cannot then be seen unless the alignment is perfect enough to produce a solar eclipse. At 2 it is at half (first quarter); at 3 it is full; and at 4 it is at half once more (last quarter). Between half and full phases the Moon is "gibbous."

The Moon takes 27.3 days to move round the barycenter – the center of gravity of the Earth–Moon system, which lies within the Earth's globe. However, the Earth is moving round the Sun. The Moon (right) is new at 1; when it has returned to 1, during the Earth's orbit of the Sun, it is still not lined up and must move on to 2 before it is new again. The "lunation," or interval between successive new moons, is therefore 29.5 days, not 27.3.

Moon: general maps

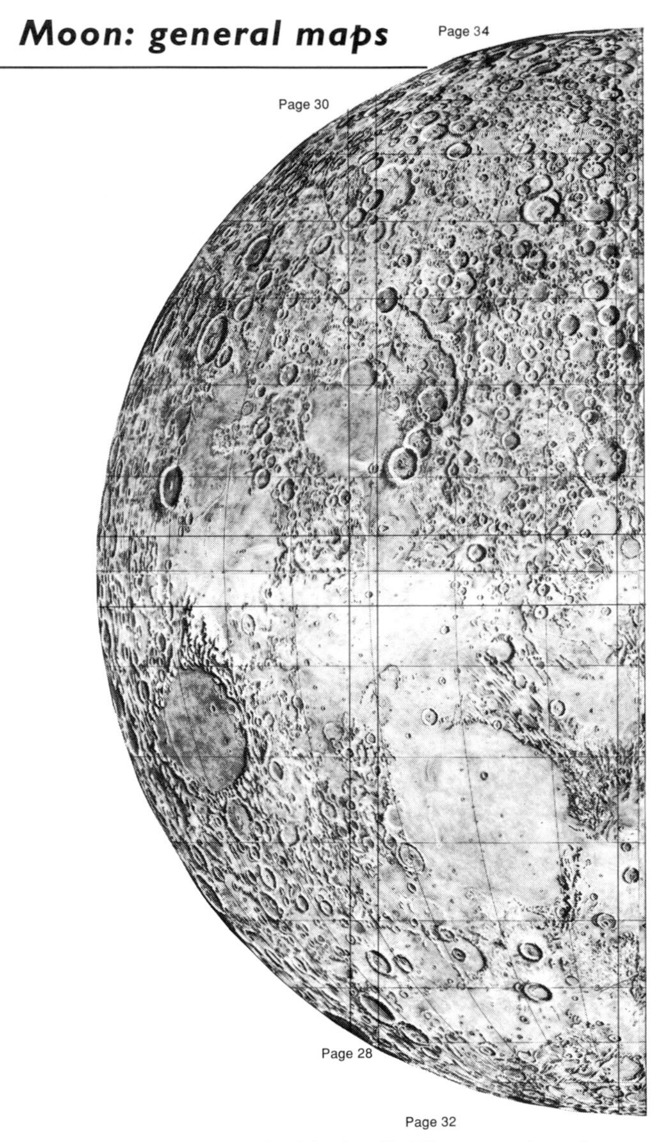

The Moon's surface is crowded with detail. There are broad, grey plains known as maria or "seas," an inappropriate name as there has never been any water in them. There are also mountains, crack-like clefts (known as rills), ridges, swellings (known as domes) and deep

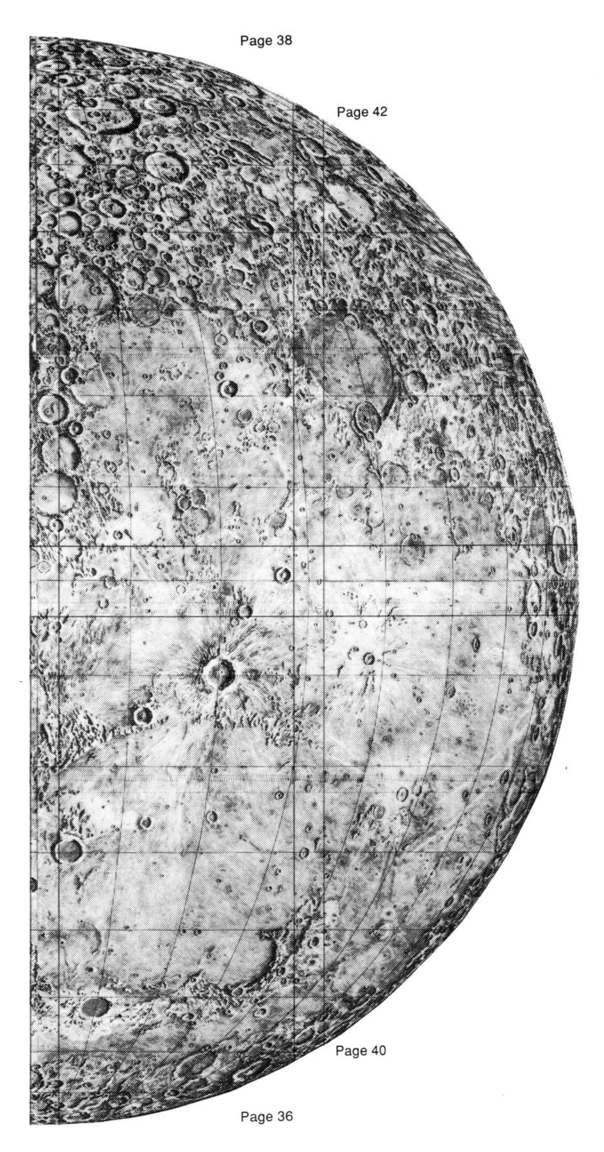

valleys. The whole surface is dominated by walled circular craters. The craters and walled plains range from enclosures that are over 200 kilometers across to tiny pits at the very limit of visibility. Some are perfectly circular, while others have been broken and distorted.

Moon: detailed maps

FAR NORTHEAST SECTOR: This region of the Moon is dominated by the Mare Crisium, which is one of the smaller lunar "seas." It is one of the most distinctive and is easily visible with the naked eye. Its appearance is deceptive, as it actually measures 560 kilometers in an east–west direction but only 448 kilometers north–south. The surface is comparatively smooth, but there are two well-marked small craters, Picard and Peirce. Outside the Mare lies Proclus, which is one of the brightest craters on the Moon and is the center of an asymmetrical system of bright rays. Cleomedes has walls with peaks rising to over 2,793 meters.

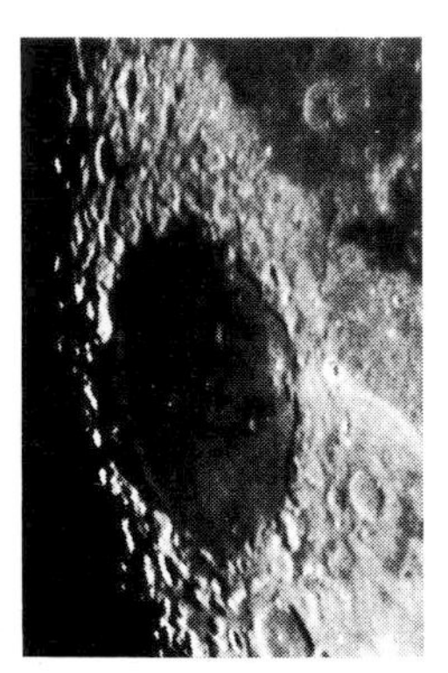

Mare Crisium area (left) at a time of unfavorable libration. Cleomedes and Tralles stand out to the south.

A more favorable libration (right). Mare Crisium is further on to the disk and is less foreshortened.

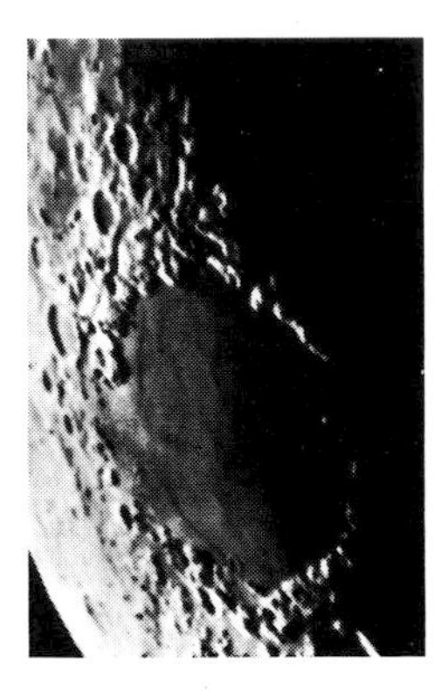

Proclus (left), the ray center to the west of the Mare. The rays are not obvious under low illumination.

Under high illumination (right) the ray system of Proclus is very clear.

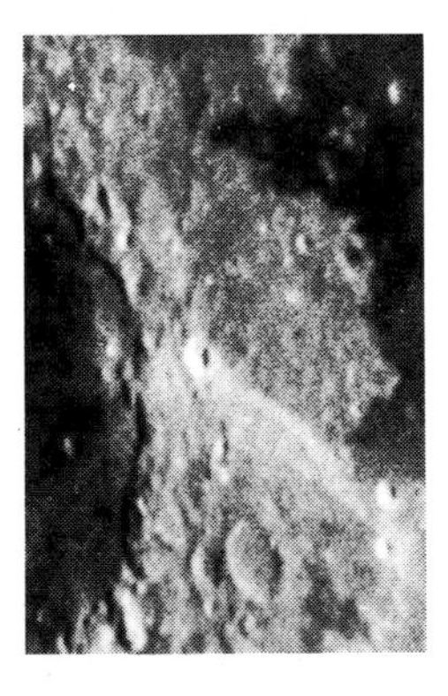

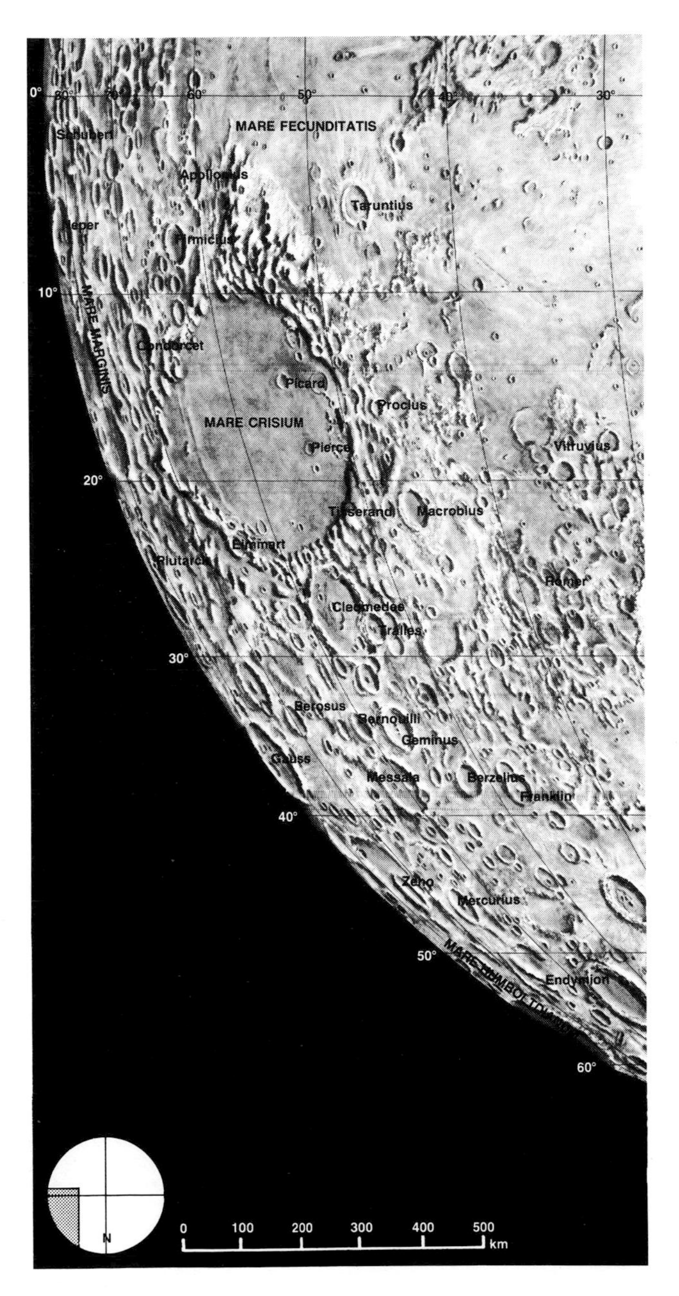
0°
80°
70°
60°
50°
40°
30°
MARE FECUNDITATIS
Schubert
Apollonius
Taruntius
Neper
Firmicius
10°
MARE MARGINIS
Condorcet
Picard
Proclus
MARE CRISIUM
Peirce
Vitruvius
20°
Tisserand
Macrobius
Eimmart
Plutarch
Römer
Cleomedes
Tralles
30°
Berosus
Bernoulli
Geminus
Gauss
Messala
Berzelius
Franklin
40°
Zeno
Mercurius
50°
MARE HUMBOLDTIANUM
Endymion
60°
N
0
100
200
300
400
500
km

FAR SOUTHEAST SECTOR: This region of the Moon contains some majestic walled plains. Langrenus is a noble formation with high, terraced walls and a central peak; Petavius is of the same type; Vendelinus is of similar size, but is less well preserved. Close to the limb are more large enclosures, such as Wilhelm Humboldt. Furnerius is another large enclosure, which is fairly well preserved. The region includes parts of the Mare Foecunditatis and the Mare Nectaris, including Fracastorius. The Messier twins have strange "comet" rays and are easy to identify and there is also the so-called Rheita Valley, which is really a crater chain. Fracastorius has a diameter of 129 kilometers. Its "seaward" wall has been so destroyed by lava that it is barely traceable, and Fracastorius has become a huge bay leading out of the Mare. Its floor, like that of the Mare, is dark grey.

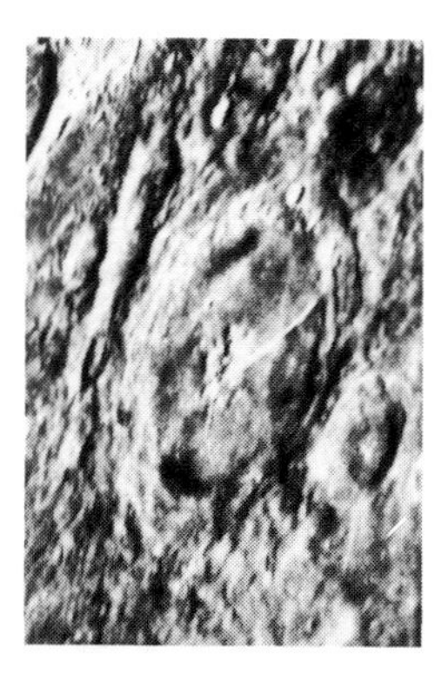

Petavius (left) is one of the finest craters on the Moon. It has a slightly convex floor and a grand central mountain group, from which a magnificent rill runs to the southwest wall.

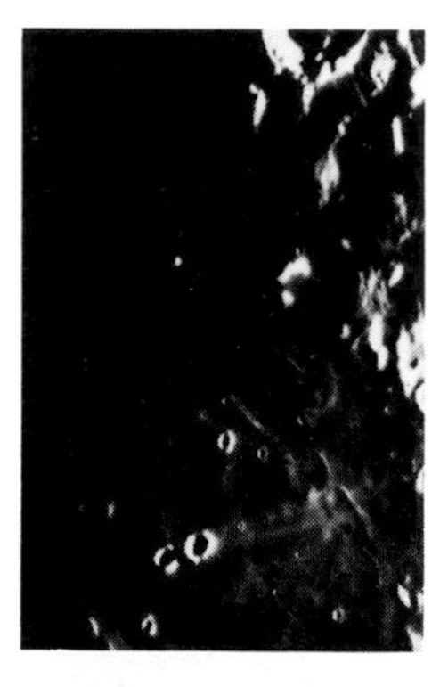

Messier and Messier A (right) are two small craters from which extends a double ray.

Fracastorius (left) is a bay at the southernmost point of Mare Nectaris.

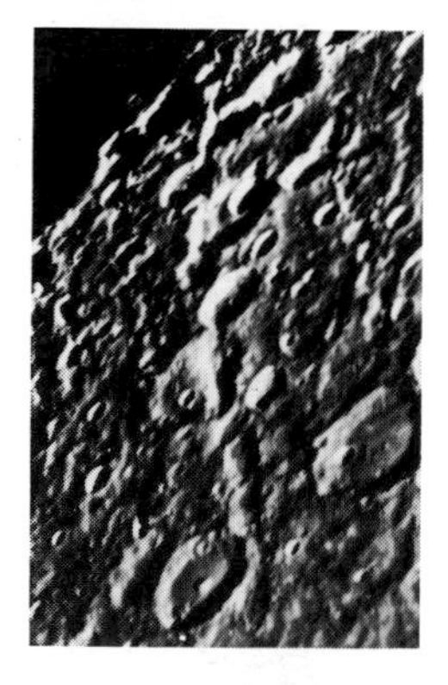

The Rheita Valley (right) is 185 km long, and 24 km across at its widest. It is not a true valley, but a crater chain.

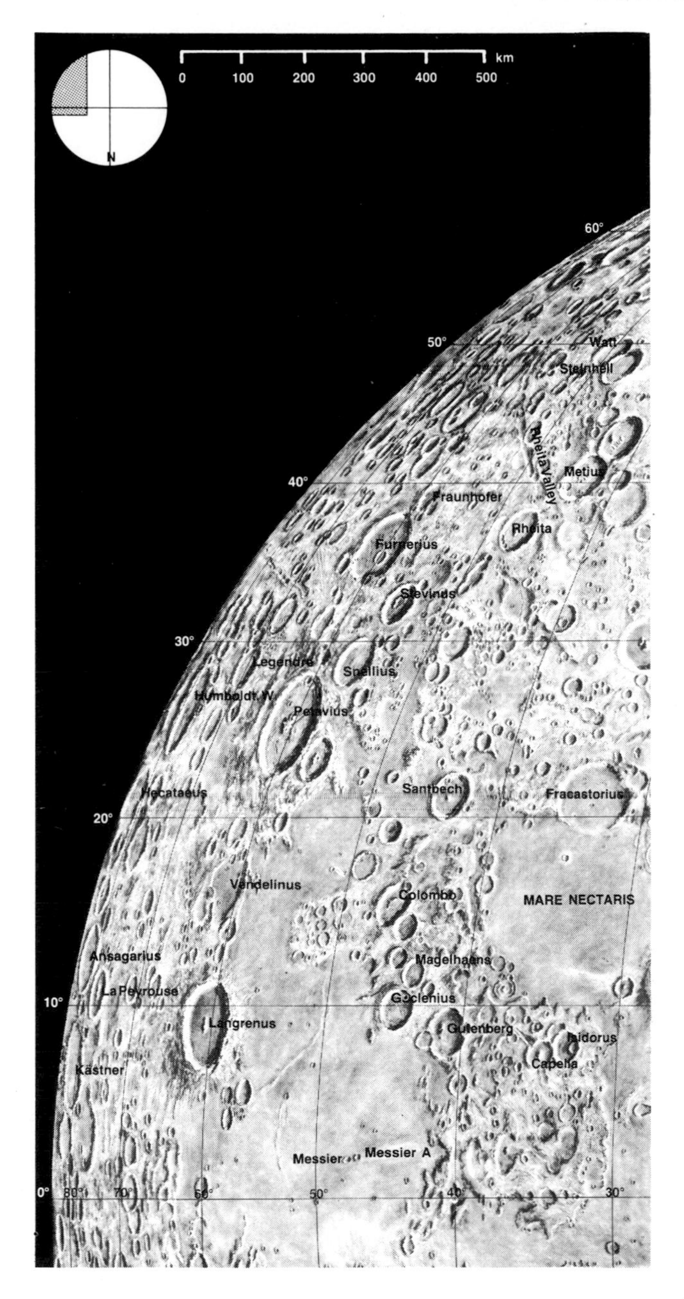
N
0
100
200
300
400
500
km
60°
50°
40°
30°
20°
10°
0°
80°
70°
60°
50°
40°
30°
Watt
Steinheil
Rheita Valley
Metius
Fraunhofer
Rheita
Furnerius
Stevinus
Legendre
Snellius
Humboldt W
Petavius
Hecataeus
Santbech
Fracastorius
Vendelinus
Colombo
MARE NECTARIS
Ansagarius
Magelhaens
La Peyrouse
Goclenius
Langrenus
Gutenberg
Isidorus
Capella
Kästner
Messier
Messier A

NORTHEAST SECTOR: This section includes part of the Mare Imbrium, including Aristillus and Autolycus, which belong to the Archimedes group. One of the most regular of all the lunar "seas" is the Mare Serenitatis, which contains no large formations, although there is one small, bright crater, Bessel, and a small craterlet, Linné. North of the Mare Serenitatis are the great walled formations Aristoteles and Eudoxus and the deep, smaller crater Bürg. To the south, Triesnecker and Hyginus, which are part of the Mare Vaporum, are visible. Boscovich and Julius Caesar are distinguishable by their dark floors and Plinius is a superb crater. It was once suggested that there had been a change in Linné, and that it had become a craterlet surrounded by a white patch instead of a deep crater such as Bessel; but it now seems certain that no real change has occurred there. Major structural alterations on the Moon belong to the remote past.

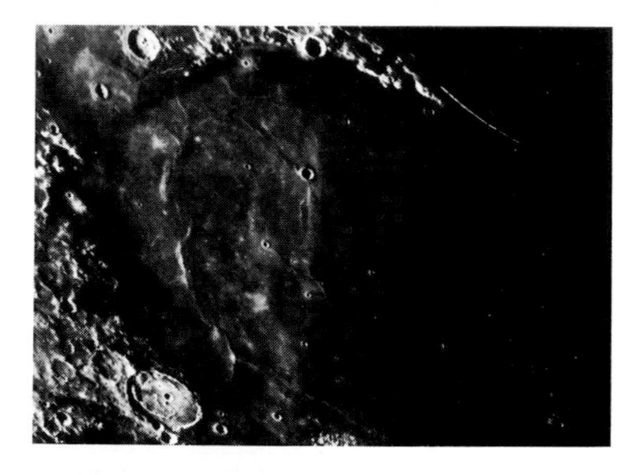

The Mare Serenitatis (left) has a relatively smooth floor but contains several craterlets. There are many prominent ridges, some of which seem to be the walls of ghost craters. A bright ray runs across the Mare.

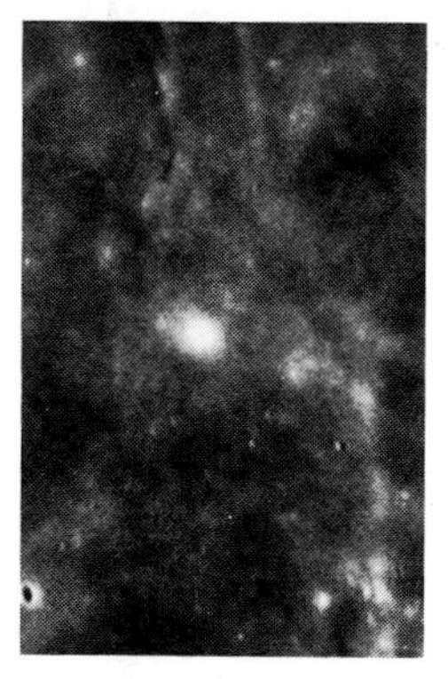

Linné (left) is a small craterlet surrounded by a light patch.

Triesnecker (right) is associated with an extensive rill system and the Hyginus "rill" (really a crater chain) is a quick and easy object to pick up under a small telescope.

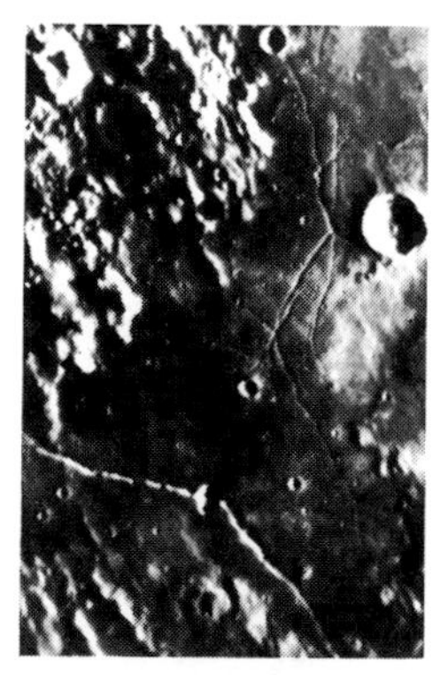

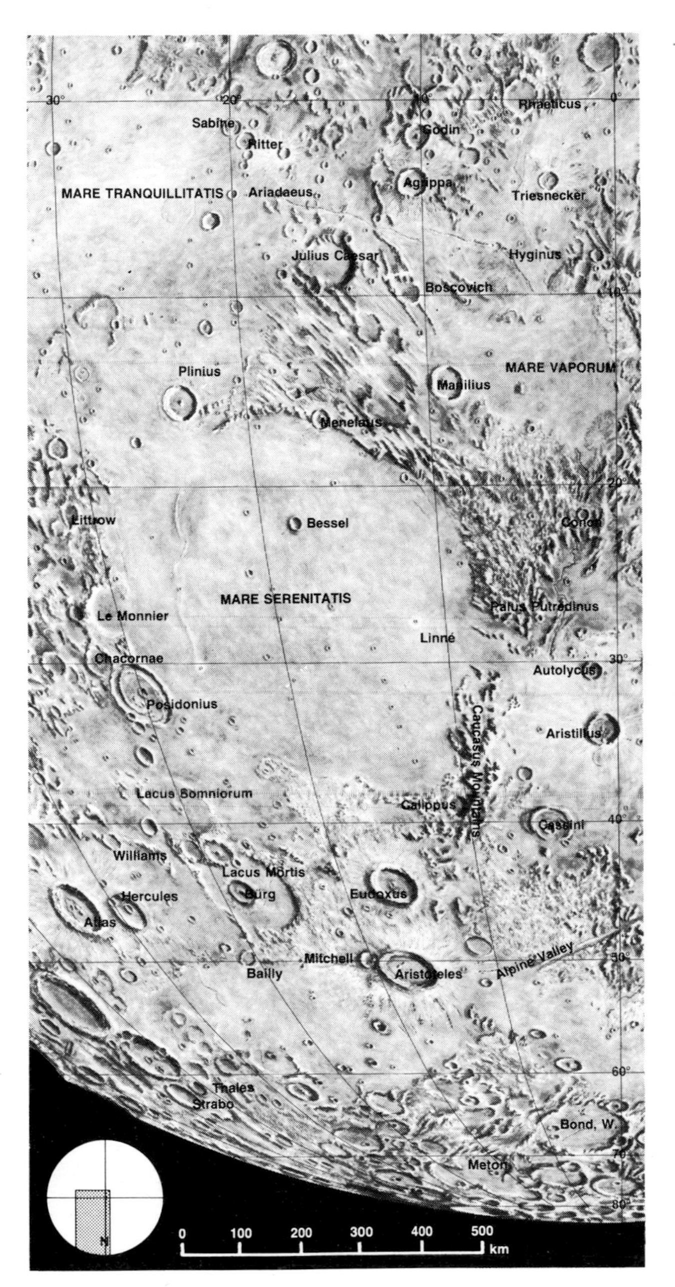
30°
20°
10°
0°
Sabine
Ritter
Godin
Rhaeticus
MARE TRANQUILLITATIS
Ariadaeus
Agrippa
Triesnecker
Julius Caesar
Hyginus
Boscovich
10°
Plinius
MARE VAPORUM
Manilius
Menelaus
20°
Littrow
Bessel
Conon
MARE SERENITATIS
Palus Putredinus
Le Monnier
Linné
Chacornae
30°
Autolycus
Posidonius
Caucasus Montains
Aristillus
Lacus Somniorum
Calippus
Cassini
40°
Williams
Lacus Mortis
Hercules
Burg
Eudoxus
Atlas
Alpine Valley
Mitchell
50°
Bailly
Aristoteles
60°
Thales
Strabo
Bond, W.
70°
Meton
80°
N
0
100
200
300
400
500
km

SOUTHEAST SECTOR: This section contains few "seas" and most of the area is occupied by rugged uplands. A variety of craters and walled plains may be seen, including the huge ruined Janssen, with a diameter of over 160 kilometers, and superb formations such as Theophilus, which is the northern member of a chain of three great walled plains and actually intrudes into its neighbor Cyrillus. There are numerous small rings and countless craterlets. Lofty mountain ranges are absent, but there is an interesting feature known as the Altai Scarp running northwest from Piccolomini. The Scarp rises to about 1,800 meters above mean ground level. The great chain of formations including Theophilus, Cyrillus and Catharina is one of the most imposing on the Moon. There are however many other chains – it is apparent that the distribution of the craters on the lunar surface is not random, and in many cases the formations in any particular chain are of obviously different ages.

Albategnius (left) is a vast crater with a few peaks over 3,000 m in height. It forms a notable pair with Hipparchus.

One of the grandest craters (right) on the Moon is Theophilus, with walls rising to over 5,000 m. It has a great multipeaked central mountain mass.

The Altai Scarp (left) is part of the ring system of the Mare Nectaris.

Stöfler (right), an enclosure that is easy to identify because of its darkish floor, is 145 km in diameter.

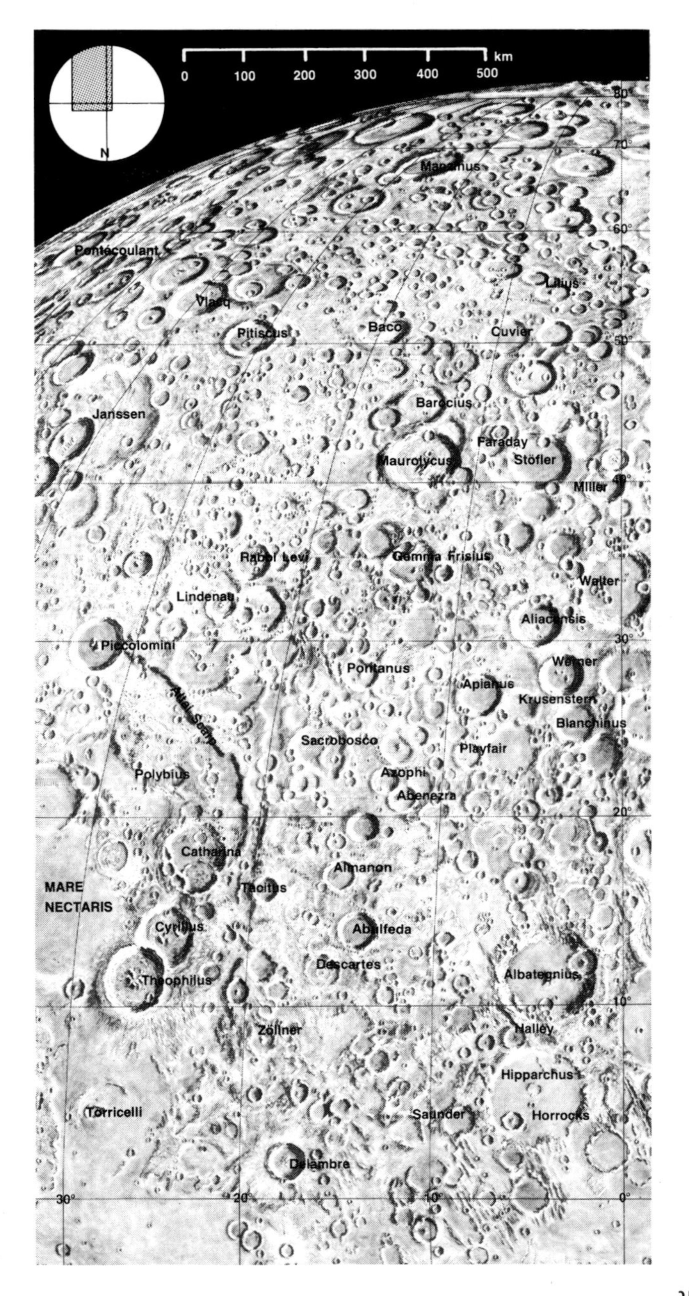
km
0
100
200
300
400
500
N
80°
70°
60°
50°
40°
30°
20°
10°
0°
30°
20°
10°
Manzinus
Pontécoulant
Lilius
Vlacq
Pitiscus
Baco
Cuvier
Janssen
Barocius
Faraday
Maurolycus
Stöfler
Miller
Rabbi Levi
Gemma Frisius
Walter
Lindenau
Aliacensis
Piccolomini
Werner
Pontanus
Apianus
Altai Scarp
Krusenstern
Blanchinus
Sacrobosco
Playfair
Polybius
Azophi
Abenezra
Catharina
Almanon
MARE
NECTARIS
Tacitus
Cyrillus
Abulfeda
Descartes
Albategnius
Theophilus
Zöllner
Halley
Hipparchus
Torricelli
Saunder
Horrocks
Delambre

NORTHWEST SECTOR: This is one of the most interesting regions on the Moon. It contains most of the Mare Imbrium as well as the superbly beautiful Sinus Iridum. The Apennines, ending near the deep, prominent crater Eratosthenes, make up part of the border of the Mare. The Carpathians are rather less lofty than the Apennines. On the Mare Imbrium itself are many small formations, as well as some major enclosures, of which Archimedes is preeminent. The regular formation Plato is in the region of the Alps. To the south of the area lies Copernicus, and near it is the famous ghost crater Stadius. Other interesting features are Beer and Feuillée, and the mountains Pico and Piton. Stadius is probably the Moon's best example of a "ghost ring." Its walls have been almost leveled, though originally it must have been a most imposing formation, comparable with Copernicus and Eratosthenes. Its surface is pitted with craterlets.

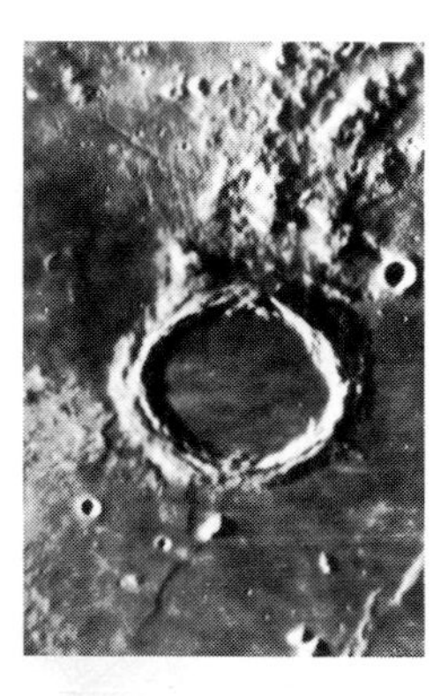

Archimedes (left) is very regular and is 80 km in diameter, with fairly low walls. The sunken floor is darkish and very smooth.

Copernicus (right) is 90 km across and has massive, terraced walls. Its rays dominate this part of the surface.

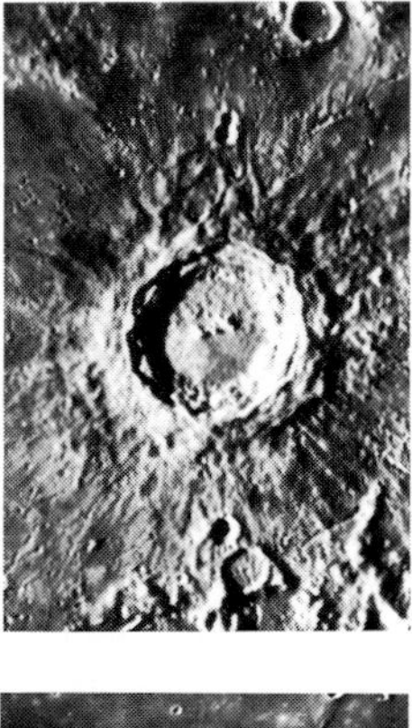

Eratosthenes (left) ranks as one of the most perfect craters on the Moon. It is 61 km across.

Sinus Iridum (right), the lovely Bay of Rainbows, is one of the most striking sights on the Moon at sunrise, when its interior is shadowed.

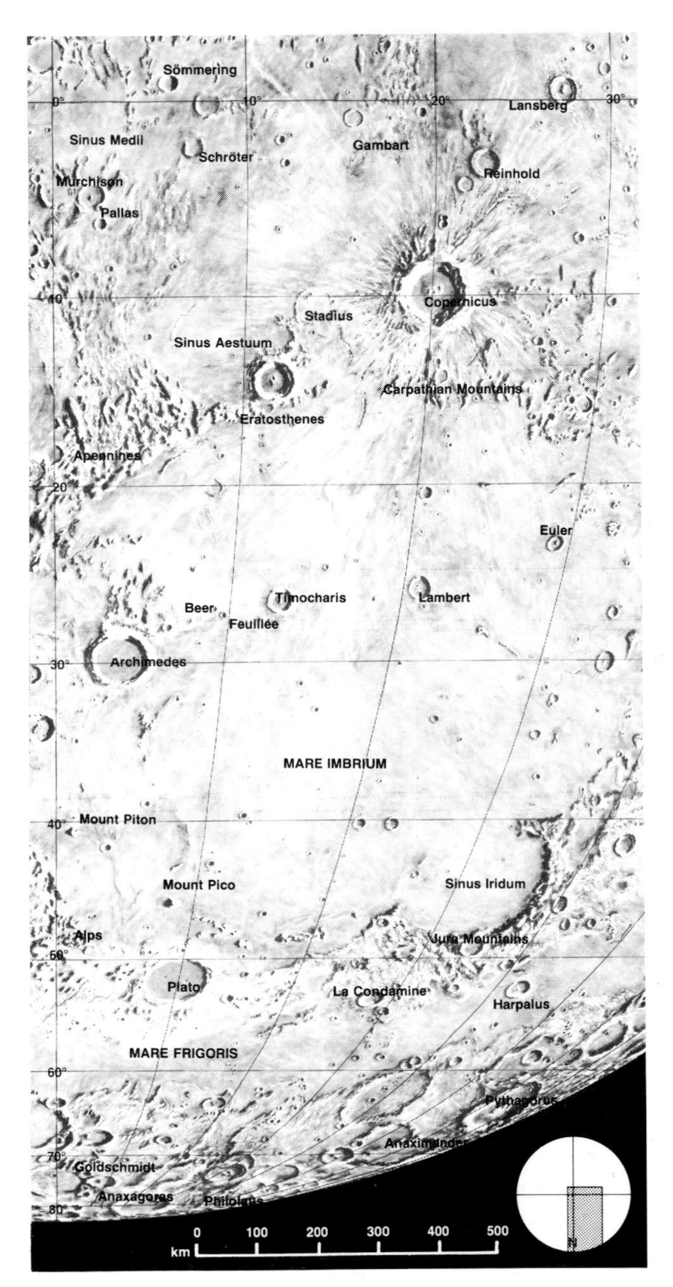
Sömmering
0°
10°
20°
30°
Lansberg
Sinus Medii
Schröter
Gambart
Reinhold
Murchison
Pallas
10°
Copernicus
Stadius
Sinus Aestuum
Carpathian Mountains
Eratosthenes
Apennines
20°
Euler
Timocharis
Lambert
Beer
Feuillée
Archimedes
30°
MARE IMBRIUM
40°
Mount Piton
Mount Pico
Sinus Iridum
Alps
Jura Mountains
50°
Plato
La Condamine
Harpalus
MARE FRIGORIS
60°
Pythagoras
Anaximander
70°
Goldschmidt
Anaxagoras
Philolaus
80°
0
100
200
300
400
500
km
N

SOUTHWEST SECTOR: Some of the most famous features of the Moon are to be found in this sector. Tycho in the uplands is well formed, and there are many other large structures, such as Clavius, Maginus, Longomontanus and the compound Schiller. To the north lies much of the Mare Nubium, on which there is the semiruined Fra Mauro group. Bullialdus, near the edge of the Mare, is a well-formed crater, and at the far side is Ptolemaeus. Thebit is a perfect example of a multiple crater, and most famous of all lunar faults is the Straight Wall, near Birt. Purbach, on the edge of the region, is associated with two more major formations, Regiomontanus and Walter. Much of this sector is occupied by the relatively smooth Mare Nubium, but the south polar area is very rough; it is not easy to study, because it is so foreshortened. Near the limb foreshortening becomes so severe that it is hard to distinguish a crater from a ridge.

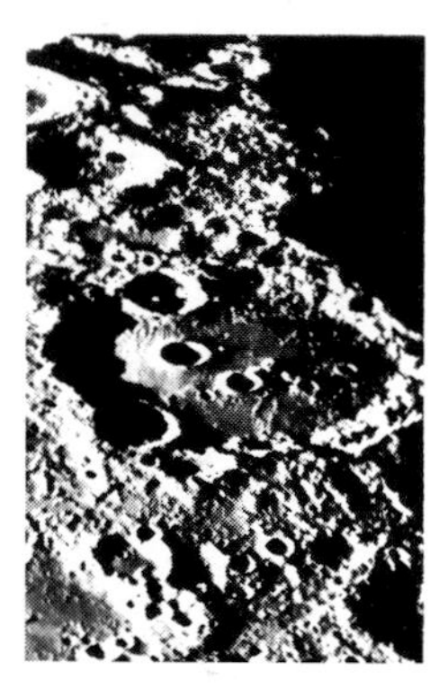

Clavius (left) is 233 km in diameter and has walls rising to over 3,600 m.

Tycho (right) lies at the center of the greatest ray system on the Moon. It is 87 km across, with high terraced walls.

Ptolemaeus, Alphonsus and Arzachel (left) form a great chain of walled plains.

The Straight Wall (right) appears dark before full Moon, because its face is in shadow, and bright after full Moon, when its face is illuminated.

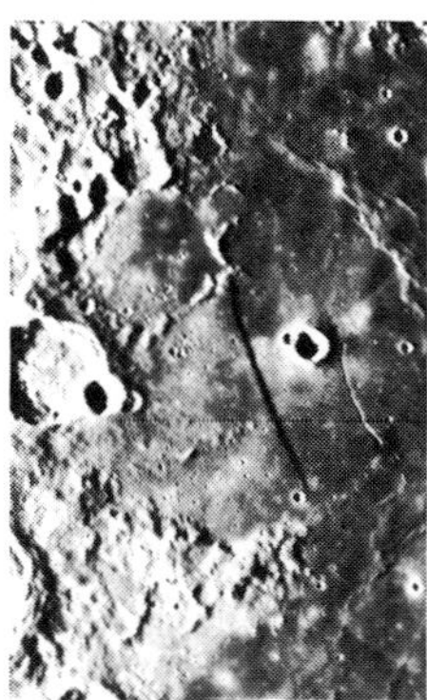

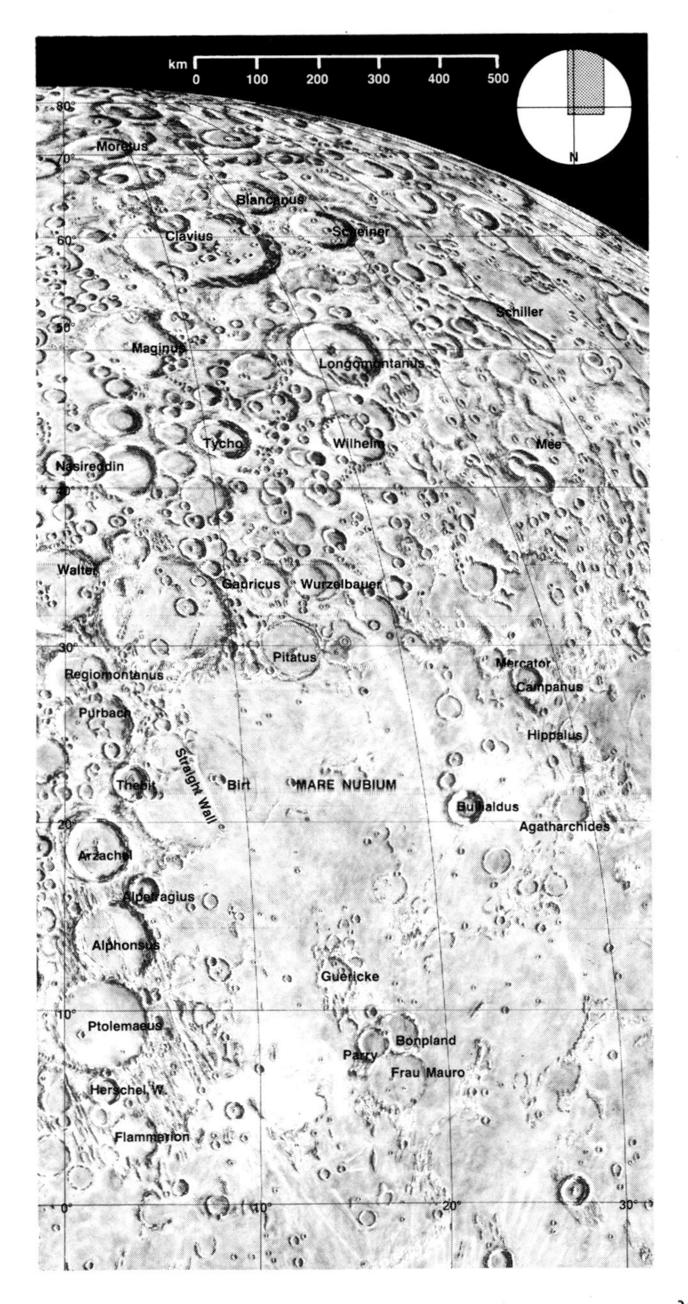
km
0
100
200
300
400
500
N
80°
70°
60°
50°
40°
30°
20°
10°
0°
10°
20°
30°
Moretus
Blancanus
Clavius
Scheiner
Schiller
Maginus
Longomontanus
Tycho
Wilhelm
Mee
Nasireddin
Walter
Gauricus
Wurzelbauer
Pitatus
Mercator
Regiomontanus
Campanus
Purbach
Hippalus
Straight Wall
Thebit
Birt
MARE NUBIUM
Bullialdus
Agatharchides
Arzachel
Alpetragius
Alphonsus
Guericke
Ptolemaeus
Bonpland
Parry
Frau Mauro
Herschel, W.
Flammarion

FAR NORTHWEST SECTOR: This area of the Moon is dominated by Oceanus Procellarum, which is much the largest, although by no means the best-defined, of the lunar "seas." The most interesting group is undoubtedly that of Aristarchus. Adjoining it is Herodotus, of similar size but much less brilliant, and extending from Herodotus is the magnificent Schröter's Valley. This whole region is subject to the elusive glows known as transient lunar phenomena, or TLPs. The Harbinger Mountains, near Aristarchus, are more like clumps of hills than a major range. Along the limb are some large, generally rather low-walled formations, and the area also includes Hevelius. Aristarchus is so bright that it is often prominent even when illuminated only by Earthshine; at times such as these, unwary observers have even taken it for an erupting volcano.

Kepler (left) has a central mountain and heavily terraced walls.

Hevelius (right), in the Grimaldi chain, has a convex floor containing a central elevation and a system of rills.

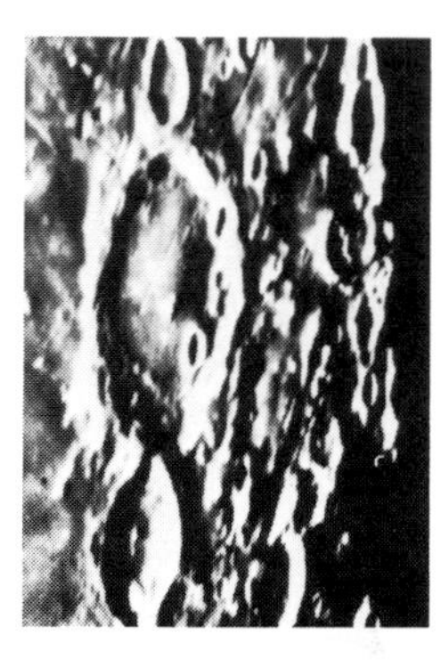

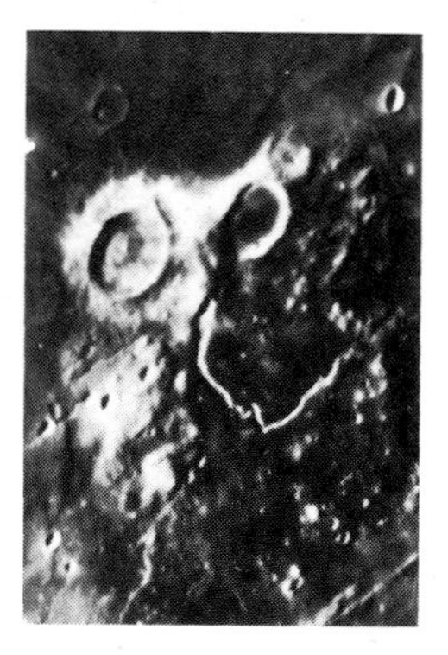

Aristarchus (left) is the most "active" and brightest formation on the Moon. Herodotus has a dark floor, in contrast to its neighbor.

Otto Struve (right) is a vast enclosure made up of two old rings, each 160 km across, which have coalesced.

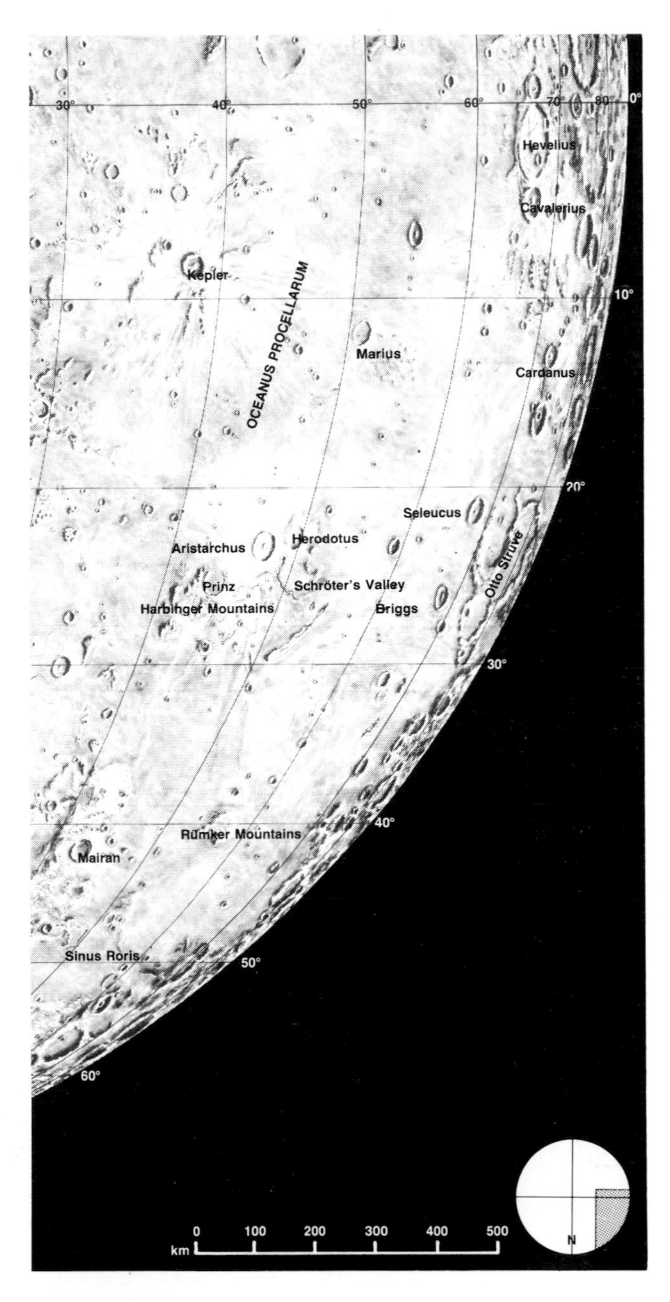
30°
40°
50°
60°
70°
80°
0°
Hevelius
Cavalerius
Kepler
OCEANUS PROCELLARUM
10°
Marius
Cardanus
20°
Seleucus
Herodotus
Aristarchus
Otto Struve
Prinz
Schröter's Valley
Harbinger Mountains
Briggs
30°
40°
Rümker Mountains
Mairan
Sinus Roris
50°
60°
0
100
200
300
400
500
km
N

FAR SOUTHWEST SECTOR: The most famous walled formation in this region is probably Grimaldi; adjoining it is Riccioli. Not far from Grimaldi is Sirsalis, a double crater associated with one of the finest rills on the Moon. The only "sea" area in the region is the Mare Humorum; at its north border is Gassendi, which has a low "seaward" wall. Doppelmayer, to the south of the Mare Humorum, is in even worse repair, the "wall" facing the Mare being barely traceable. To the far south lies the majestic Schickard and the celebrated plateau Wargentin, which is close to the limb and very foreshortened, but it is not hard to identify with a small telescope. Grimaldi is the darkest formation on the Moon, and is always easy to identify when it is in sunlight. It is a member of a major chain that also includes Riccioli, Hevel and Lohrmann. Schickard is of much the same size as Grimaldi, but has a much lighter floor.

The Mare Humorum (left) is a superb minor circular "sea." There are many rills and bays in and near it.

Gassendi (right) is 88 km in diameter. The floor includes a central peak and a magnificent system of rills.

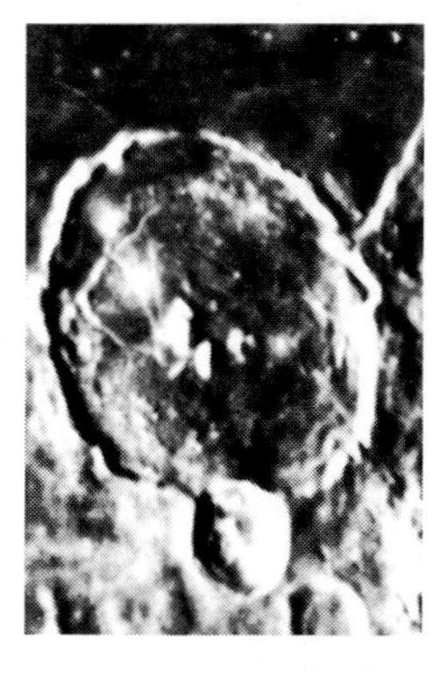

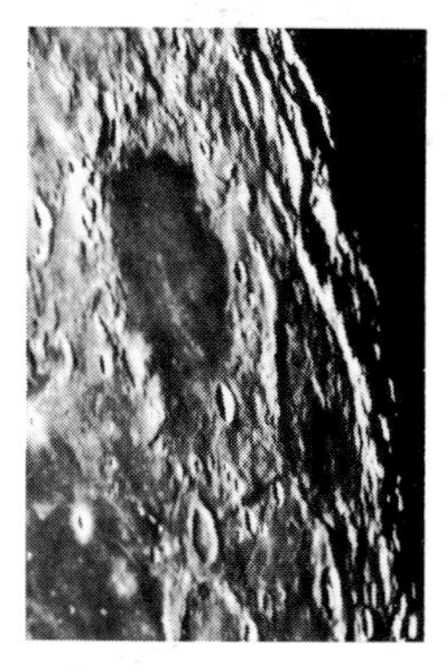

Grimaldi (left) is one of the grandest enclosures on the Moon; so too is its neighbor, Riccioli. Both have dark patches on their floors.

Wargentin (right) represents the best example of a lunar plateau; it is 88 km in diameter, and filled with lava.

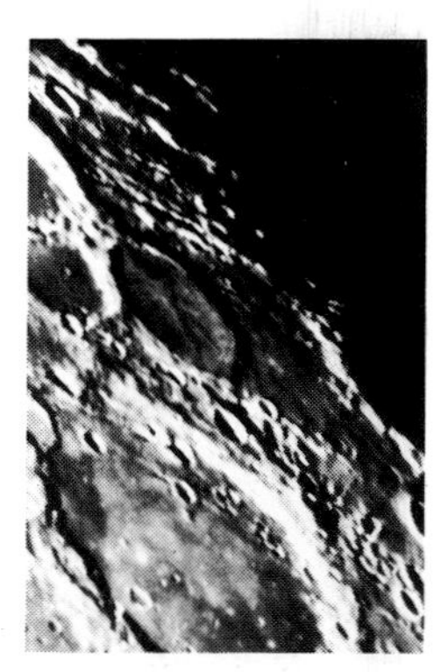

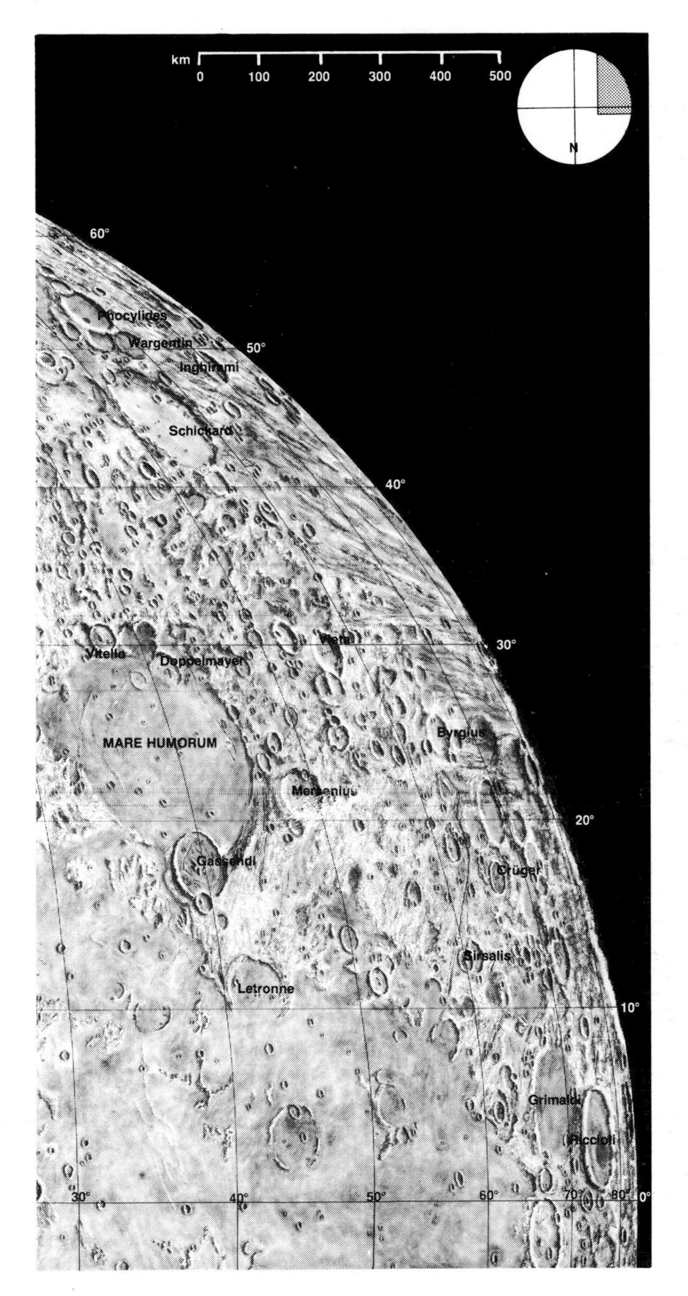
km
0
100
200
300
400
500
N
60°
Phocylides
Wargentin
50°
Inghirami
Schickard
40°
Vitello
Doppelmayer
Vieta
30°
MARE HUMORUM
Byrgius
Mersenius
20°
Gassendi
Crüger
Sirsalis
Letronne
10°
Grimaldi
Riccioli
30°
40°
50°
60°
70°
80°
0°

Moon maps index

Feature	Page	Position	Feature	Page	Position
Legendre	31	29°S 70°E	**Pontécoulant**	35	59°S 66°E
Le Monnier	33	26°N 31°E	**Posidonius**	33	32°N 30°E
Letronne	43	10°S 43°W	**Prinz**	41	26°N 44°W
Lilius	35	54°S 6°E	**Proclus**	29	16°N 47°E
Lindenau	35	32°S 25°E	**Ptolemaeus**	39	9°S 2°W
Linné	33	28°N 12°E	**Purbach**	39	25°S 2°W
Littrow	33	22°N 31°E	**Pythagoras**	37	63°N 62°W
Longomontanus	39	50°S 21°W	**Rabbi Levi**	35	35°S 24°E
Macrobius	29	21°N 46°E	**Regiomontanus**	39	28°S 0°
Maginus	39	50°S 6°W	**Reinhold**	37	3°N 23°W
Magelhaens	31	12°S 44°E	**Rhaeticus**	33	0° 5°E
Mairan	41	42°N 43°W	**Rheita**	31	37°S 47°E
Manilius	33	15°N 9°E	**Rheita Valley**	31	38°S 46°E
Mare Crisium	29	17°N 60°E	**Riccioli**	43	3°S 75°W
Mare Foecunditatis	29	2°S 50°E	**Ritter**	33	2°N 19°E
Mare Frigoris	37	58°N 10°W	**Römer**	29	25°N 37°E
Mare Humboldtianum	29	55°N 75°E	**Rümker Mt**	41	41°N 58°W
Mare Humorum	43	26°S 38°W	**Sabine**	33	2°N 20°E
Mare Imbrium	37	35°N 20°W	**Sacrobosco**	35	24°S 17°E
Mare Marginis	29	10°N 90°E	**Santbech**	31	21°S 44°E
Mare Nectaris	31	12°S 30°E	**Saunder**	35	4°S 9°E
Mare Nubium	39	20°S 10°W	**Scheiner**	39	60°S 28°W
Mare Serenitatis	33	24°N 20°E	**Schickard**	43	44°S 55°W
Mare Tranquillitatis	33	7°N 22°E	**Schiller**	39	52°S 39°W
Mare Vaporum	33	13°N 4°E	**Schröter**	37	3°N 7°W
Marius	41	12°N 51°W	**Schröter's Valley**	41	25°N 50°W
Maurolycus	35	42°S 14°E	**Schubert**	29	3°N 81°E
Manzinus	35	68°S 27°E	**Seleucus**	41	21°N 66°W
Mee	39	44°S 35°W	**Sinus Aestuum**	37	15°N 9°W
Menelaus	33	16°N 16°E	**Sinus Iridum**	37	45°N 25°W
Mercator	39	29°S 26°W	**Sinus Medll**	37	2°N 0°
Mercurius	29	46°N 65°E	**Sinus Roris**	41	48°N 60°W
Mersenius	43	21°S 49°W	**Sirsalis**	43	13°S 60°W
Messala	29	39°N 60°E	**Snellius**	31	29°S 56°E
Messier	31	2°S 48°E	**Sömmering**	37	0° 7°W
Messier A	31	2°S 47°E	**Stadius**	37	11°N 14°W
Metius	31	40°S 44°E	**Steinheil**	31	50°S 46°E
Meton	33	74°N 19°E	**Stevinus**	31	33°S 54°E
Miller	35	39°S 1°E	**Stöfler**	35	41°S 6°E
Mitchell	33	50°N 20°E	**Strabo**	33	62°N 54°E
Moretus	39	70°S 8°W	**Tacitus**	35	16°S 19°E
Murchison	37	5°N 0°	**Taruntius**	29	6°N 46°E
Nasireddin	39	41°S 0°	**Thales**	33	62°N 50°E
Neper	29	7°N 83°E	**Thebit**	39	22°S 4°W
Oceanus Procellarum	41	10°N 46°W	**Theophilus**	35	11°S 26°E
Otto Struve	41	25°N 75°W	**Timocharis**	37	27°N 13°W
Pallas	37	5°N 2°W	**Tisserand**	29	21°N 48°E
Palus Putredinus	33	26°N 0°	**Torricelli**	35	5°S 29°E
Parry	39	8°S 16°W	**Tralles**	29	28°N 53°E
Petavius	31	25°S 61°E	**Triesnecker**	33	4°N 4°E
Philolaus	37	72°N 32°W	**Tycho**	39	43°S 11°W
Phocylides	43	54°S 58°W	**Vendelinus**	31	16°S 62°E
Picard	29	15°N 55°E	**Vieta**	43	29°S 57°W
Piccolomini	35	30°S 32°E	**Vitello**	43	30°S 38°W
Pico Mt	37	46°N 9°W	**Vitruvius**	29	18°N 31°E
Pierce	29	18°N 54°E	**Vlacq**	35	53°S 39°E
Pitatus	39	30°S 14°W	**Walter**	35	33°S 0°
Pitiscus	35	51°S 31°E	**Wargentin**	43	50°S 60°W
Piton Mt	37	41°N 1°W	**Watt**	31	50°S 49°E
Plato	37	51°N 9°W	**Werner**	35	28°S 3°E
Playfair	35	23°S 9°E	**Wilhelm**	39	43°S 20°W
Plinius	33	15°N 24 E	**Williams**	33	42°N 37°E
Plutarch	29	25°N 79°E	**Wurzelbauer**	39	34°S 16°W
Polybius	35	22°S 26°E	**Zeno**	29	45°N 70°E
Pontanus	35	28°S 15°E	**Zöllner**	35	8°S 19°E

4

THE SUN

The Sun seen from Earth

The Sun is the only star that is close enough to us to be studied in detail. For this reason astronomers regard it as vitally important to understand the processes that generate its heat and light.

The Sun is a star of no particular importance; indeed, astronomers relegate it to the status of a yellow dwarf. Yet, compared to the Earth, it is a vast globe. Its diameter is 1,392,000 kilometers and its volume is well over a million times that of the Earth. It is a typical member of the Galaxy, and takes as much as 225 million years to make one complete revolution of the galactic center, along with the Earth and all other members of the Solar System.

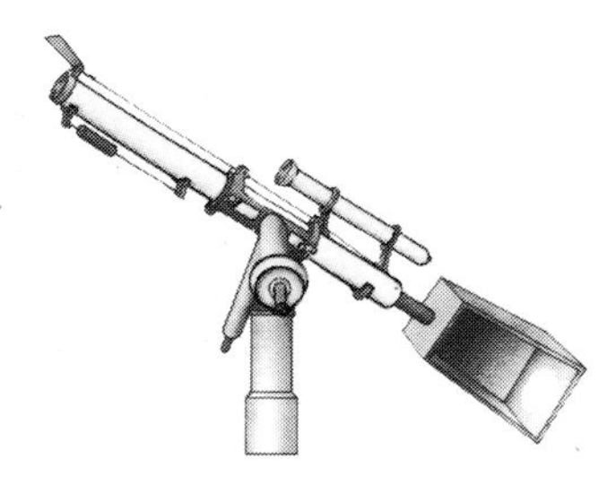

To project the Sun's image safely (left), point the telescope sunward, without looking through the eyepiece. The Sun's image may be thrown on to a screen inside a box or on to a sheet of paper.

Eclipses of the Sun (below) occur when the Earth, Moon and Sun are aligned. A total eclipse is seen at places along a narrow band, the path of totality. Over a wide area outside this band a partial eclipse is seen.

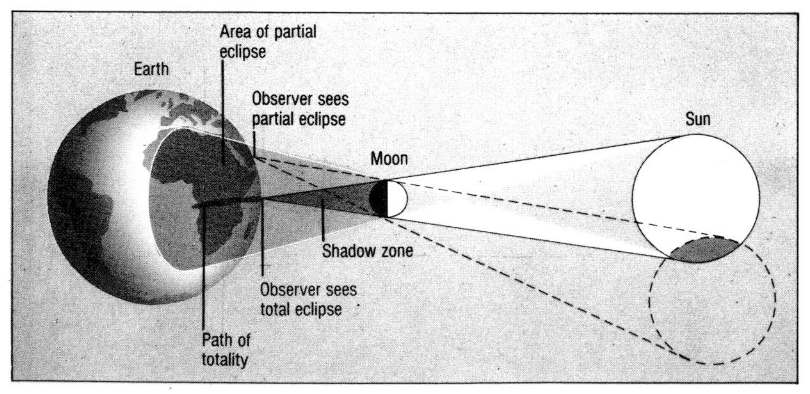

ECLIPSES OF THE SUN				
Date	***UT (h)***	***Type***	***Max. duration (or cover if partial)***	***Area***
1998 Feb 26	17	Total	4m 8s	Pacific, Atlantic
1998 Aug 22	02	Annular	3m 14s	Indian Ocean, E. Indies, Pacific
1999 Feb 16	07	Annular	1m 19s	Indian Ocean, Australia, Pacific
1999 Aug 11	11	Total	2m 23s	England, France, Turkey, India
2000 Feb 5	13	Partial	56%	Antarctic
2000 Jul 31	02	Partial	60%	Arctic
2000 Dec 25	18	Partial	72%	Arctic
2001 Jun 21	12	Total	4m 56s	Atlantic, S. Africa
2001 Dec 14	31	Annular	3m 54s	Central America, Pacific
2002 Jun 10	24	Annular	1m 13s	Pacific

The surface temperature of the Sun is 6,000 degrees centigrade. Near the core, the temperature rises to at least 14 million degrees centigrade. The Sun is not "burning" in the conventional sense. Deep inside its globe, nuclear transformations are taking place. Hydrogen is being converted into helium, and each time a new helium nucleus is formed from four hydrogen nuclei a little energy is released and a little mass is lost. This energy keeps the Sun shining, while the mass-loss amounts to four million tonnes per second. Although the Sun is halfway through its lifetime, it still has sufficient fuel for another five thousand million years.

The Sun is a fascinating, although dangerous, object to study. *An observer must never look directly at the Sun through a telescope or even binoculars.* The inevitable result will be permanent blindness. Nor is it safe to look directly at the Sun merely using a dark filter over the eyepiece. The only sensible way to view the Sun's surface is to project it through a telescope. Refractors are ideal for all solar work, although reflectors may also be used.

The most obvious features on the solar surface are the sunspots, which appear as dark patches. A typical large sunspot has an umbra, or central area, surrounded by a lighter area known as the penumbra. No sunspot can last for more than a few months, and many have lifetimes of only a few days; when the Sun is at its most active many groups can be seen at the same time.

When the Moon passes between the Sun and the Earth, there is a solar eclipse. If the eclipse is total, the Sun's outer surroundings flash into view and the appearance is magnificent. Unfortunately solar eclipses do not occur every month because the Moon's orbit is inclined to that of the Earth and, at most new Moons, the Moon passes either "above" or "below" the Sun in the sky, thereby avoiding eclipse.

The surface of the Sun

The visible surface of the Sun is known as the photosphere. This emits virtually all the Sun's radiation, and has a temperature of almost 6,000 degrees centigrade. Through a small telescope the photosphere will have a mottled appearance; larger telescopes refine this to reveal granulation – the effect of convection in the outer layers of the Sun.

Interesting and easily observable phenomena to study on the Sun are dark patches known as sunspots, which appear dark in comparison with the photosphere only because they have a lower temperature. Although they are not completely understood, sunspots are known to be associated with magnetic fields.

The solar cycle (below) has an average period of 11 years. The graph is based on an index of spot activity called the Zürich Relative Sunspot Number. At times of minimum activity no spots may be visible for some weeks, while at solar maximum there may be many sunspot groups. Maximum sunspot activity is often reached 4.5 years from a minimum; at this time flares may occur and eject charged particles. Some of these can reach the Earth and cause magnetic disturbances high in the ionosphere.

The aurora borealis (above) shimmers in the upper reaches of the Earth's atmosphere. These "northern lights," like their southern equivalent, the aurora australis, are caused by electrically charged particles from the Sun. When the particles collide with atoms of oxygen, nitrogen and other gases in our atmosphere, they knock electrons from them. When the electrons recombine with the atoms, light is given out.

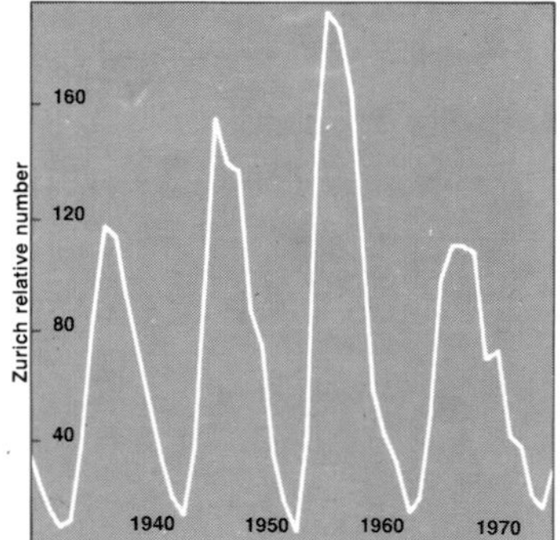

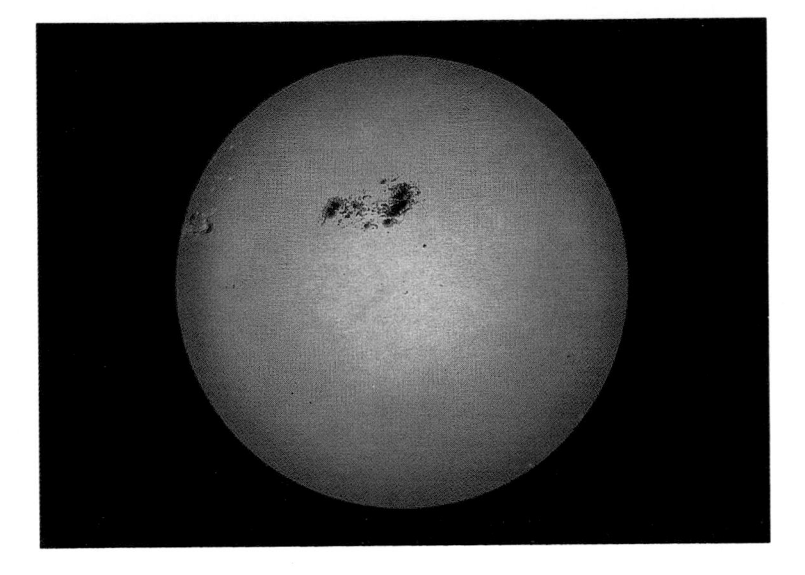

An enormous group of sunspots dominates the Sun's surface in this picture. The entire Earth could easily be swallowed by one of these spots.

The Sun's rotation period is 26 days at the equator, but 36 days at the poles, because it does not rotate in the manner of a solid body. The lines of magnetic force, therefore, become twisted, causing a loop of magnetic energy to burst through the photosphere and form a sunspot pair. Single sunspots are not uncommon, but generally spots appear in groups between 10 degrees and 30 degrees latitude north and south. The bright patches that are usually associated with sunspots are known as faculae.

In addition to sending the Earth virtually all its light and heat, the Sun has much less obvious effects. Of special importance is the so-called solar wind, made up of charged particles streaming continually outward from the Sun in all directions. As it passes the Earth it is moving at about 600 kilometers per second. The solar wind affects the magnetosphere of the Earth (the area inside which the Earth's magnetic field is dominant). The particles also enter the radiation zones around the Earth, the Van Allen belts. At times of great activity there is interference with radio reception because of disturbances in the ionosphere (the region of the Earth's upper air in which radio waves are reflected). These effects are most noticeable when the Sun is near the maximum of its cycle. Apart from visible light and heat, the Sun is also a source of radio waves, ultraviolet radiation and X-rays.

Flares and prominences

Simple telescopic observations of thc Sun show only the bright surface, or photosphere, and features such as spots and faculae. The photosphere is surrounded by the chromosphere, or "colored sphere" – so called because it consists of glowing red hydrogen gas. In the chromosphere violent explosions called flares occur; these emit charged particles and short-wave radiation. Less violent are prominences,

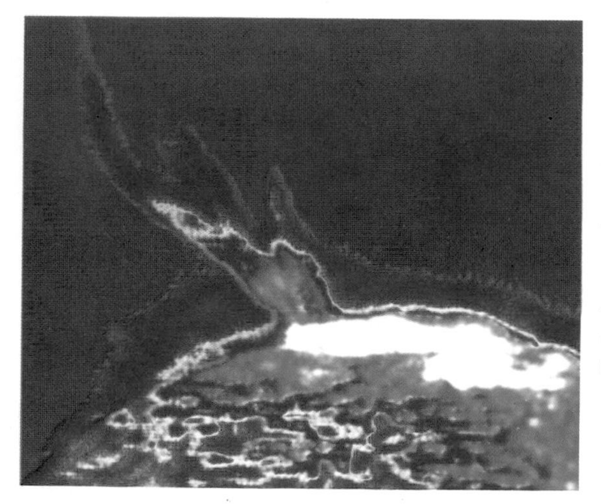

Solar flares are sudden short-lived outbreaks in the Sun's chromosphere and are associated with sunspots. This false-color picture of a huge flare was taken from the Skylab space station. The image was made with very short-wavelength ultraviolet light produced by the eruption.

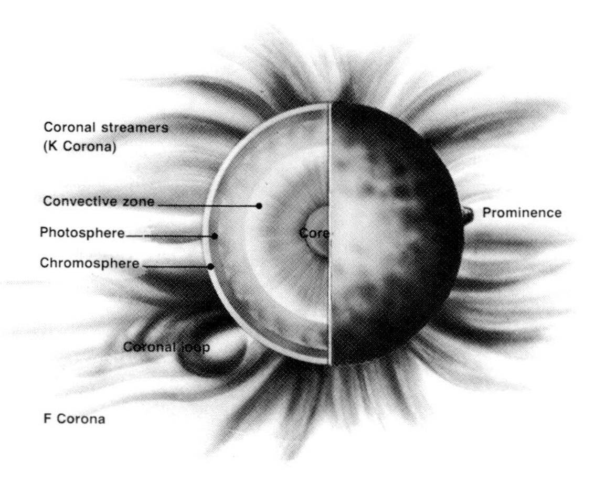

Energy is generated in the Sun's core by the transformation of hydrogen into helium. In the inner Sun the energy travels as radiation; in the convective zone the circulation of hot gases carries the energy to the surface. The white-hot visible disk of the Sun is the photosphere; the chromosphere and corona are invisible to the naked eye, except during total eclipses.

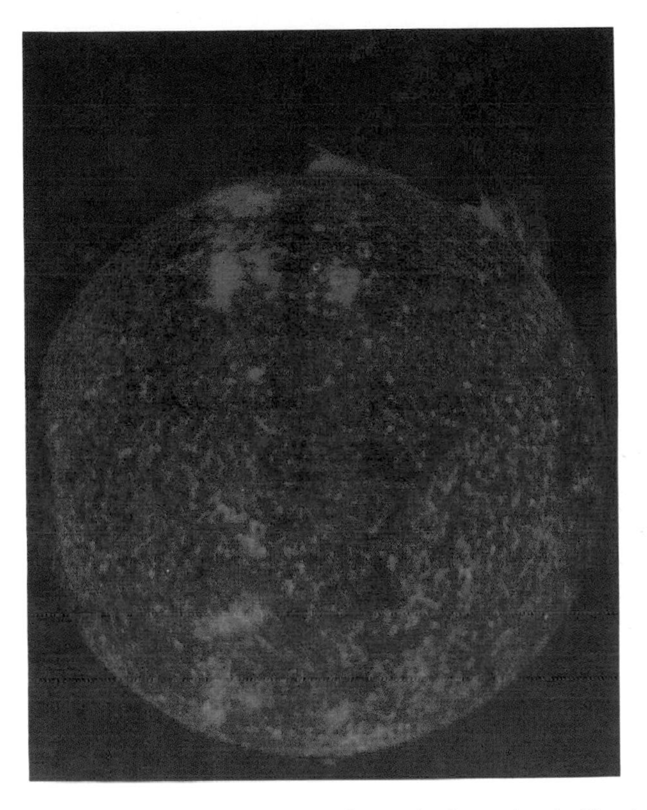

A solar prominence of incandescent hydrogen arches 400,000 km above the Sun – a height equal to the Moon's distance from the Earth.

outbursts of glowing hydrogen gas. "Quiescent" prominences are relatively stable and may persist for days or months before dispersing. "Eruptive" prominences are more violent and can reach to at least 2 million kilometers from the Sun.

Beyond the chromosphere comes the inner K corona, made of extremely hot but tenuous gas. It glows with a pearly white light. Both the chromosphere and the K corona become visible when the dazzling light of the photosphere is cut off during a total eclipse. At other times they can be observed with the aid of the spectroheliograph; this is an instrument based on the principle of the spectroscope, with which the Sun can be observed in the light of a single element – usually hydrogen or calcium. The outer F corona extends far into space and is difficult to observe; it has no well-defined boundary.

5

THE PLANETS

Nine major planets circle the Sun, in company with thousands of minor planets, or asteroids. The inner planets are rocky worlds; the outer ones are gas giants, apart from tiny Pluto at the very edge of the system.

Mercury

Mercury is the smallest of the principal planets (with the exception of Pluto) and, although it can become brighter than any star, it is not very easy to see with the naked eye. It is never visible against a dark background, and always keeps close to the Sun in the sky, so that, without optical aid, it is only seen low in the west after sunset or low in the east before sunrise. Telescopically, all that can be made out is the

The phases of Mercury may be followed with a small telescope. When "full" (1) the planet is on the far side of the Sun, and when "new" (2) it is invisible (except during transit). So it is best seen as a crescent (3), at half-phase (4), or when gibbous (5). It is unwise to search for it with the Sun above the horizon, unless your telescope has accurate setting circles.

Mercury is seen in transit when it passes directly between the Sun and the Earth. It appears as a black disk against the brilliant solar surface; at these times it is obviously much darker than a sunspot. During transits Mercury cannot be seen with the naked eye and the best way to observe it is by projection with a telescope. The next transit is due on 15 November 1999.

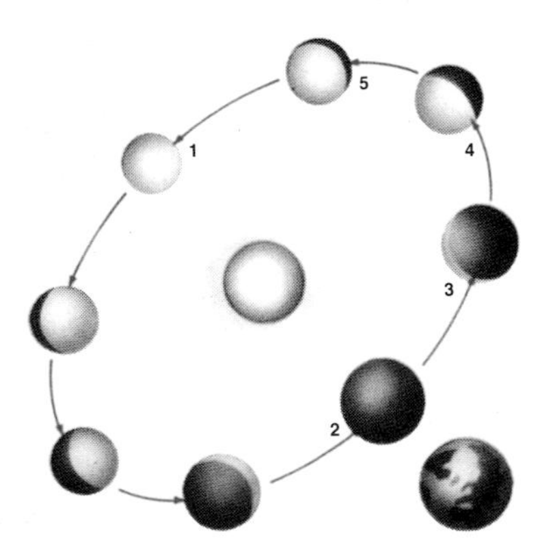

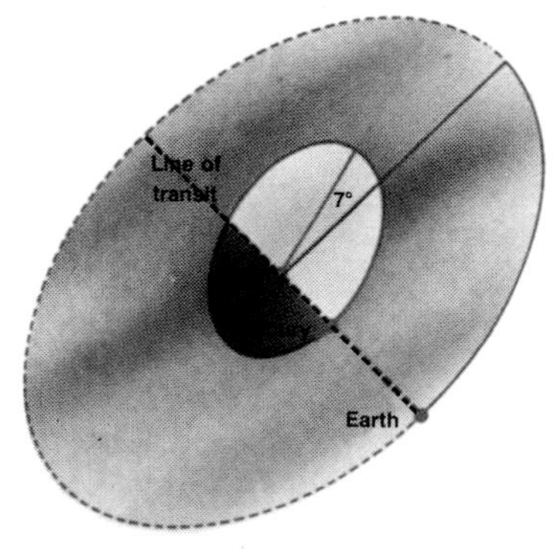

characteristic phase. Before the flight of Mariner 10 in 1974 little was known of its surface.

Mariner 10, which made three active passes of Mercury during 1974 and 1975, showed that the surface is heavily cratered, looking superficially very much like that of the Moon. There are mountains, valleys, scarps and ridges. There are also depressed basins, of which one (the Caloris Basin) is 1,300 kilometers in diameter. Mariner 10 photographed only part of the surface but there is no reason to doubt that the remainder is essentially similar.

Mercury has only an extremely tenuous atmosphere, corresponding to a laboratory vacuum. There is a relatively large, heavy core and a weak magnetic field. Life there is certainly out of the question.

The cratered surface of Mercury (right) presents a decidedly "lunar" landscape in this mosaic of images from the Mariner 10 spacecraft. The craters are probably formed in the same way as those on the Moon.

Mercury rotates slowly on its axis (below). It used to be thought that the rotation period was the same as the planet's "year" (88 Earth days). In this case one half would be in permanent daylight and the other in permanent night. In fact its period is 58.6 days, two-thirds of Mercury's "year." When best placed for observation, the same face is always turned toward us.

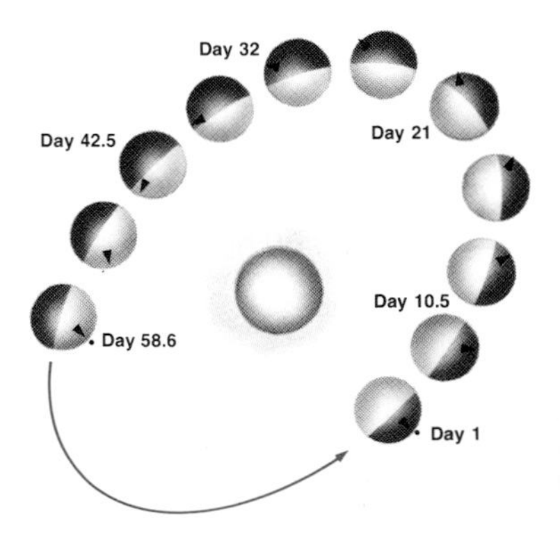

Venus

Venus is the most brilliant of the planets, but through a telescope little detail can be seen apart from the phase; the surface is permanently hidden by layers of cloud.

Before the space age very little was known about the surface conditions on Venus, but in 1962 Mariner 2 bypassed the planet and showed that the surface temperature is very high – thereby disproving an older theory that there were oceans on the planet. Since then probes have landed on Venus; the soft-landers Venera 9 and Venera 10 even sent back one picture each before being put out of action by the intensely hostile conditions.

When Venus (left) is closest to the Earth, at inferior conjunction, it is new, with the dark side turned toward Earth. As the phase increases the diameter appears to shrink. When full (at superior conjunction) Venus is on the far side of the Sun. Greatest brilliancy occurs when Venus is a crescent, and it is then at its best for the naked-eye observer.

The swirling winds of Venus (left) are revealed in a false-color image from the Mariner 10 spacecraft. The picture was made with ultraviolet light: in ordinary light the disk appears featureless. These clouds are largely composed of sulfuric acid droplets.

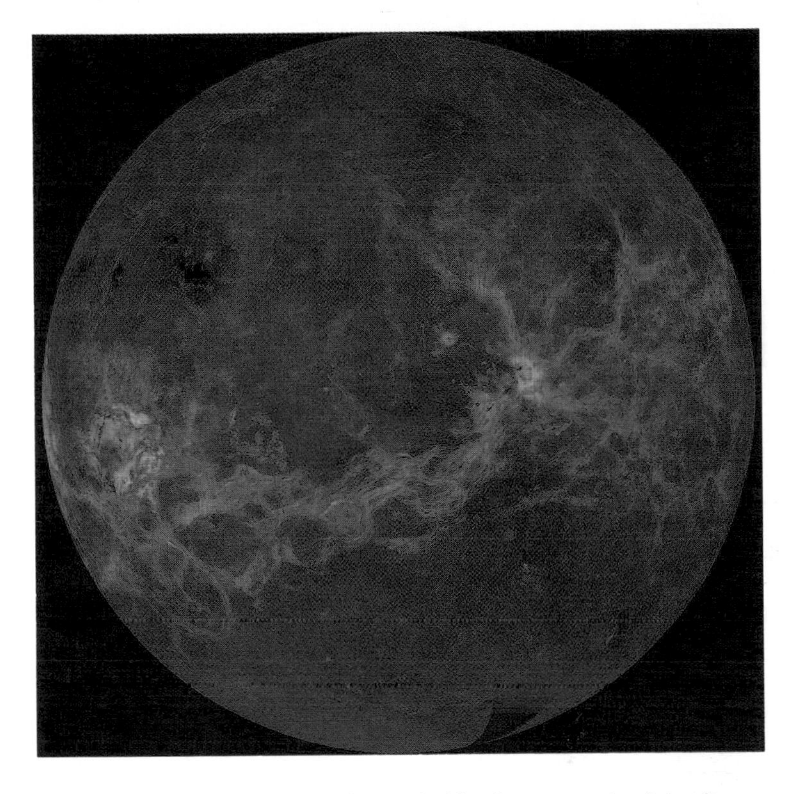

The surface of Venus, stripped of clouds, is seen in a radar mosaic from the Magellan space probe. Aphrodite Terra is the bright band at the equator.

By now we have been able to obtain accurate maps of Venus, thanks to the space probes, the latest of which, Magellan, sent back magnificent radar images. Venus is a world of plains, lowlands and two highland areas, Ishtar and Aphrodite. A huge rolling plain covers 65 per cent of the surface; the highest peaks, the Maxwell Mountains adjoining Ishtar Terra, rise to over 8 kilometers above the adjacent surface. There are craters, valleys and extensive lava flows. One smaller highland area, Beta Regio, includes two shield volcanoes, Rhea Mons and Theia Mons, which are probably active. The surface seems to have been shaped by the same sort of tectonic forces that power the drifting of the continents on the Earth.

It is possible that at an early stage in the story of the Solar System, Venus and the Earth began to evolve along similar lines, with oceans and similar atmospheres. However, the Sun became steadily more

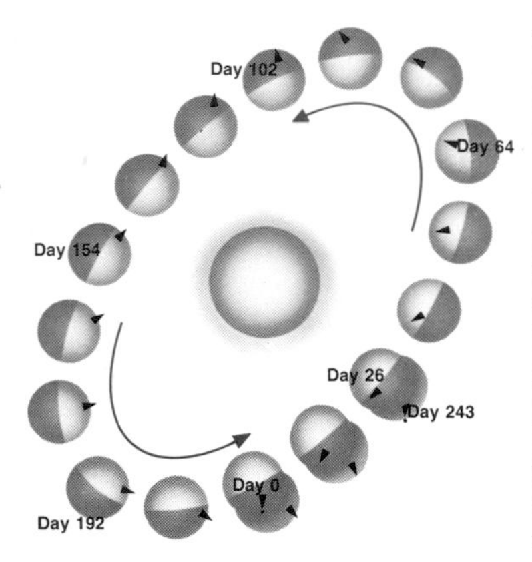

PHENOMENA OF VENUS, 1997–2003	
Eastern elongation	***Inferior conjunction***
1997 Nov 6	1998 Jan 16
1999 Jun 11	1999 Aug 20
2001 Jan 17	2001 Mar 30
Western elongation	***Superior conjunction***
1998 Mar 27	1998 Oct 30
1999 Oct 30	2000 Jun 11
2001 Jun 8	2002 Jan 14
2003 Jan 11	2003 Aug 18

Venus is unique (above) in having an axial rotation period longer than the planet's period of revolution around the Sun. The Venus "year" is 224.7 Earth days; the axial rotation period is 243 Earth days. Furthermore, Venus rotates in a retrograde or backward direction: east to west instead of west to east. To an observer on Venus, the interval between one sunrise and the next would be 118 Earth days.

The volcano Sapas Mons (below) is about 400 km across and 1.5 km high. Its sides are covered with overlapping lava flows. Smooth areas are dark in this radar image. A lava-covered impact crater lies above it and to the right.

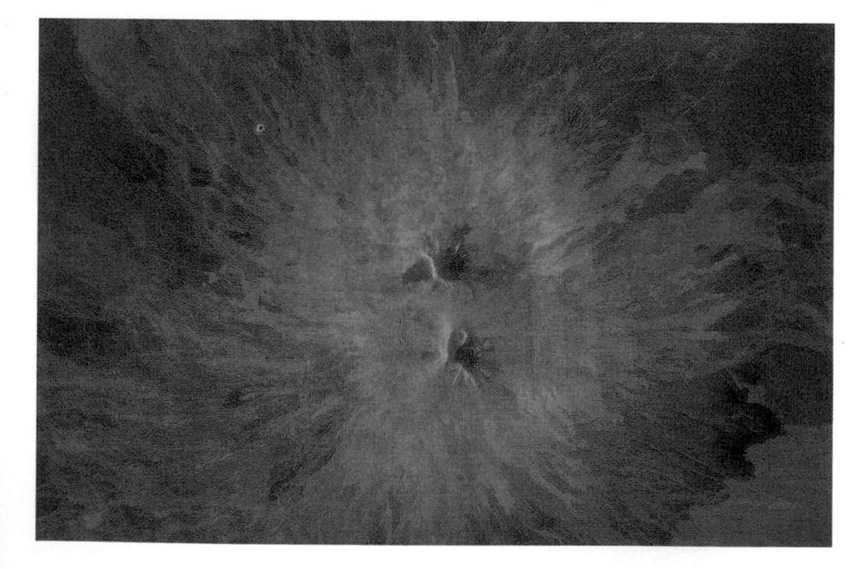

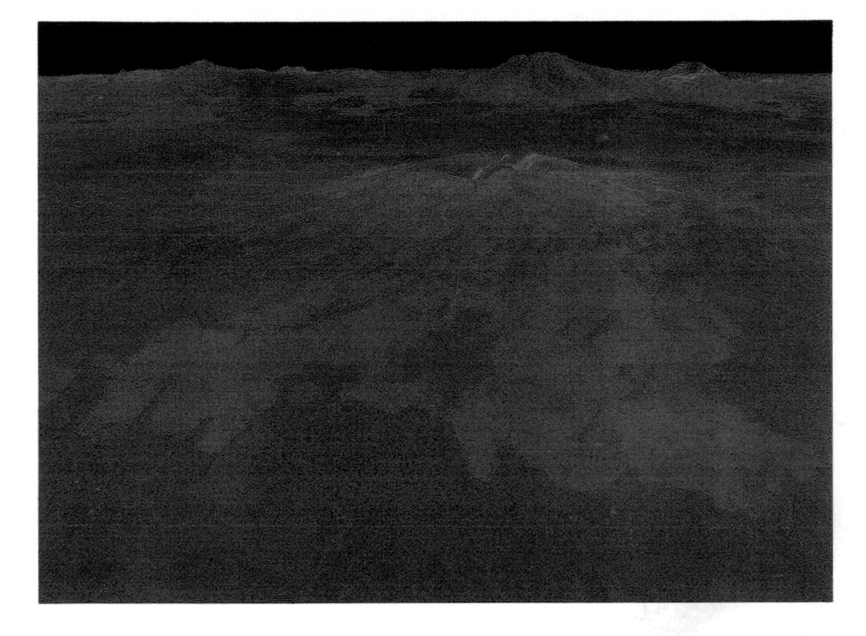

The landscape of Venus stretches into the distance. In the foreground is Sapas Mons; in the background is the largest volcano on Venus, Maat Mons.

luminous, and Venus, so much closer to the Sun than we are, suffered a runaway greenhouse effect; the oceans boiled away, the carbonates were driven out of the rocks, and in a short time, cosmically speaking, Venus changed into the furnace-like environment of today. Life there is certainly out of the question. The surface temperature is almost 500 degrees centigrade; the atmosphere is mainly carbon dioxide with a surface pressure 90 times that of the Earth's air at sea level, and the clouds are rich in sulfuric acid. Venus, far from being the pleasant, welcoming world so often envisaged in science fiction, approximates much more closely to the conventional idea of hell.

But little sign of the extreme conditions beneath the clouds can be seen by the Earthbound observer. It has to be admitted that Venus is a disappointing object telescopically. No well-marked surface details can be seen, and the only visible markings are hazy cloud shadings. These do not even reveal the fact, discovered by the space probes, that the upper clouds rotate in only four days. All that can really be seen is the characteristic phase. Transits of Venus across the face of the Sun are rare. The last occurred in 1882, and the next is not due until 2004.

Mars

Mars, the first planet beyond the Earth in the Solar System, is easy to recognize because of the strong red color which led to it being named in honor of the God of War. At its brightest it outshines every other planet apart from Venus. However, at minimum it may fall to the 2nd magnitude and is then easy to mistake for a star.

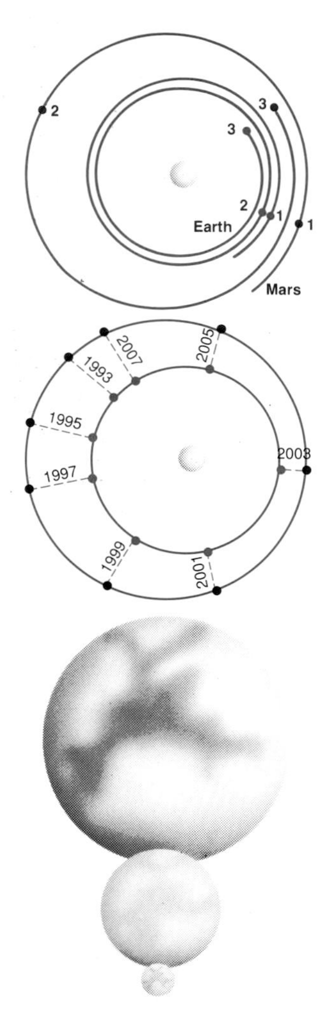

OPPOSITIONS OF MARS, 1999–2005

Date	Min. distance (million km)	Mag.	Constell'n
1999 Apr 24	87	−1.5	Virgo
2001 Jun 13	67	−2.1	Sagittarius
2003 Aug 28	56	−2.7	Capricornus
2005 Nov 7	43	−2.1	Aries

An opposition of Earth and Mars (top left) is shown at (1) in the diagram. A year later the Earth has returned to position (1) while Mars has traveled just over halfway around its orbit (2). The Earth "catches it up" at the next opposition (3), roughly 780 days later. Opposition distance may vary from 56 million km to 101 million km (center). The apparent diameter of Mars (bottom left) ranges between 25″.7 and 3″.5.

Clouds of ice crystals (below) fill the canyons of Noctis Labyrinthus, a huge system of valleys that lies in a high plateau region of Mars.

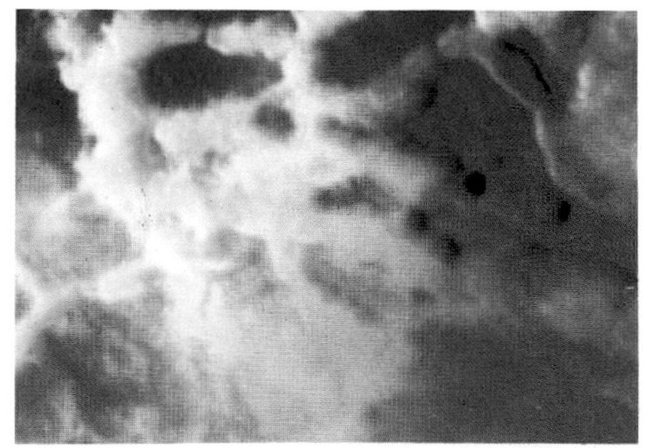

Mars at crescent phase, a sight never seen from the Earth. This spectacular picture was sent back from the Viking 2 spacecraft. The volcano Ascraeus Mons is conspicuous in the upper part of the daylight area. Closer views disclose the extensive cratering of the surface – a discovery that was made by the earliest space probes and that was completely unforeseen by Earth-based astronomers.

Mars in general is not easy to study with a small telescope, but under good conditions, adequate instruments show dark, well-defined features and white polar caps. The dark areas were once thought to be old seabeds filled with vegetation but this attractive idea has now been disproved and as yet there is no firm evidence of any life on Mars. Some of the dark regions, such as Syrtis Major, are elevated plateaux.

The Martian "day" is about half an hour longer than that of the Earth. The apparent drift of features across the disk is therefore obvious even over a short period of observation. Generally the atmosphere is transparent, except for occasional dust storms which may hide the surface features.

The first reasonably accurate maps of Mars were drawn in the 19th century. In 1877 G. V. Schiaparelli, from Milan, compiled a chart and named the main features. These names have been generally retained, although recently modified in view of the space-probe discoveries. Schiaparelli also drew a network of fine, sharp lines which

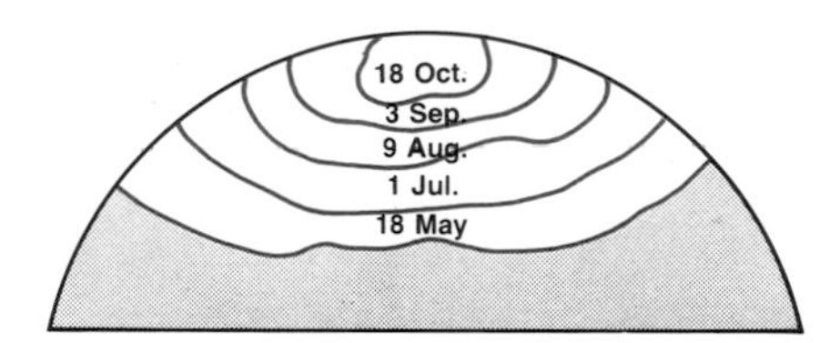

The southern polar cap (left) shrinks dramatically during the Martian summer. Its changes in the course of 1956 are shown in the diagram, based on observations made with a 25 cm reflector.

The axial inclination of Mars (right) is 23° 58′, almost identical to the Earth's 23° 27′. The seasons are of the same type but of much longer duration.

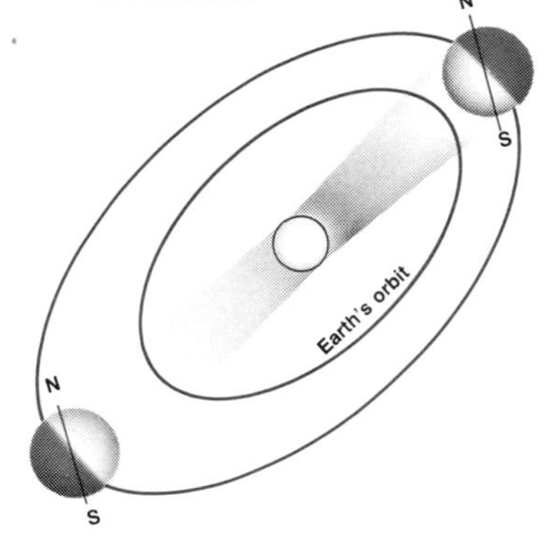

Mars has been mapped (below) in stunning detail from spacecraft images. Terms used in names are: planitia – low-lying plain; planum – upland plain; mons – mountain or volcano; patera – volcano; fossa – groove.

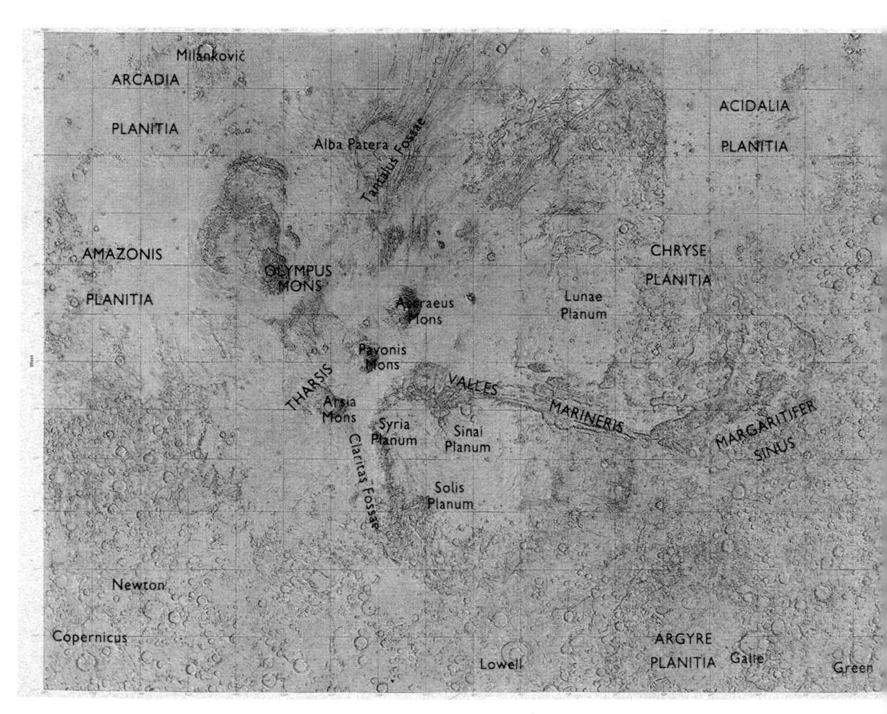

became known as canals and were regarded by some astronomers as being of artificial origin. It is now known that the canal network does not exist in any form and that it was purely illusory.

The most prominent dark markings are Syrtis Major, a V-shaped feature close to the Martian equator, and Acidalia Planitia in the north. Both of these are very easy to see with a small telescope when Mars is suitably placed near opposition. South of Syrtis Major is Hellas, a circular basin which can at times be so bright that it is easy to confuse with a polar cap. Little further detail can be seen when Mars is well away from opposition.

The thin atmosphere is made up chiefly of carbon dioxide and the ground pressure is below 10 millibars, so that no surface water can exist. Clouds are frequent, but extensive dust storms are less common.

Present knowledge of Mars has been obtained mainly by means of space probes. Mariner 9, which reached the neighborhood of the planet in November 1971 and was put into a closed path around Mars, sent back thousands of high-quality pictures, showing craters, valleys and towering volcanoes. The highest of these is Olympus Mons, the highest known mountain in the Solar System. Its summit lies

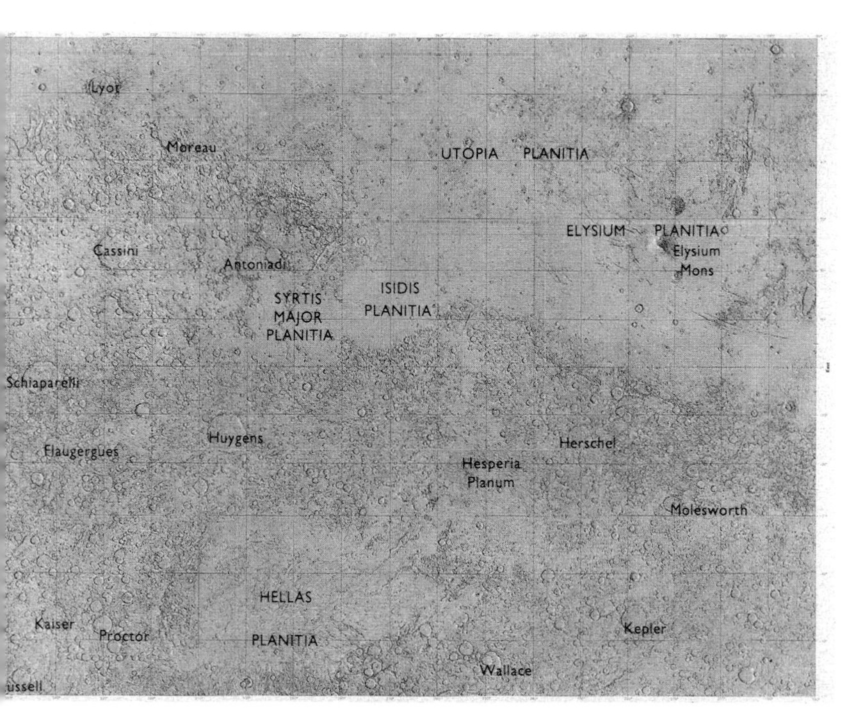

SATELLITES OF MARS				
	Mean dist. from center of Mars (km)	***Diameter (km)***	***Magnitude***	***Orbital period (days)***
Phobos	9,270	17 × 14 × 11	11.6	0.32
Deimos	23,400	9 × 7 × 6	12.8	1.26

Both the satellites of Mars are very small, and are probably captured asteroids. Both were imaged from the Vikings; they proved to be dark, irregular in shape, and cratered. The largest crater on Phobos, named Stickney, is huge in relation to the satellite: 5 km across.

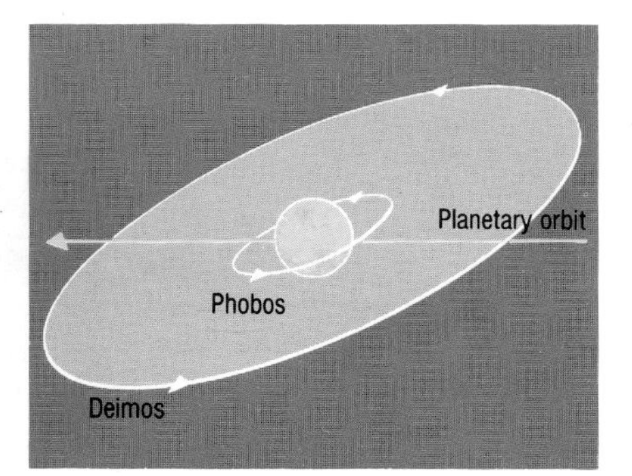

The two moons of Mars orbit relatively close to the parent body. Since Phobos takes less than a Martian day to go once round the planet, it would be seen from the surface to rise in the west and set in the east 5.5 hours later. Deimos, by contrast, orbits in slightly more than the Martian day, and would seem to move extremely slowly through the sky.

25 kilometers above the general level of the Martian surface, and is topped by a multiringed caldera 65 kilometers in diameter.

Another spectacular feature is the system of canyons named Valles Marineris, after the Mariner probes that discovered them. Many of these valleys dwarf the Grand Canyon on Earth: they reach a maximum depth of 7 kilometers.

In 1976 two Viking probes made soft landings on the surface, Viking 1 in Chryse and Viking 2 in Utopia. Samples from the surface were analysed in a search for life, but no trace of organic matter was found. Abundant evidence of past water activity was confirmed when the orbiting sections of the Vikings picked up the existence of deep canyons and features which are almost certainly dry river beds. In 1996 it was claimed that a meteorite found in Antarctica came from Mars, and was blasted off the Martian surface by the impact of a comet or asteroid long ago; it was also claimed that this meteorite contained indications of past very primitive life forms. This is certainly a fascinating possibility, though as yet the evidence is very far from conclusive. We will no doubt find out when it becomes possible to use an unmanned probe to bring back Martian samples, as should become practicable within the next few years.

The asteroids

The minor planets, or asteroids, are dwarf worlds. Only one (Ceres) is as much as 1,000 kilometers in diameter and only Vesta is ever visible to the naked eye. Most of the asteroids move in the region between the orbits of Mars and Jupiter. It has been suggested that they are the fragments either of a former planet that broke up or of a potential planet that failed to form properly. Most astronomers now believe that the asteroids never formed part of a large body; their combined mass is much less than that of the Moon.

The first four asteroids (Ceres, Pallas, Juno and Vesta) were discovered between 1801 and 1807. Today several thousand asteroids are known but most are very faint. Some have orbits that swing them away from the main group. Thus Icarus (a mere 1 kilometer across) moves closer to the Sun than the orbit of Mercury, while the so-called Trojans move in the same orbit as Jupiter. Chiron, discovered in 1977, moves mainly between the orbits of Saturn and Uranus, and even more remote asteroidal bodies have now been found.

Another asteroid with a highly eccentric orbit is Hidalgo, which travels inside the path of Mars at its closest to the Sun, and out to the orbit of Saturn at its furthest. Some asteroids are "Earth-grazers": in 1991 a tiny one passed by at less than half the distance of the Moon.

Many asteroids are discovered in time-exposure photographs. When the telescope carrying the camera follows the sky's movement, the stars in the photograph appear as circular spots of light. Asteroids (and also comets) move appreciably in relation to the background of stars, and appear as short, easily noticeable trails of light.

The main asteroid zone lies between the orbits of Mars and Jupiter. A number of gaps in the zone are caused by the cumulative perturbations from Jupiter's gravitational pull. These are known as Kirkwood gaps, named after their discoverer. Some satellites move in orbits outside the main zone. For example, two groups, the Trojan asteroids, share the orbit of Jupiter: one moves 60° ahead of the planet, the other an equal distance behind.

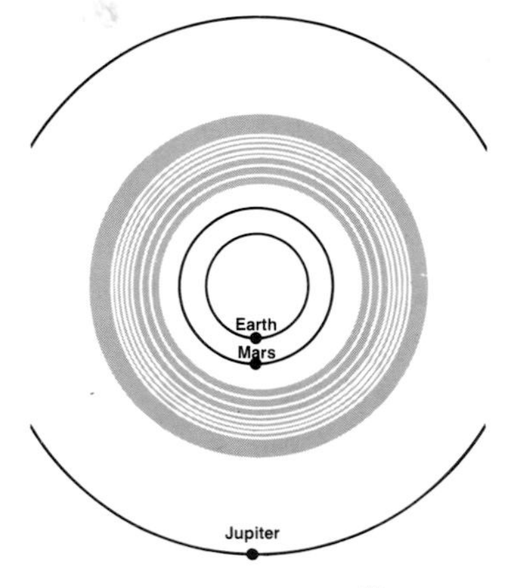

Jupiter

Jupiter comes to opposition each year and is a brilliant object. Its yellowish disk is markedly flattened because of its rapid axial rotation, and there are dark belts, bright zones, spots, wisps and festoons to be seen over the gaseous surface, which is always changing.

The belts are regions where material is descending; the bright zones represent upcurrents. There are two prominent belts, one to either side of the equator. Of other features, the most interesting is the Great Red Spot, now known to be a kind of whirling storm – a phenomenon of Jupiter's meteorology, which is generally very violent. It is thought that Jupiter has a small solid core, overlain by liquid hydrogen, above which is the gaseous "atmosphere." A flat ring of particles also exists, but is too thin and too faint to be seen from Earth.

Jupiter has been surveyed by several spacecraft, notably the Voyagers in 1979. In 1995 the Galileo probe reached Jupiter; the entry section plunged into the clouds, sending back surprising information, notably the fact that Jupiter's atmosphere contains much less water than had been expected. The four large satellites were surveyed. Callisto and Ganymede are icy and cratered; Ganymede has been found to have a magnetic field. Europa is also icy, but has a surface which is almost smooth; there are few craters, and there are many shallow cracks. Io was the real surprise. It is red, and has a sulfurous surface with volcanoes which are erupting all the time. The volcanic vents are hot, though the general surface is bitterly cold; moreover, the satellite is connected to Jupiter by a strong electric current. It moves in the midst of the strong radiation zones associated with Jupiter, and must be just about the most dangerous world in the entire Solar System.

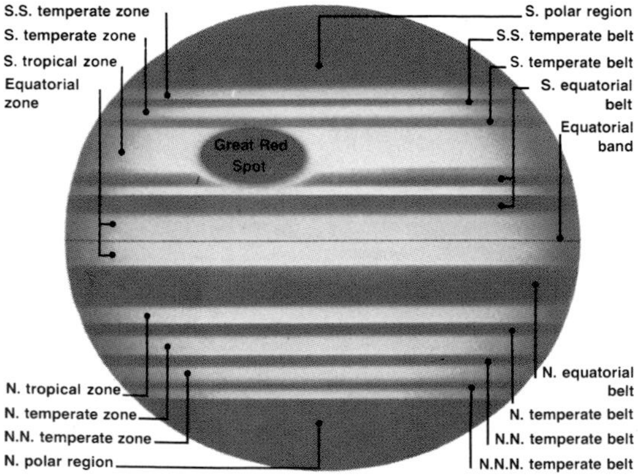

The nomenclature of Jupiter's surface. The belts (dark areas) and the Great Red Spot do not change appreciably in latitude. The rotation period of the equatorial zone is about 5 minutes shorter than that of the rest of the planet. Various features, including the Great Red Spot, drift in longitude over the surface.

The Great Red Spot is an enormous storm that has been raging essentially unchanged since it was first seen through the telescopes of 17th-century observers. It is about 25,000 km long.

OPPOSITIONS OF JUPITER, 1998–2004		
Date	***Magnitude***	***Constellation***
1998 Sep 16	−2.5	Aquarius
1999 Oct 23	−2.5	Pisces
2000 Nov 28	−2.4	Taurus
2002 Jan 1	−2.3	Gemini
2003 Feb 2	−2.0	Cancer
2004 Mar 4	−2.0	Leo

Jupiter's four largest satellites were observed telescopically by Galileo in 1610, and since then they have been known as the Galileans. All are planet-sized and all can be seen with any optical aid; a few keen-sighted people can even glimpse them with the naked eye. The remaining 12 satellites are extremely small and faint. The first probe to bypass Jupiter was Pioneer 10, in 1973; Pioneer 11 followed in 1974, and Voyagers 1 and 2 in 1979. However, the most elaborate Jovian probe so far has been Galileo, of 1995–6.

When viewed from Earth all four Galileans appear to keep almost in a straight line, as they lie in the plane of Jupiter's equator. But they change their positions in a few hours, since they orbit the planet so rapidly. A Galilean passing in front of Jupiter may appear in transit, with or without its shadow. It may also be occulted (hidden) by the disk of Jupiter, or it may be eclipsed as it enters the planet's shadow.

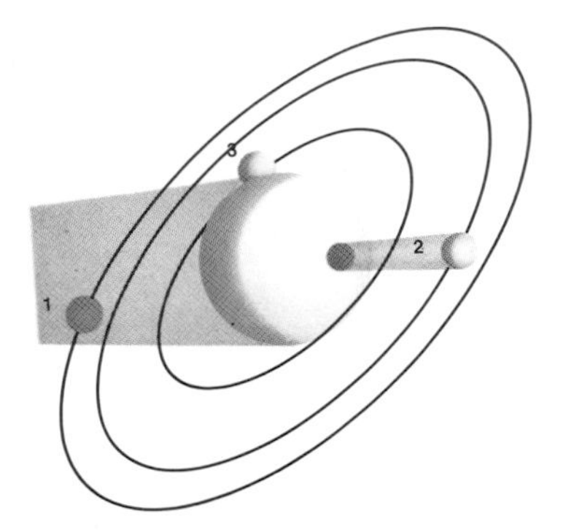

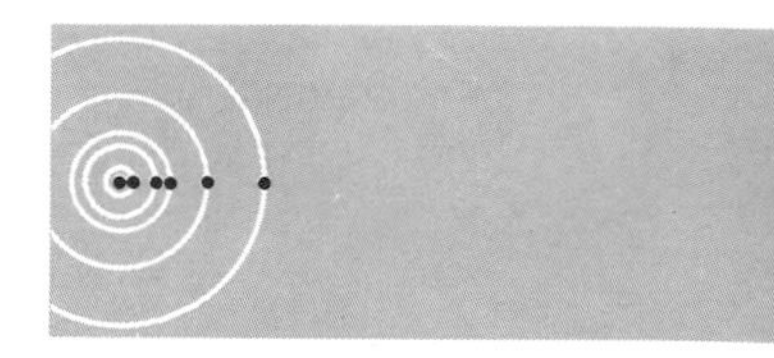

Eclipses, occultations and transits (left) of the Galilean satellites. A satellite can be eclipsed by Jupiter's shadow (1), can appear in transit, together with its shadow (2), or be occulted by Jupiter (3).

The reddish moon Io stands out against the giant globe of Jupiter in this Voyager 1 image. Io is the innermost of the four Galilean satellites and is stretched and squeezed by the planet's huge gravity. This stress creates the internal heating that drives the tiny world's volcanoes.

Jupiter's satellites (left) fall into several groups. The outer moons are asteroidal in nature, and the four outermost have retrograde motion.

A volcanic plume (right) arches 300 km above the horizon of Io. The volcano, called Pele, throws out material consisting of sulfur and sulfur compounds. This image was made in March 1979; by July this eruption had stopped.

SATELLITES OF JUPITER				
Name	*Mean distance from Jupiter (km)*	*Orbital period (days)*	*Diameter (km)*	*Magnitude*
Metis	127,960	0.295	40	17.4
Adrastea	128,980	0.298	26 × 20 × 16	18.9
Amalthea	181,300	0.498	262 × 146 × 143	14.1
Thebe	221,900	0.675	68 × 56	15.5
Io	421,600	1.769	3660 × 3637 × 3631	5.0
Europa	670,900	3.551	3130	5.3
Ganymede	1,070,000	7.155	5268	4.6
Callisto	1,880,000	16.689	4806	5.6
Leda	11,094,000	238.7	10	20.2
Himalia	11,480,000	250.6	170	14.8
Lysithea	11,720,000	259.2	24	18.4
Elara	11,737,000	259.7	80	16.7
Ananke	21,200,000	631*	20	18.9
Carme	22,600,000	692*	30	18.0
Pasiphaë	23,500,000	735*	36	17.7
Sinope	23,700,000	758*	28	18.3
(* = retrograde)				

Saturn

There can be little doubt that, when seen through a telescope, Saturn is the most beautiful object in the entire sky. Even a small telescope is adequate to show the ring system when it is well placed, and there is nothing else like it. The globe is essentially of the same type as that of Jupiter, though it is less active; there are cloud belts, and occasional spots – such as the bright white spots seen in 1933 and 1990, which were, however, temporary. The globe, like that of Jupiter, is clearly flattened, because of the quick rotation. Here, too, the equatorial regions have a faster rotation period than those of higher latitudes.

The rings are made up of icy particles, spinning round Saturn in the manner of dwarf satellites. The main system consists of three rings: two bright (A and B) and a third semitransparent (C, or the Crêpe Ring). Between Rings A and B there is a gap, known as Cassini's Division, and there is a narrower gap, Encke's Division, in Ring A.

Much of our knowledge of Saturn has come from the Voyager probes, which bypassed the planet in 1980 and 1981 respectively. It was found that the rings are more complicated than had been thought, with hundreds of ringlets and narrow divisions. Ring B also contains curious "spokes," probably consisting of particles elevated from the ring-plane by magnetic forces. New rings were found outside the main

The rings of Saturn are very broad, measuring over 270,000 km overall.

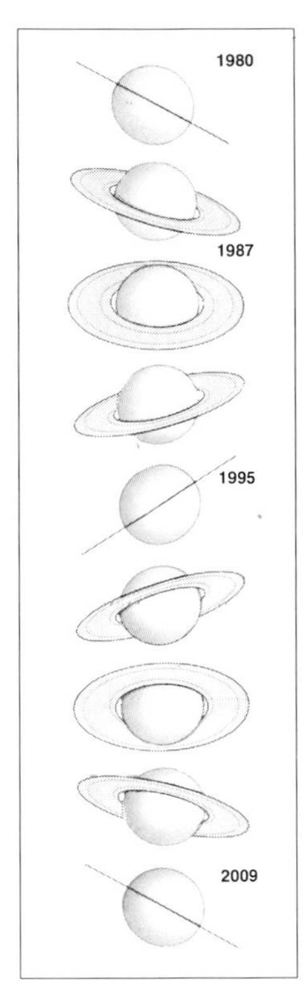

Changing aspects of Saturn's rings (above). Roughly every 15 years the rings appear edge-on to us, becoming almost invisible. Between these times they seem to "open out," and then "close up" again. At their maximum they present a truly magnificent spectacle to the telescopic observer.

This plan view of the ring system (right), as it appears through telescopes on Earth, is drawn to scale. The divisions in the rings are caused by the gravity of Saturn's inner satellites.

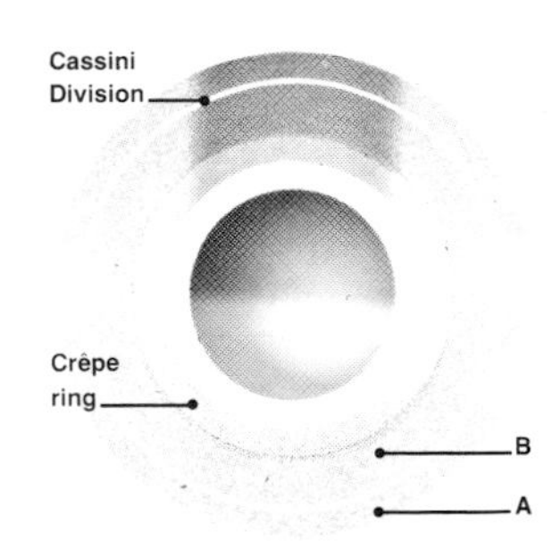

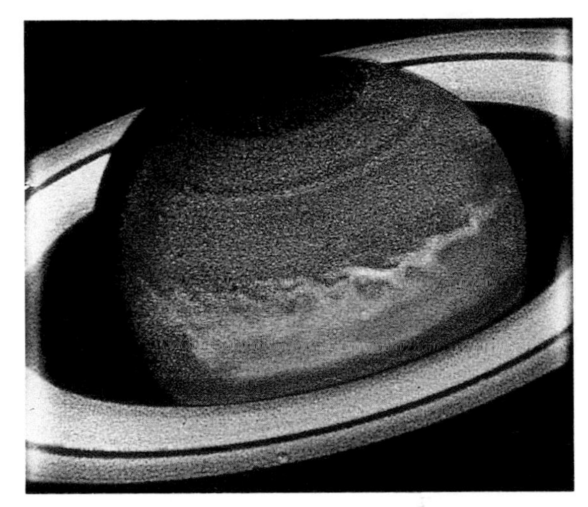

White clouds in Saturn's atmosphere (above) are the remains of a storm that began as a white spot in 1990 and then dispersed.

OPPOSITIONS OF SATURN, 1997–2002		
Date	***Magnitude***	***Constellation***
1997 Oct 10	+0.4	Pisces
1998 Oct 23	+0.2	Pisces
1999 Nov 6	0.0	Aries
2000 Nov 19	−0.1	Aries
2001 Dec 3	−0.3	Taurus
2002 Dec 7	−0.3	Taurus

system, and there are narrow rings even within the main divisions. It was confirmed that the rings are very thin – less than a kilometer thick – so that when they are edge-on to the Earth, as in 1995, they almost disappear.

Nine satellites were known before the Voyager passes. Of these the largest is Titan, which was known to have an atmosphere and to be comparable in size with the planet Mercury; of the rest Iapetus and Rhea were fairly easy objects, and Dione and Tethys could be seen with a moderate telescope, while the rest (Mimas, Enceladus, Hyperion and Phoebe) were fainter. All these were studied from the Voyagers. Most are icy and cratered; remarkably, Iapetus has one bright and one dark hemisphere. Titan's atmosphere is mainly composed of nitrogen, with a ground density 1.5 times that of the Earth's air at sea level. Phoebe

Saturn's rings (left), seen from Voyager 2, display "grooves" and also narrow rings in the Cassini Division.

Titan, Saturn's largest moon (below), has an atmosphere of nitrogen and methane, with thick orange clouds.

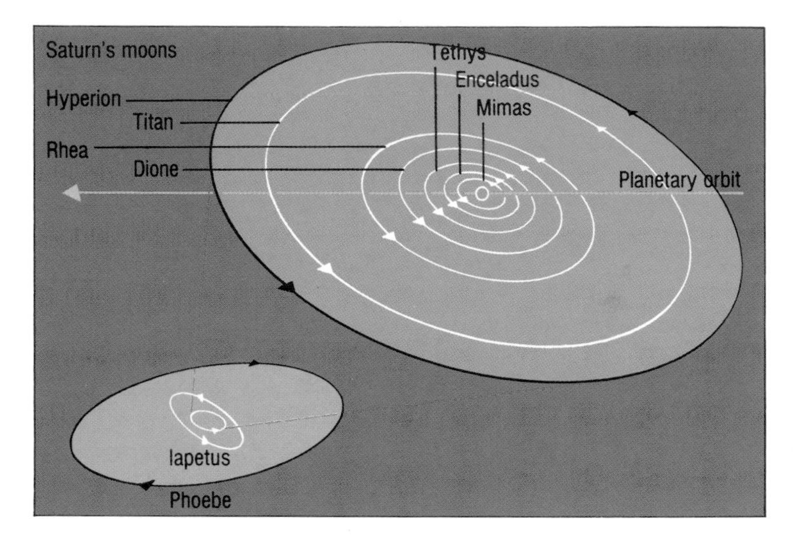

Orbits of Saturn's satellites (above). The paths of the nine main satellites, known before the Voyager missions, are shown here. Iapetus and Phoebe are so far from Saturn that they have to be shown separately; moreover, Phoebe has retrograde motion, and is probably a captured body rather than a bona-fide satellite.

Dione is heavily cratered (left), like several other satellites of Saturn. These features are mostly due to intense meteoric bombardment in the distant past. But there is also evidence of past geological activity.

SATELLITES OF SATURN				
Name	***Mean dist. from Saturn (km)***	***Orbital period (days)***	***Diameter (km)***	***Magnitude***
Pan	133,600	0.57	12	21
Atlas	137,670	0.602	37 × 34 × 27	18.1
Prometheus	139,350	0.613	148 × 100 × 68	16.5
Pandora	141,700	0.629	110 × 88 × 62	16.3
Epimetheus	151,420	0.694	194 × 190 × 154	15.5
Janus	151,470	0.695	138 × 110 × 110	14.5
Mimas	185,540	0.942	421 × 395 × 385	12.9
Enceladus	238,040	1.370	512× 495 × 488	11.8
Tethys	294,670	1.888	1,046	10.3
Telesto	294,670	1.888	30 × 25 × 15	19.0
Calypso	294,670	1.888	30 × 16 × 16	18.5
Helene	377,410	2.737	35	18.5
Dione	377,420	2.737	1,120	10.4
Rhea	527,040	4.518	1,528	9.7
Titan	1,221,860	15.945	5,150	8.4
Hyperion	1,481,100	21.277	360 × 280 × 225	14.2
Iapetus	3,651,300	79.331	1,436	8.6–11.5
Phoebe	12,954,000	550.4*	230 × 220 × 210	16.5
(* = retrograde)				

has retrograde motion, and is probably a captured asteroid rather than a bona-fide satellite. Several new satellites were discovered by the Voyager probes, but all are small. Some of these have extraordinary orbits. Telesto and Calypso, for example, share the orbit of Tethys, while Helene shares that of Dione. Prometheus orbits just inside the thin F-ring, while Pandora moves outside it: their gravitational influences "shepherd" the bodies that make up the ring.

Like Jupiter, Saturn is now thought to have a solid core, surrounded by layers of liquid hydrogen, which are in turn overlaid by the hydrogen-rich atmosphere. There is an appreciable magnetic field, though it is much weaker than Jupiter's, and Saturn, like Jupiter, is a source of radio emissions.

A new space mission to Saturn will undoubtedly extend our knowledge of the ringed planet greatly. Called Cassini after the great Italian astronomer, it may also make a controlled landing on Titan, which may well have oceans of methane or ethane on its surface. If all goes well, the mission should reach its target in the year 2004.

Uranus

Uranus, the first planet to be found in near-modern times, was discovered by William Herschel in the year 1781. It is just visible with the naked eye, and telescopes can show its pale, greenish disk. It had in fact been seen on a number of occasions before 1781, but had always been mistaken for a star.

Uranus proved to be a giant planet, though its diameter (51,118 kilometers) is less than half that of Saturn. It is 14.6 times as massive as the Earth. It is very remote – its mean distance from the Sun is 2,870 million kilometers – and it has an orbital period of 84 years.

No Earth-based telescope will show much on its disk, but it was soon found that the axial inclination amounts to 98 degrees; this is more than a right-angle, so that the rotation of Uranus is technically retrograde. The "seasons" are strange: during the course of one revolution around the Sun each pole will have a night lasting for 21 Earth years, and a "day" of equal length.

One space probe has bypassed Uranus: Voyager 2 in 1986. Little detail was seen on the disk, but the existence of a system of thin, dark

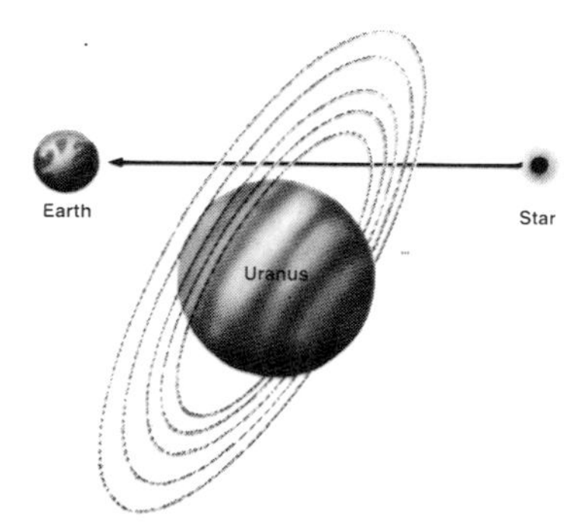

The Uranian ring system (above) was discovered when it occulted a star, causing it to "wink" repeatedly. There are nine known rings. They are too thin and dark to be observed directly by Earth-bound astronomers.

Uranus is a blue crescent (right) in this Voyager image. The blue is caused by methane in the planet's atmosphere.

SATELLITES OF URANUS				
Name	***Mean dist. from Uranus (km)***	***Orbital period (days)***	***Diameter (km)***	***Magnitude***
Cordelia	49,471	0.330	26	
Ophelia	53,796	0.372	30	
Bianca	59,173	0.433	42	
Cressida	61,777	0.463	62	
Desdemona	62,676	0.475	54	
Juliet	64,372	0.493	84	
Portia	66,085	0.513	108	
Rosalind	69,941	0.558	54	
Belinda	75,258	0.622	66	
Puck	86,000	0.762	154	
Miranda	129,400	1.414	472	16.5
Ariel	191,000	2.520	1,158	14.4
Umbriel	266,300	4.144	1,169	15.3
Titania	435,000	8.706	1,578	14.0
Oberon	583,500	13.463	1,523	14.2

rings, previously detected from Earth, was confirmed. They had revealed themselves to astronomers when they caused a star to wink before and after it was occulted (hidden) by the planet.

Voyager also showed that Uranus has a magnetic field, and is a radio source. Surprisingly, the magnetic axis is inclined to the axis of rotation by almost 60 degrees, and does not even pass through the center of the globe.

Uranus differs in composition from Jupiter and Saturn. Below the clouds there is a dense region in which gases are mixed with "ices" – that is, substances that would be frozen out at the low temperature of the clouds. It seems that water is the main constituent, and that water, ammonia and methane condense in that order to form thick layers of cloud, consisting of ice crystals. Methane forms the top layer; this absorbs red light, which explains the bluish-green color of Uranus. Alone among the giant planets, Uranus lacks a strong internal heat source.

Five satellites were known before Voyager: Miranda, Ariel, Umbriel, Titania and Oberon. All were imaged by the space probe and found to be icy and cratered; Miranda has a strangely varied surface. Voyager found 10 new satellites, all small and close to the planet.

Neptune

After Uranus had been under observation for some time, it was found to be moving in a somewhat irregular manner. Two mathematicians, Urbain Le Verrier in France and John Couch Adams in England, independently decided that the cause must be perturbation by a planet moving at a greater distance from the Sun. In 1846 Johann Galle and Heinrich D'Arrest at Berlin Observatory – working according to the position given by Le Verrier – identified the new planet, later named Neptune. It proved to be very slightly smaller than Uranus, but appreciably more massive. Binoculars show it as a bluish, starlike point of about magnitude 7.7.

Voyager 2 flew by Neptune in 1989 and found it to be a much more active world than Uranus. Its composition is similar, but it has a marked internal heat source, and it does not share Uranus' strong axial tilt. The magnetic axis is sharply inclined to the axis of rotation.

The main surface feature discovered by Voyager 2 was the Great Dark Spot, a huge whirling storm. Other spots were also seen, and Neptune was found to have very strong winds. The surface temperature is similar to that of Uranus, even though Neptune is so much further from the Sun.

Two satellites, Triton and Nereid, were known before the Voyager

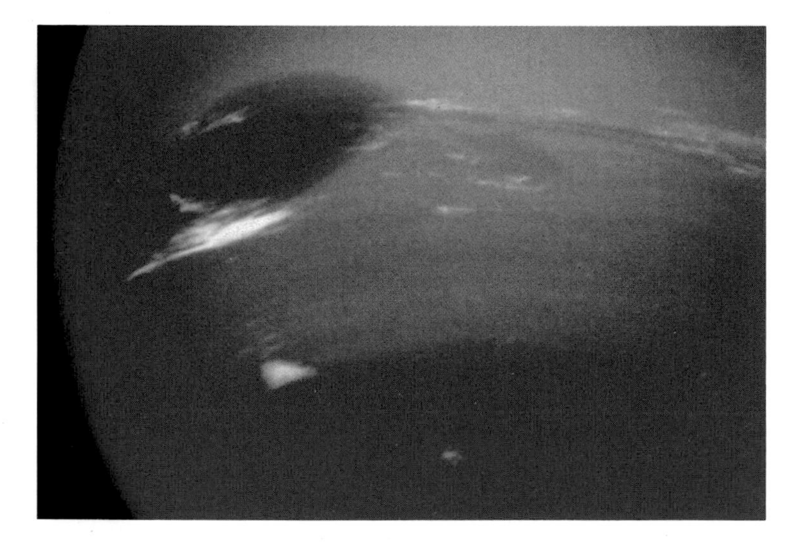

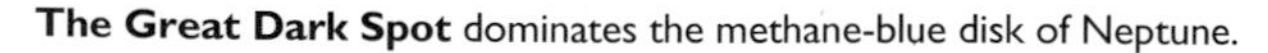

The Great Dark Spot dominates the methane-blue disk of Neptune.

Neptune looms in the sky of Triton in this artist's impression.

SATELLITES OF NEPTUNE

Name	*Mean dist. from Neptune (km)*	*Orbital period (days)*	*Diameter (km)*	*Magnitude*
Naiad	48,000	0.296	54	25
Thalassa	50,000	0.312	80	24
Despina	52,500	0.333	180	23
Galatea	62,000	0.429	150	23
Larissa	73,600	0.554	192	21
Proteus	117,600	1.121	416	20
Triton	354,800	5.877*	2,705	13.6
Nereid	1,345,500–9,688,500	360.16	240	18.7
(* = retrograde)				

mission. Triton is unique among large satellites in having retrograde motion; Voyager showed that it has pink nitrogen snow at its poles, and active nitrogen geysers. Nereid, not well imaged from Voyager, has a very eccentric orbit. The six new satellites found by Voyager are all small and close in. Neptune also has a system of thin, dark rings. They are not uniform but show "clumping" at some points, an effect that is produced by the gravitational influence of the inner satellites.

Pluto

Pluto was discovered by Clyde Tombaugh in 1930 from calculations made by Percival Lowell. It is very much of a problem: it is smaller and less massive than the Moon; its orbit is highly eccentric, bringing it at perihelion closer to the Sun than Neptune; and it has a companion, Charon, more than half its own diameter. It has a very thin atmosphere, and is probably composed of a mixture of rock and ice. It is much too faint – magnitude 14 – to be seen without a telescope, and it seems too small to be classed as a true planet. Lowell had inferred the existence of a "Planet X" because the gravitation of Neptune could explain only part of the perturbations of Uranus. But Pluto is so tiny that there may well be yet another planet beyond it, awaiting discovery.

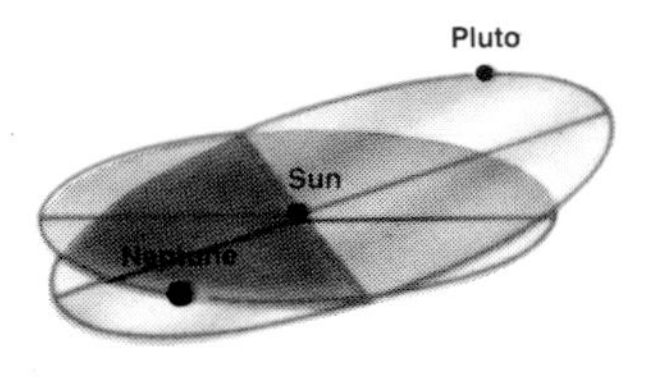

SATELLITE OF PLUTO (CHARON)	
Distance from Pluto	19,640 km
Orbital period	6.387 days
Orbital eccentricity	0
Diameter	1,212 km
Magnitude	16.8

Pluto has an exceptional orbit (above) that is highly eccentric and highly inclined, being tilted fully 17° to the ecliptic.

Charon and Pluto (below) are clearly distinguished by the Space Telescope. When Charon was discovered in 1978, it looked like nothing more than a bump on Pluto's image. The satellite's revolution time immediately revealed the mass of Pluto, long a mystery.

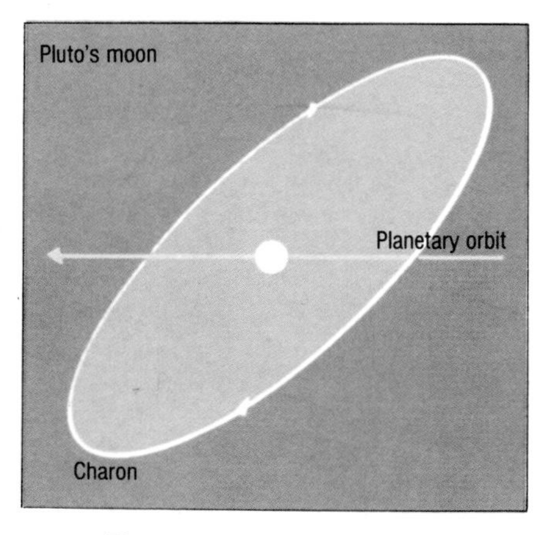

Charon's orbit (left) is tilted in relation to Pluto's. The two bodies occulted each other regularly in the late 1980s.

6

DEBRIS OF THE SOLAR SYSTEM

Comets

Though they comprise a negligible part of the total mass of the bodies that orbit the Sun, comets, meteors and meteorites yield a great deal of information about the origin of the Solar System.

A brilliant comet with a shining head and a tail stretching halfway across the sky is an awesome sight. Indeed, comets caused considerable alarm in ancient times, as they were often regarded as harbingers of doom. Yet in comparison with a planet, a comet is flimsy and insubstantial: on several occasions the Earth has been known to pass unharmed through a comet's tail.

Comets have been aptly described as "dirty ice-balls." The main nucleus is composed of rocky fragments held together with various frozen materials, such as methane, ammonia, carbon dioxide and water. Surrounding the nucleus is the coma, or head, which is made up of dust and gas. The Sun makes the comet visible by floodlighting the coma. When the comet nears the Sun, the radiation pressure of the sunlight and the solar wind drive the dust and gases away from the coma. This produces a tail that usually consists of both dust and gas. However, many faint comets do not develop a tail.

A number of comets move round the Sun in fairly small, generally

A comet's tail makes a magnificent spectacle in a long-exposure photograph. Yet it contains less matter than a typical laboratory vacuum on Earth. Heating by sunlight causes gases to be driven from the nucleus of the comet as it nears the Sun. The gases are then pushed away from the comet by the radiation pressure of the sunlight, forming a tail.

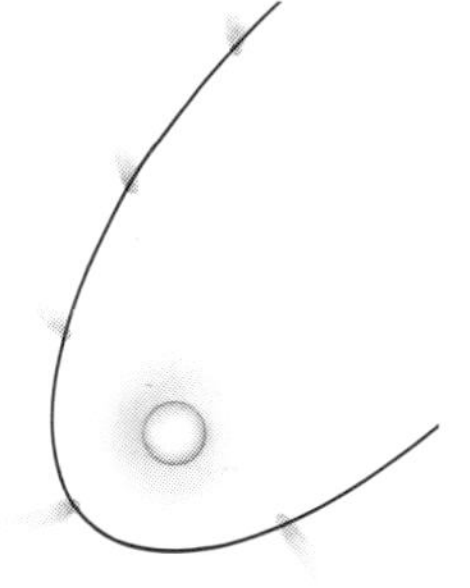

A bright comet (above) usually consists of three principal parts: the compact nucleus, the surrounding gas and dust cloud (the coma) and the tenuous tail. Most comets move in highly elliptical orbits (left) and become conspicuous only near perihelion. At this point one or more tails begin to develop. Tails always point more or less directly away from the Sun. As the comet recedes from the Sun, the tails decline.

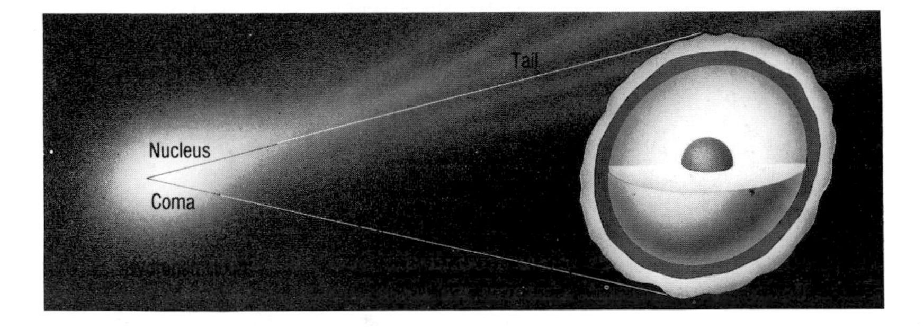

The nucleus of a comet consists of a mixture of rock and ice. The Sun's warmth drives gas and dust from the nucleus to form the coma and tail.

SOME PERIODIC COMETS

Name	*Period (yrs)*	*Dist. from Sun (a.u.)*		*Orbital inclination (°)*
		min	*max*	
Encke	3.3	0.34	4.09	12.0
Grigg–Skjellerup	5.1	1.00	4.94	21.1
Tempel 2	5.3	1.36	4.68	12.5
D'Arrest	6.2	1.17	5.61	16.7
Pons–Winnecke	6.3	1.25	5.61	22.3
Giacobini–Zinner	6.5	0.99	5.98	31.7
Finlay	6.9	1.10	6.19	2.6
Faye	7.4	1.62	5.98	9.1
Tuttle	13.8	1.02	10.46	54.4
Halley	76.1	0.59	35.33	162.2
Swift–Tuttle	133	0.96	51.7	113.4

elliptical orbits in periods of a few years; others have longer periods. However, most really brilliant comets (except Halley's) have orbital periods of thousands or even millions of years, which means that the times of their appearance cannot be predicted. Few have been seen in the 20th century. Halley's Comet returned in 1986 and a small flotilla of probes from several countries was sent to it. The most spectacular results came from the European Space Agency's Giotto craft, which sent back pictures as it flew close by the nucleus. The flow of data ceased when the craft was damaged by the storm of fine dust particles from the comet. The last bright naked-eye comets were Hyakutake (1996) and Hale–Bopp (1997). Hyakutake will return in about 15,000 years, while Hale–Bopp will be back in about 3,500 years.

Meteors and meteorites

The streaks of luminosity that we call shooting stars, or meteors, are caused by tiny grains of dust – properly called meteoroids – which dash into the upper air from space, moving at speeds up to 45 kilometers per second. They become intensely hot from friction with the air and burn up.

Many meteoroids move around the Sun in swarms; each time the Earth passes through a swarm we see a shower of shooting stars. The

A Perseid meteor (left) streaks across the sky in this time-exposure photograph. The trail ends with several flare-ups in brightness as intense heating makes the body disintegrate.

A meteor trail (right) stands out against the starry background of the Milky Way. Meteor images are quite often caught by chance when some other object is being photographed.

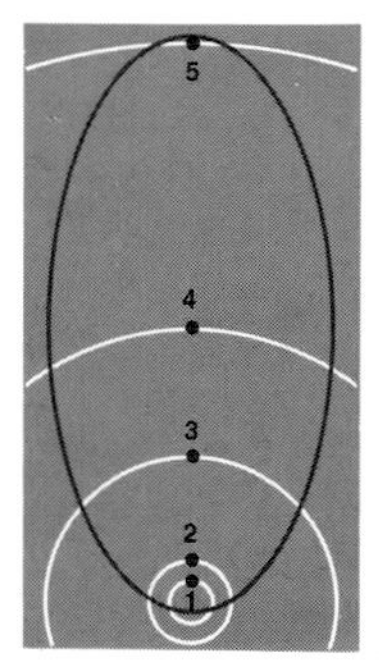

The orbit of the Leonids (left) intersects the orbits of the Earth (1), Mars (2), Jupiter (3), Saturn (4) and Uranus (5).

swarms are remains of dead comets. The most spectacular annual shower is known as the Perseids (25 July to 18 August). In addition to meteor showers, there are also non-shower or sporadic meteors, which may come from any direction at any moment.

Meteorites are fragments of meteoritic rock that succeed in reaching the ground. They are thought to be associated with fragmented asteroids. They are of two main types: irons (siderites) and stones (aerolites), although there is no hard-and-fast distinction, and many intermediate types are known.

Most museums have meteorite collections. The largest specimen on public view is the Ahnighito (Tent), which weighs more than 30 tonnes. It was found in Greenland by Robert Peary and is now in the Hayden Planetarium, New York. This is dwarfed by the Hoba West Meteorite, still lying where it fell at Grootfontein, Namibia, in prehistoric times. It weighs at least 60 tonnes. In recent years many meteorites have been discovered in Antarctica, where they have lain undisturbed for long periods.

Many meteorite falls have been recorded, but major falls are rare and there is no reliable record of any human casualty caused by a meteorite. It was formerly believed that meteoric bodies would prove a major hazard in space research but it is now clear that the danger is less than was originally thought.

METEOR SHOWERS

Name	*Date of return*	*Maximum*	*Radiant*	*Comments*
Quadrantids	1–6 Jan	4 Jan	Boo	Quite fast, blue
Corona Australids	14–18 Mar	16 Mar	CrA	
Lyrids (April)	19–24 Apr	21 Apr	Lyr	Fast, brilliant
η-Aquarids	1–8 May	5 May	Aqr	Fast, persistent
Lyrids (June)	10–21 Jun	15 Jun	Lyr	Blue
Ophiuchids	17–26 Jun	20 Jun	Oph	
Capricornids	10 Jul–15 Aug	25 Jul	Cap	Yellow, very slow
δ-Aquarids	15 Jul–15 Aug	28 Jul	Aqr	Slow, long paths
Pisces Australids	15 Jul–20 Aug	30 Jul	PsA	
Capricornids	15 Jul–25 Aug	1 Aug	Cap	Yellow
ι-Aquarids	15 Jul–25 Aug	6 Aug	Aqr	
Perseids	25 Jul–18 Aug	12 Aug	Per	Fast, fragmenting
Cygnids	18–22 Aug	20 Aug	Cyg	Bright, exploding
Orionids	16–26 Oct	21 Oct	Ori	Fast, persistent
Taurids	10 Oct–30 Nov	1 Nov	Tau	Slow, brilliant
Leonids	15–19 Nov	17 Nov	Leo	Fast, persistent
Phoenicids	4 Dec	–	Phe	
Geminids	7–15 Dec	14 Dec	Gem	White
Ursids	17–24 Dec	22 Dec	UMi	

7
THE STARS

Stars and their evolution

The stars are suns, and our Sun is only an average star. It takes an effort of the imagination to realise that some stars are thousands or even millions of times more luminous than the Sun.

The stars show a tremendous range in size, luminosity and surface temperature. The hottest have temperatures of at least 80,000 degrees centigrade; the coolest, less than 3,000 degrees. They are divided into certain classes according to their spectra: types O, B and A are very hot, and white or bluish-white; types F and G are yellow; K, orange; and M, R, N and S, orange–red.

A star begins its career when it condenses from a nebula, a cloud of dust and gas consisting mostly of hydrogen. A star of mass less than one tenth that of the Sun will shrink, because of gravitation, but will never become hot enough for nuclear reactions to begin. It will fade, becoming a dim, red star, before finally growing cold.

A star of 0.1–1.4 solar mass will behave very differently. When the core temperature has reached about 10 million degrees centigrade, nuclear reactions will begin; hydrogen will be converted to helium, and the star will settle down to a long period of stable existence. Eventually no more hydrogen will be available. Different reactions will begin and the star's outer layers will expand and cool, so that the star will become a red giant. When the nuclear energy is exhausted, the star will collapse into a small, superdense object known as a white dwarf.

A star that is more massive still will run through its life-span much more quickly and will die in a more spectacular manner. There will be a tremendous outburst, known as a supernova explosion, and much of the star's material will be blown away, leaving an expanding gas cloud and a stellar remnant made up of neutrons – atomic particles with no electrical charge. A cupful of neutron-star material would weigh thousands of millions of tonnes.

Finally, there are stars of even greater mass and shorter lives. Once the final collapse starts, it is so rapid and violent that nothing can stop it. The end-product is a remnant so small, and pulling so strongly, that not even light can escape from it, and the old star is surrounded by an area virtually cut off from the rest of the universe – a black hole.

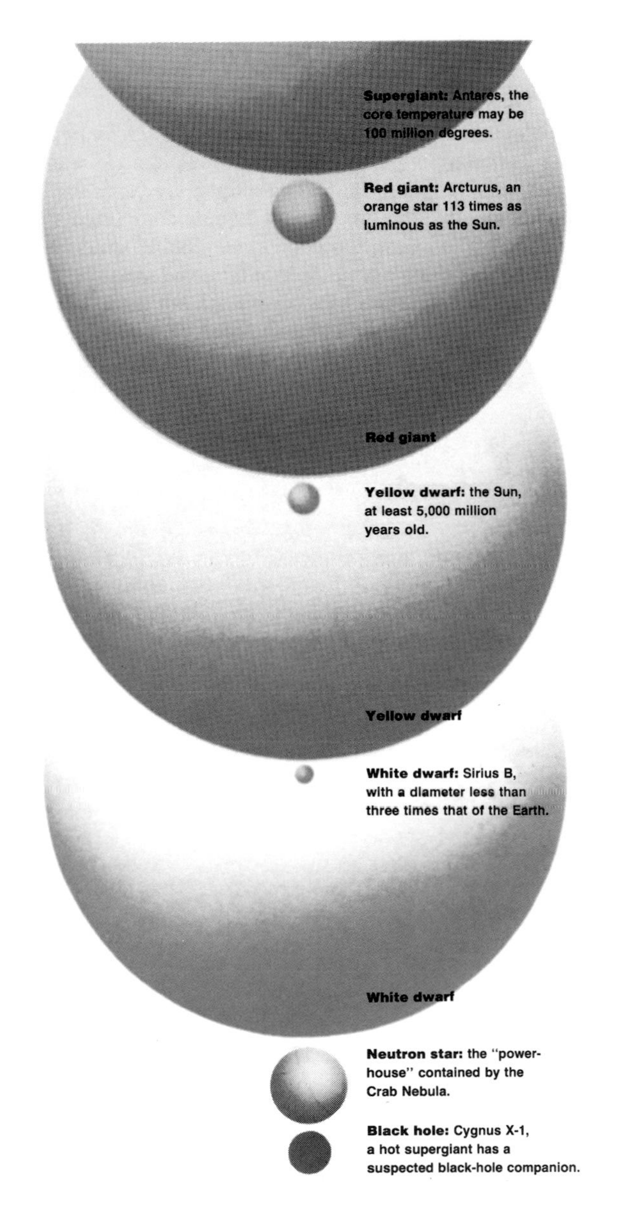

The enormous range in the sizes of stars is illustrated here, with a named example given for each.

Double stars

Not all stars are solitary wanderers in space. Many may be attended by families of planets, but there are also many double stars and even multiple systems. For example, Castor (Alpha Geminorum) is revealed in a small telescope to be a multiple with two components – but each of these is shown by the spectroscope to be double. And also included in the Castor system is a separate pair of faint red stars.

In some cases the two stars of a couple are not genuinely associated; one component merely lies almost in front of the other by chance. Rather surprisingly, however, optical pairs of this type are not nearly so numerous as physically associated pairs of binary systems. In a binary, the two components move around their common center of gravity. Periods may range from less than half an hour for ultraclose pairs to millions of years. For example, Arich, in Virgo, has almost equal components; the revolution period is 171 years.

Some doubles are separable with the naked eye. Mizar, in Ursa Major, has a naked-eye companion (Alcor), and in a small telescope Mizar itself is seen to be double. Other wide, easy pairs are Alpha Centauri and Alpha Crucis, two of the most brilliant stars of the southern hemisphere. Sometimes the components are quite unequal; thus the brilliant Sirius (Alpha Canis Majoris) has a dim white companion,

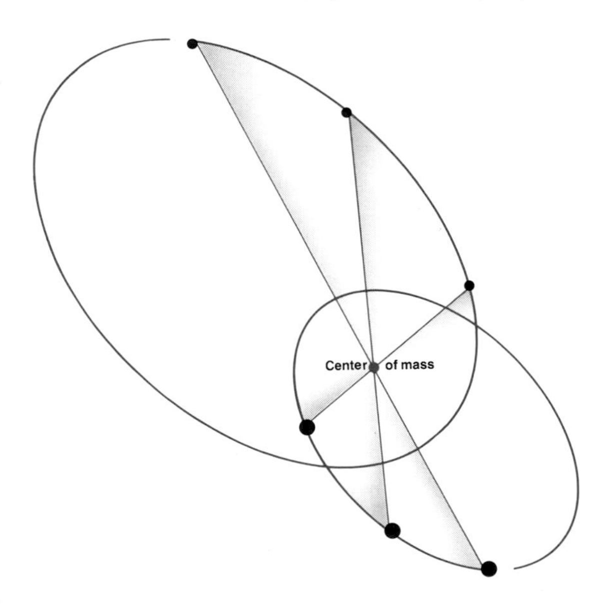

Orbits of Sirius and its white dwarf companion. The companion is only 1/10,000th as luminous as its primary, but it is extremely dense and the discrepancy in mass is not nearly as great as might be expected. The companion is difficult to see, as it is overpowered by Sirius itself. In any such binary system the less massive star has the larger orbit.

Mizar and Alcor comprise a famous naked-eye double in the tail of the Great Bear. Mizar is of the second magnitude, while Alcor is of the fifth. In this photograph the eighth-magnitude Sidus Ludovicianum is just visible, slightly to the left of the line joining the two brighter stars.

a white dwarf (see page 84) with only one ten-thousandth of the luminosity of Sirius itself.

It used to be thought that a binary pair resulted from the breakup of a formerly single star, but it is now thought more likely that the components of a binary were formed from the same cloud of material in the same region of space.

Variable stars, as their name implies, brighten and fade over relatively short periods (see pages 88–89). Yet not all such stars are actually changing. With eclipsing binaries, such as Algol in Perseus, the variations result from one component passing in front of the other and cutting out some or all of its light. Algol "winks" regularly every two and a half days, falling from magnitude 2 to below magnitude 3.

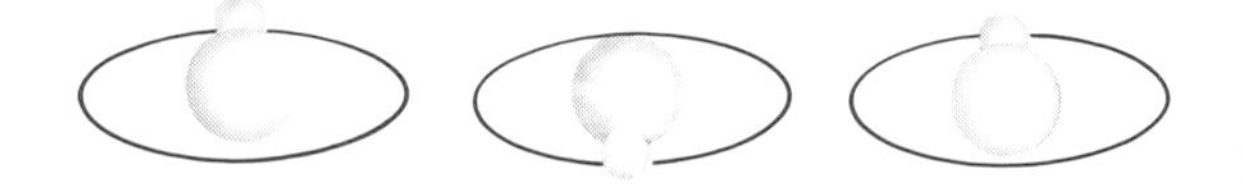

An eclipsing binary. When the larger, fainter component passes in front of the smaller, brighter one, there is a marked reduction in the light reaching us; when the fainter is eclipsed, the fall in the system's overall brightness is slight.

Variable stars

A variable star is one that changes in brightness. Variable star work is a very important branch of modern astronomy – and amateurs can make very valuable contributions, because there are so many variables that professional astronomers cannot hope to monitor them all.

There are numerous types of variables. With a true variable, the changes are intrinsic. In an eclipsing variable, the changes are simply due to the effects of the movements of two stars that are orbiting each other (see page 87). The different kinds of true variable include:

(1) Mira stars, named after the brightest and best-known member of the class, Mira in Cetus (the Whale). These are old red giants, which have become unstable and are pulsating. Their periods range from many weeks to several years, though neither the periods nor the amplitudes are constant from one cycle to another. Mira itself ranges in magnitude from 1.7 to below 10, though at some maxima it never exceeds magnitude 4. Mira-type stars are very common indeed.

(2) Semiregular variables. These also are mainly red, but have smaller amplitudes than the Mira stars, and their periods vary very widely.

(3) Cepheids. These are named after the prototype star, Delta Cephei. Their periods are short – a few days – and the amplitudes fairly small; Delta Cephei itself ranges from magnitude 3.4 to 5.1 in a period of just over five days. The Cepheids are of special importance because their period gives a key to their real luminosities; the longer the period, the more luminous the star. It is therefore possible to tell the luminosity of a Cepheid, and hence its distance, merely by observing it, and these stars act as "standard candles" in space. They are highly luminous, and can be seen in external galaxies, thereby providing a

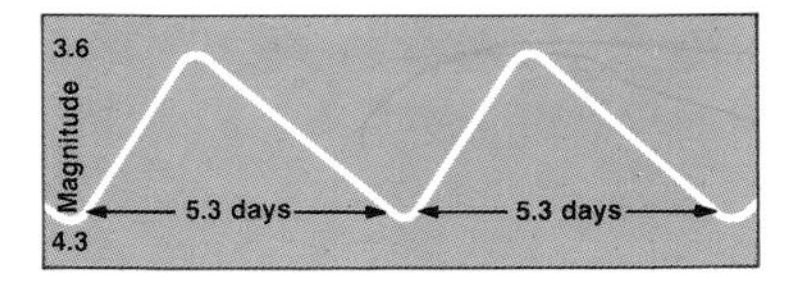

The light-curve of Delta Cephei shows regular fluctuations. In general the absolute brightness of Cepheid variables is greater the longer the period of brightness variation.

The light-curve of Nova Aquilae 1918 shows an abrupt rise, marking the explosive disruption of a previously obscure star, followed by a sharp fall, and then a more gradual decline.

valuable link carrying the astronomer's distance scale beyond our own Galaxy. RR Lyrae stars are also short-period variables, but their periods are shorter – usually less than a day – and all are about 90 times as luminous as the sun.

(4) Eruptive variables. These are irregular in behavior. For instance, T Tauri stars are very young, and have not yet settled down to a sober, steady existence, so that their fluctuations are unpredictable. T Coronae stars show sudden fades, caused by the accumulation of clouds of soot in their atmospheres.

(5) Cataclysmic variables. These show violent outbursts. Of special importance are novae. The name means "new stars," but they are not new at all. A nova is a binary system; the fainter, highly condensed component draws material away from its companion, becomes unstable and flares up, so that the star increases to many times its normal brilliance. Supernovae are even more spectacular; such an outburst involves the virtual destruction of a star. The last supernova to be seen in our Galaxy appeared as long ago as 1604, though many have been seen in other systems. In 1987 a supernova flared up in the Greater Magellanic Cloud. Since this irregular galaxy, a satellite to our own, is only 170,000 light-years away, the supernova was virtually "on our doorstep" and is still yielding a fund of valuable information.

The magnitude of a variable star is gauged by comparing it with nearby stars that do not vary. It is here that amateurs with patience and experience can carry out such valuable work.

The extraordinary variable Eta Carinae lies at the heart of this nebula in Carina (the Keel). In the 19th century it grew to become the brightest star in the sky after Sirius. Then it faded until now it is just below naked-eye visibility. But at infra-red wavelengths it is as bright now as it was at visible wavelengths in the last century. Eta Carinae is one of the most massive stars known, and also one of the most luminous.

Clusters

Some of the most spectacular objects in the Galaxy are the star clusters, which are of two main types: open and globular. Some, such as the Pleiades, or Seven Sisters, are visible with the naked eye.

Open or loose clusters have no particular shape, and may contain several dozen to several hundred stars. It may be assumed that the stars in a cluster are of about the same age and have a common origin. In some cases the prominent stars are hot and white, with considerable nebular material spread between them, as in the Pleiades. Other clusters are much more advanced in their evolution. An open cluster may eventually be so affected by adjacent non-cluster stars that it will disperse and cease to be recognizable.

Globular clusters are different. They are symmetrical systems, containing up to a million stars, surrounding the Galaxy in what is termed the "galactic halo." Out of the hundred known globular clusters, only three are clearly visible with the naked eye. Globulars are commonest in the southern hemisphere of the sky, giving the first definite proof that the Sun lies well away from the galactic center. They contain Cepheids, and so their distances may be measured. Most of them are extremely remote, being at least 20,000 light-years away. If the Sun were a member of a globular cluster, the stars would indeed make the sky a glorious sight. There would be many stars bright enough to cast shadows and there would be no real darkness at night.

A globular cluster is a ball of thousands of stars, packed into a space perhaps 150 light-years across. The colors of this photograph are misleading: the stars in a globular cluster are reddish, being among the oldest known.

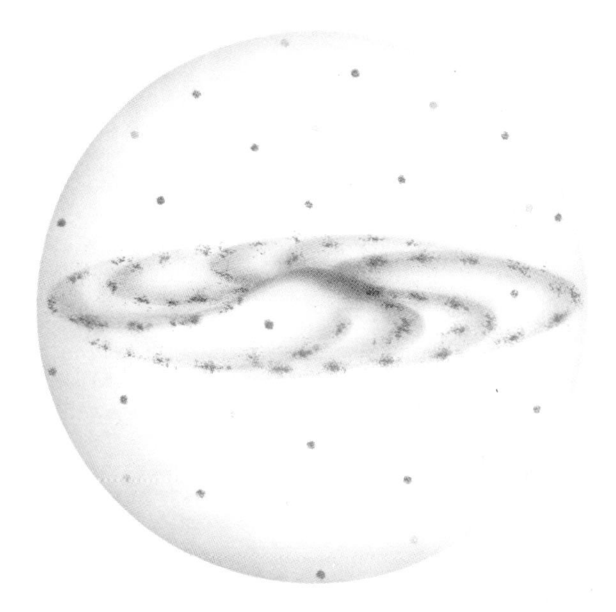

The distribution of clusters in our galactic system. Globular clusters form a spherical "halo," while open clusters are distributed throughout the disk of the Galaxy. Many consist of young stars and are associated with the lanes of gas and dust that are the birthplaces of stars.

The Pleiades (below) are intensely hot young stars, swathed in remnants of the interstellar gas from which they were born.

Nebulae

Gaseous or galactic nebulae (Latin *nebula,* a cloud) are stellar birthplaces: fresh stars are slowly condensing out of the nebular material. The nebulae shine because of the stars in or close to them. If the intermixed stars are extremely hot, they cause the nebular material to emit a certain amount of self-luminosity and it is described as an emission nebula; if the stars are cooler, the nebula shines only by reflection and is called a reflection nebula. If there are no suitable stars available, the nebula cannot shine at all and appears as a dark mass blotting out the light of objects beyond. But a nebula that appears dark to one observer might look bright from another direction.

The material in a nebula is highly rarefied. A good example is the famous nebula in Orion's Sword, which is 30 light-years in diameter. If it were possible to take a 2.5-centimeter core sample right through it, the total mass of material collected would be less than that of a small coin. Astronomers have also found that many bright nebulae are merely the visible parts of much larger clouds.

The so-called planetary nebulae are neither planets nor nebulae in the true sense. They are called planetary because in the telescope they resemble the disks of planets. They are really shells of gas surrounding hot stars that are evolving to the white dwarf stage.

Also of interest are the supernova remnants, of which the most famous example is the Crab Nebula. Just as an ordinary nebula is the material from which a star first forms, so a supernova remnant is the matter left behind after a massive star's explosive death.

The Great Nebula in Orion (left) is shown in its glory in photographs.

The Veil Nebula (opposite, top) is part of a ring of gas expanding from a supernova that exploded about 30,000 years ago.

The center of the Rosette Nebula (opposite) is swept clear of matter by radiation from young stars within.

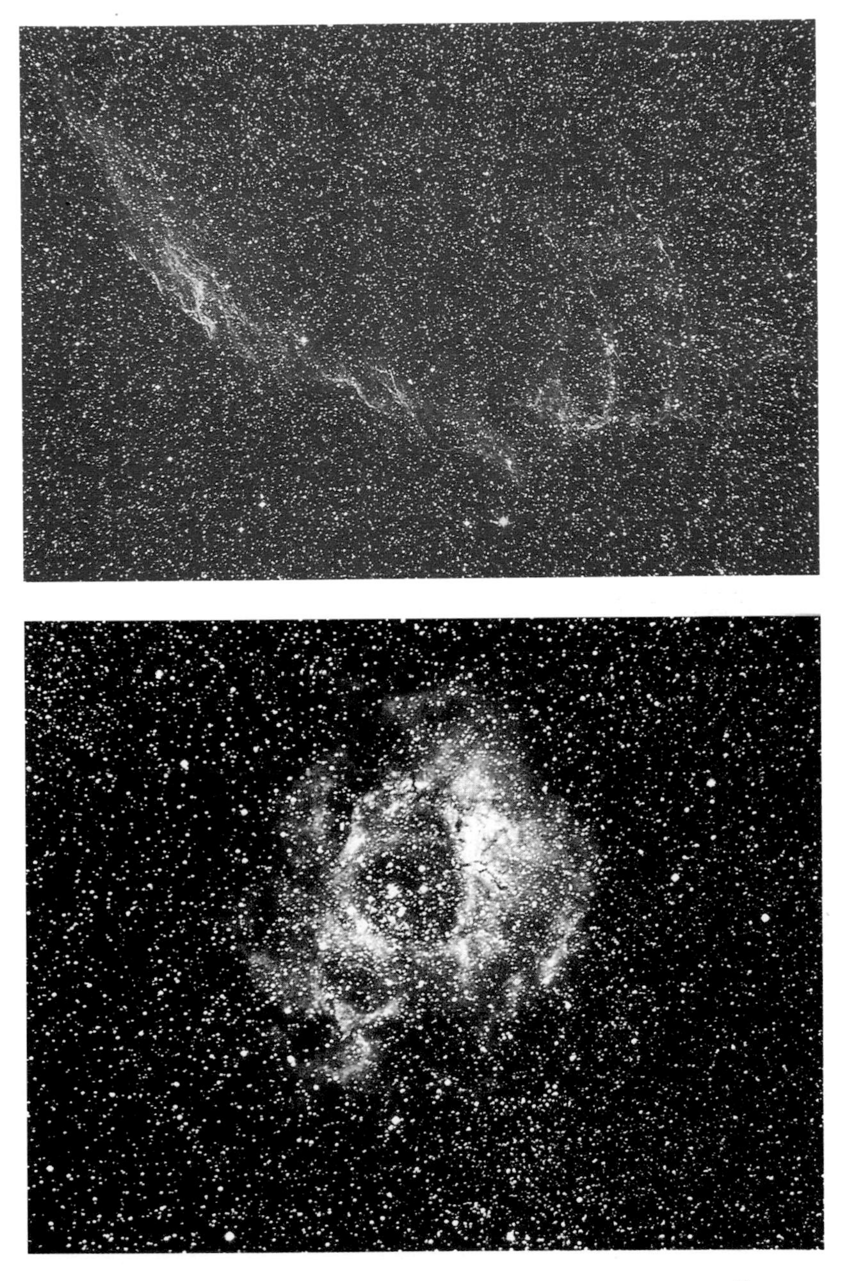

8

THE GALAXIES

The galaxies are like great cities of stars, and like cities, they occur in a great range of sizes and forms. Some are irregular in shape; some are ellipsoidal; and some, like our own, are beautiful spirals.

Our Galaxy

Our Galaxy, the star system of which the Sun is a member, contains about 100,000 million stars, as well as nebulae and much interstellar dust. It has an overall diameter of perhaps 100,000 light-years (although some astronomers believe this to be an overestimate) and the central "bulge" has a thickness of about 20,000 light-years.

The Galaxy is a flattened system. It has been compared, graphically if unromantically, to two fried eggs clapped together back to back. It is this shape that gives rise to the Milky Way – that glorious band of radiance stretching right round the sky, the subject of innumerable legends. Any telescope, or even binoculars, will show that the Milky Way is made up of stars, and at first glance it might seem that they are

The true shape of the Galaxy (below). In plan, it is a spiral. From the side, it would appear flattish with a central bulge. Its diameter is about 100,000 light-years.

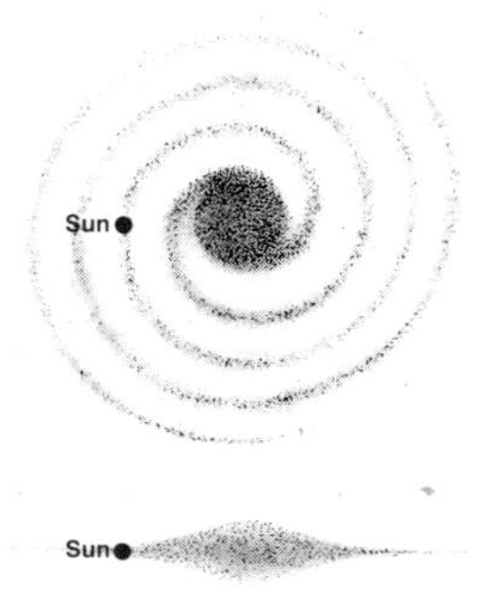

The Milky Way (below) bisects this photographic mosaic of the sky, compiled by the Lund Observatory.

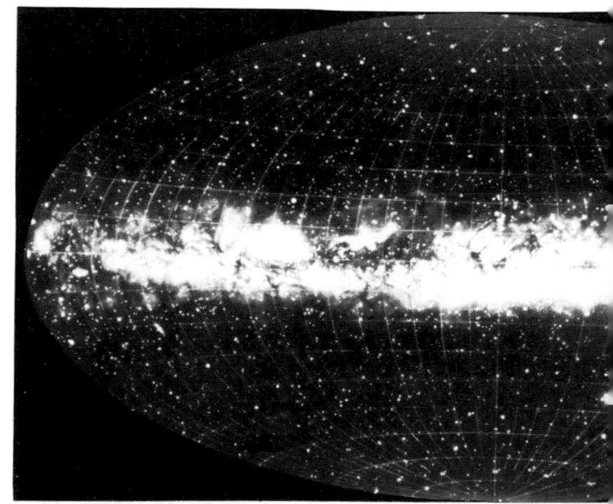

crowded together. Once again, appearances are deceptive. An observer looking along the main plane of the Galaxy sees many stars in almost the same direction, and this accounts for the Milky Way. It is nothing more than a line-of-sight effect. Dark "rifts" may be seen here and there in the Milky Way band, indicating not an absence of stars but rather the presence of dark, obscuring material.

Seen in plan, the Galaxy is a spiral, rather like a huge whirlpool. This is not surprising, because many other galaxies are spiral in form. Like them, our Galaxy also rotates. The Sun, at its distance of some 32,000 light-years from the center of the system, takes 225 million years to complete one revolution – a period sometimes nicknamed the "cosmic year."

Unfortunately it is impossible to see right through to the center of the Galaxy, because there is too much obscuring matter in the way, but radio waves can pass through this material. The galactic center lies beyond the lovely star clouds seen in the constellation of Sagittarius.

What lies at the actual center of the Galaxy? We have to admit that even today we are not certain. We cannot see through to it at visible wavelengths, because there is too much obscuring matter in the way, so that we have to depend mainly upon radio emissions. There is certainly a major radio source, known as Sagittarius A* (pronounced Sagittarius A-star), and there may well be a massive black hole, swallowing matter that pours out energy before it disappears. However, this is speculative as yet, and the presence of a black hole remains to be confirmed.

The Andromeda Galaxy (right), our twin.

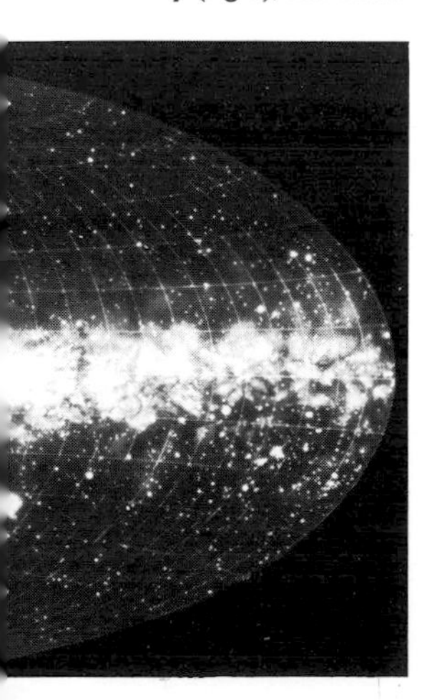

Other galaxies

Our Galaxy is a typical spiral, and is certainly not unique. Many other spirals are known, and formerly they were incorrectly termed spiral nebulae. But not all galaxies take this form. Some are irregular, while others are elliptical.

The Local Group, of which our Galaxy is a member, contains a few large systems and more than twenty smaller ones. The Andromeda Spiral, at a distance of 2.2 million light-years, is a member; so are the Clouds of Magellan, a spiral in Triangulum, and another massive system, Maffei 1, which is hard to see because it is so heavily obscured by material in the plane of our Galaxy. Beyond the Local Group the distances become immense and astronomers know of galaxies that are more than 13,000 million light-years away. Distances of the nearer galaxies can be measured by the Cepheid method. With more remote systems the Cepheids fade into the background blur, and the measurement of distances becomes much more difficult. Apart from the normal

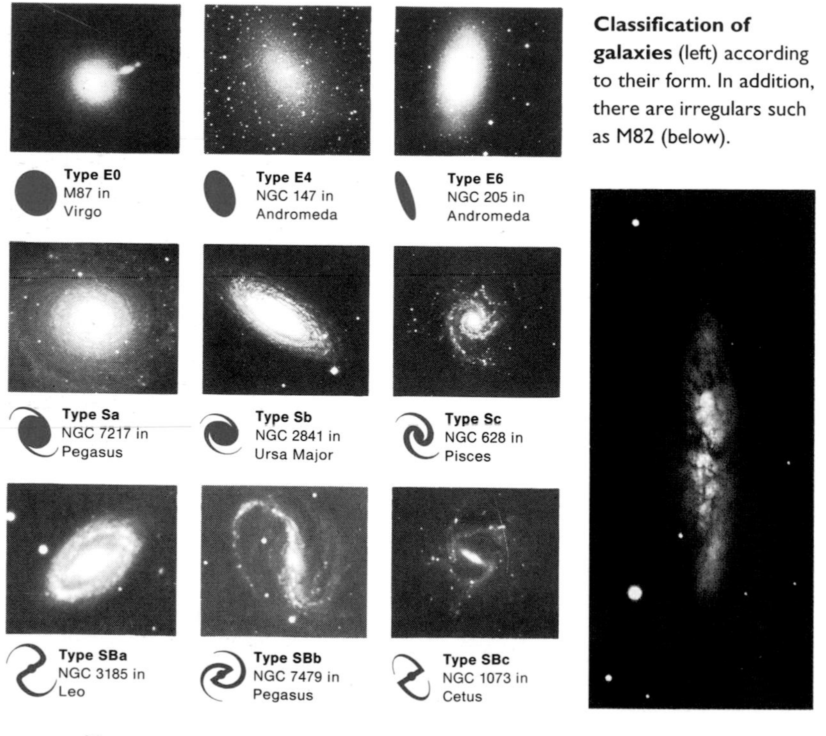

Classification of galaxies (left) according to their form. In addition, there are irregulars such as M82 (below).

spirals, there are "barred spirals" in which the arms extend from the ends of a bar through the main plane. There are also elliptical and irregular systems. Some galaxies emit strong ultraviolet or radio waves – for example, the Seyfert systems, which have condensed active nuclei and inconspicuous spiral arms. Then there are quasars, which are believed to be the nuclei of very active galaxies.

Galaxies tend to occur in clusters (not to be confused with star clusters), some of which are much more populous than the Local Group. The Virgo cluster, for example, at about 65 million light-years, contains hundreds of systems. In remote groups even a massive galaxy will appear as nothing more than a tiny smudge of light. Apart from the members of the Local Group, all the galaxies are racing away from Earth, so that the entire universe is expanding.

In 1781 the French astronomer Charles Messier drew up a list of more than 100 clusters, nebulae and galaxies and his numbers, called "M" numbers, are still used; the Andromeda Spiral, for example, is M31. Other systems of nomenclature have since been introduced but most of the brightest objects were included in Messier's list.

Chart of the Local Group (below). Fainter galaxies are identified by their catalog numbers; Nubecula Minor and Major are the Magellanic Clouds. The Andromeda Spiral is M31.

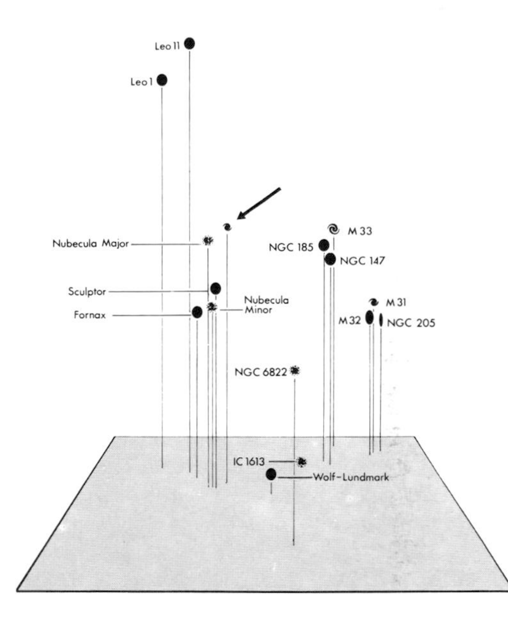

The heart of the Andromeda Galaxy (above) is relatively free of gas and dust and therefore of young stars. New stars are born in the arms, which are unseen at this exposure.

Cosmology

How did the universe begin? How is it evolving, and how will it die – assuming that it will come to an end? The first great breakthrough in tackling these questions was made in 1923 when Edwin Hubble, at the Mount Wilson Observatory, proved that the "spiral nebulae" are independent galaxies. He was able to obtain the distances of some by observing Cepheids in them, and it became clear that our Galaxy was only one of many.

Hubble also confirmed that, apart from the members of the Local Group, all the galaxies appear to be receding from each other and that the further away they are, the faster they go. The essential clue came from the Doppler effect. If a light-source approaches Earth, the wavelength of the light appears to be shortened and the light seems to be "too blue"; if it recedes, the wavelength is increased, and the light appears to be "too red." If the dark lines in the spectrum of a galaxy are moved toward the red end of the rainbow band, then the object must be receding. The amount of red-shift is a key to the speed of recession.

Astronomers are now aware of galaxies and quasars that are moving away at appreciable fractions of the velocity of light. If the law of higher velocity with greater distance holds good indefinitely, then a point must be reached at which a galaxy is moving away at the velocity of light and cannot be observed There will be a limit to the observable universe, although not necessarily to that of the universe itself.

The recession of galaxies is interpreted as a general expansion of the universe, which seems to have begun about 15,000 million years ago. When astronomers begin to consider the origin of the universe certain difficulties immediately arise. According to the generally held Big Bang theory, everything was created at one moment in a kind of "primeval atom"; expansion began and galaxies, and then stars, were formed. One important piece of evidence for the theory has been the discovery of "background radiation" coming from all directions, and possibly a remnant of the Big Bang.

It is possible that the present phase of expansion will be followed by a period of contraction. The galaxies will come together again and there will be another Big Bang. This is the cyclic or oscillating universe theory. If it is correct there should be a Big Bang every 80,000 million years or so. Everything depends upon whether the density of material in the universe as a whole is great enough for its gravitational effect to prevent the expansion from continuing indefinitely. Present research indicates that the density is below this critical point, though it is widely believed that most of the mass of the universe is made up of "dark matter," which may be quite unlike anything in our present experience.

Looking toward the edge of the universe (right) is looking backward in time, because light takes time to travel. Quasars seem to have formed about 1,000 million years after the Big Bang.

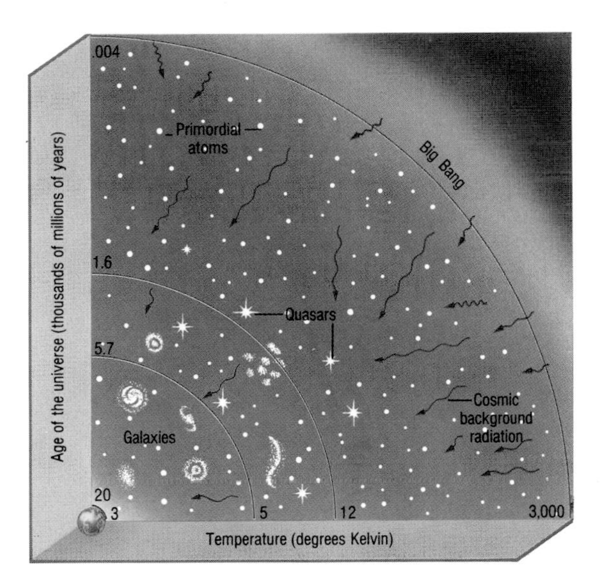

The universe shows structure (below) on an enormous scale. The Local Supercluster comprises clusters of galaxies within a cube that is 100 million light-years on a side.

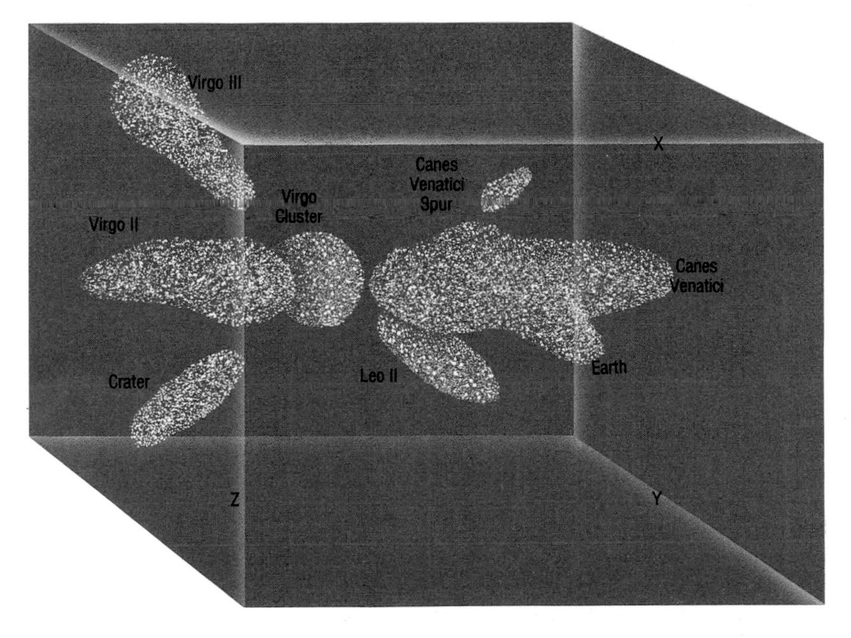

9

NEW ASTRONOMY

Extending our view of the universe

Astronomers now observe the universe at all wavelengths, from the very shortest X-rays to the longest radio waves. Novel telescopes are placed on mountain tops and above the Earth's atmosphere in space.

Most radiations coming from space are blocked by layers in the Earth's atmosphere. We are therefore confined mainly to visible light and a few "windows," such as that in the radio range. The situation can be somewhat improved by building equipment at high altitude. For example, much of the incoming infra-red radiation is absorbed by

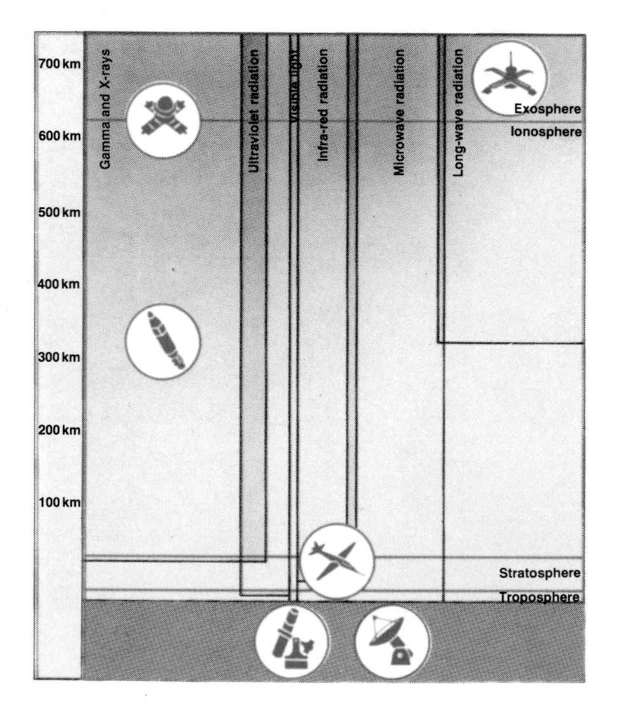

Two "windows" open on to the universe – the two wavelength bands, of light and radio, that can penetrate the Earth's atmosphere and can therefore be analysed at ground level. Other wavelengths have to be viewed by high-flying aircraft or by satellites.

atmospheric water vapor, and the world's largest infra-red telescope, UKIRT (United Kingdom Infra-Red Telescope) has been set up at the summit of Mauna Kea, in Hawaii, more than 4,000 meters above sea level.

Infra-red radiation is emitted by cool bodies, and studies of the sources can tell us a great deal about stellar evolution. In 1983 IRAS, the Infra-Red Astronomical Satellite, carried out a survey of the entire sky, and discovered over 200,000 sources – including cool material round some stars, notably Vega, Fomalhaut and the southern Beta Pictoris; this may well be indicative of planetary systems, though one must be very wary of jumping to conclusions. For studies of extraterrestrial X-rays, gamma-rays and most ultraviolet wavelengths, satellites are essential. X-rays and gamma-rays are emitted by very hot, energetic bodies, and are of special interest; for example, some X-ray sources may indicate the presence of black holes – the X-rays in such a case being emitted by material which is being intensely heated just before being sucked into the black hole.

Perhaps the most successful of all satellites has been the IUE, or International Ultraviolet Explorer, which was launched in 1978 with a life expectancy of two years – but operated excellently until 1996. It has carried out a full-sky survey in ultraviolet, providing information which could never have been obtained from ground level.

The first large optical telescope to be launched into space was the HST or Hubble Space Telescope, named in honor of the great astronomer Edwin Hubble. It was launched from Cape Canaveral, via the Space Shuttle, on 25 April 1990, and was put into an orbit at an altitude of 600 kilometers; its orbital period is 95 minutes and the inclination of its orbit to the equator is 28°.5. The total weight is 11,360 kg, the length 13.3 meters, and the diameter 12.3 meters with the solar arrays extended. The mirror is 240 centimeters in diameter. This is much smaller than many ground-based telescopes, but seeing conditions in space, beyond the atmosphere, are perfect all the time, and it was confidently expected that the HST would make notable advances.

These expectations were fully realized, and the Hubble Telescope has been an outstanding success. The photographs sent back of planets, stars and star systems are much the best ever obtained, and it has been possible to obtain images of objects in the far depths of the universe.

The telescope will continue to operate for many years, and it can be periodically overhauled; one major servicing mission has already been carried out by astronauts. No doubt there will be more space telescopes in the future, but certainly the HST has proved to be even more effective than had originally been hoped.

Radio astronomy

Radio astronomy began by accident. In 1931 the American radio engineer Karl Jansky was studying "static" on behalf of Bell Telephone Laboratories, using an improvised aerial, when he found that he was picking up radiations from the sky. These were traced to the Milky Way. Strangely, Jansky never followed up this discovery, and it was not until later that the first true radio telescopes were built.

The term "radio telescope" is in some ways misleading, because the instrument is more in the nature of an antenna – one certainly cannot look through it, and it does not produce a visible picture unless its data are processed by a computer. The "noise" from the sky that has so often been broadcast is produced inside the equipment.

The sun is a source of radio waves; so is the planet Jupiter, and the space missions have also detected radio emissions from the other giant planets. However, most sources lie far beyond the Solar System. Supernova remnants are powerful radio emitters; the Crab Nebula is the classic example. Some kinds of galaxies are also very powerful at

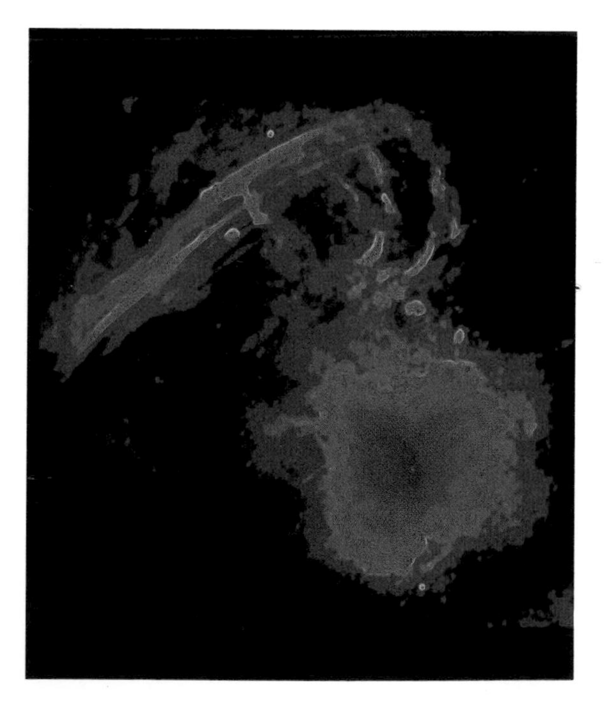

The center of our galaxy, seen in a radio image from the VLA (Very Large Array). A strange feature can be seen: the looping filaments, which probably result from a magnetic field drawing radiating gas along its field lines. The object at the center of the red area is Sagittarius A West, believed to lie at the very hub of the galaxy.

The 76-meter Lovell Telescope at Jodrell Bank, UK, was completed in 1957. For a long period it was the world's largest fully steerable radio dish.

radio wavelengths. Pulsars, which are now known to be rotating neutron stars, were discovered when Jocelyn Bell-Burnell noticed their extremely regular radio waves in 1967.

Radio telescopes working together can provide remarkable resolution, and extensive networks have been set up, such as the Merlin equipment in Britain and the Australia Telescope in the southern hemisphere. At Arecibo, in Puerto Rico, there is a 305-meter dish built into a natural hollow in the ground. The dish is not steerable, but it sweeps a band of the sky as the Earth rotates each day. In New Mexico there is the VLA, or Very Large Array, which consists of 27 linked antennae which can move on a Y-shaped railway.

Orion, showing Betelgeux (top left), Rigel (bottom right) and the Great Nebula (below the Belt).

10

THE CONSTELLATIONS

Introduction to the constellation maps

The constellation patterns we use today are based on those of the Greeks, but have no real significance, as the stars in any constellation are at very different distances from us.

Learning the constellations is not nearly so difficult as might be thought. Much the best method is to select a few groups which cannot be mistaken – notably Ursa Major, Orion and the Southern Cross – and use them as guides to the less obvious groups. Once a constellation has been identified, it will not easily be forgotten.

The maps given here are seasonal, and there are of course separate sets for the northern and southern hemispheres – remember, Ursa Major is not well seen from countries such as Australia, and in North America the Southern Cross never rises. Fortunately Orion is crossed by the celestial equator, and so can be seen from any part of the world.

In the separate constellation pages, the main stars (usually those above the third magnitude) have generally been listed, and there are also lists of interesting objects. The positions in right ascension and declination are given, to help those who have telescopes equipped with setting circles.The telescopic objects have in general been limited to those which are fairly easy for equipment used by the average amateur observer, though a few "tests" have also been included. Positions are given for epoch 2000. The distances and luminosities of stars have been taken from the authoritative Cambridge catalog – other catalogs may in some cases give rather different values – and with any but the nearest stars there is bound to be a degree of uncertainty. Moreover, the lists given here do not pretend to be in any way complete, and are not intended to be anything more than a general introduction.

Greek letters have been used. The Greek alphabet is as follows:

α Alpha	ε Epsilon	ι Iota	ν Nu	ϱ Rho	φ Phi
β Beta	ς Zeta	κ Kappa	ξ Xi	σ Sigma	χ Chi
γ Gamma	η Eta	λ Lambda	ο Omicron	τ Tau	ψ Psi
δ Delta	θ Theta	μ Mu	π Pi	υ Upsilon	ω Omega

Northern Hemisphere

COMA BERENICES
BOÖTES
Arcturus
CANES VENATICI
CORONA BOREALIS
SERPENS CAPUT
HERCULES
OPHIUCHUS
DRACO
URSA MINOR
Polaris
Vega
LYRA
CEPHEUS
Deneb
CYGNUS
VULPECULA
SAGITTA
SERPENS CAUDA
SCUTUM
AQUILA
Altair
LACERTA
DELPHINUS
PEGASUS
EQUULEUS
AQUARIUS
Magnitudes: −1 0 1 2 3 4 5

Southern Hemisphere

Magnitudes: -1 0 1 2 3 4 5

Guide to the stars (North)

The charts on this page and the two following show the general appearance of the constellations as viewed from the northern hemisphere throughout the year. The upper chart in each pair shows the skies looking southward, toward the equator; the lower one shows the skies looking toward the pole. The scales at the sides of each map show the

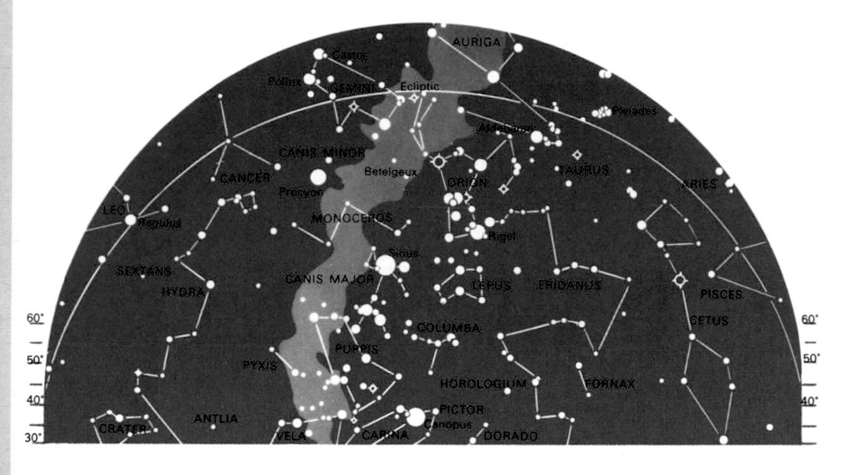

Evening
January 1 at 11.30
January 15 at 10.30
January 30 at 9.30

Morning
October 1 at 5.30
October 15 at 4.30
October 30 at 3.30

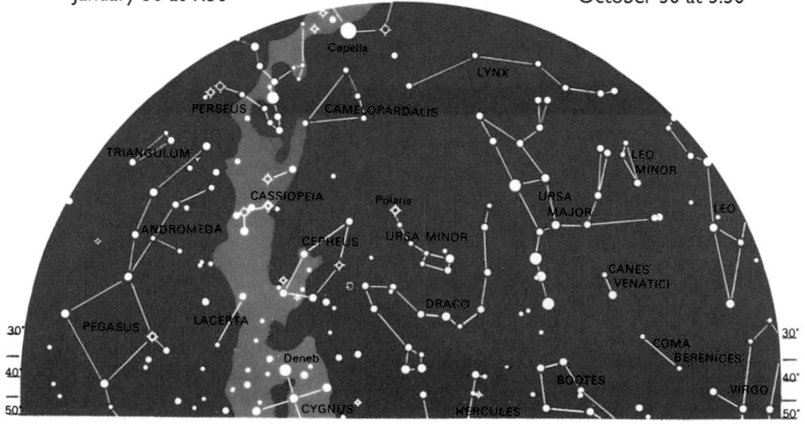

horizon for the latitude indicated. Thus an observer at 30° north can see further south of the celestial equator than one at 50°, and the celestial pole appears closer to the horizon. To be precise, he or she can see 90° − 30° = 60° south of the celestial equator, and the pole, roughly marked by Polaris, lies 30° above the horizon. Stars that are circumpolar for this observer – meaning that they never set – are those that lie less than 30° from the pole; and correspondingly for any other latitude. An observer at 50° N can see only 90° − 50° = 40° south of

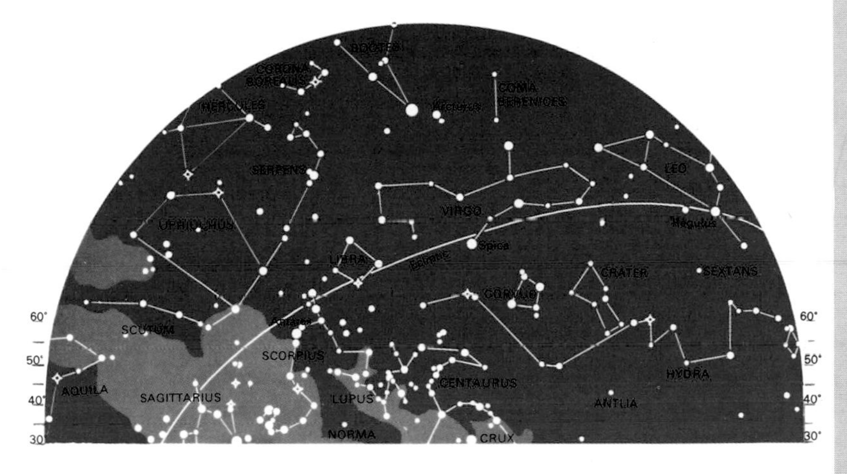

Evening
May 1 at 11.30
May 15 at 10.30
May 30 at 9.30

Morning
January 15 at 5.30
February 1 at 4.30
February 14 at 3.30

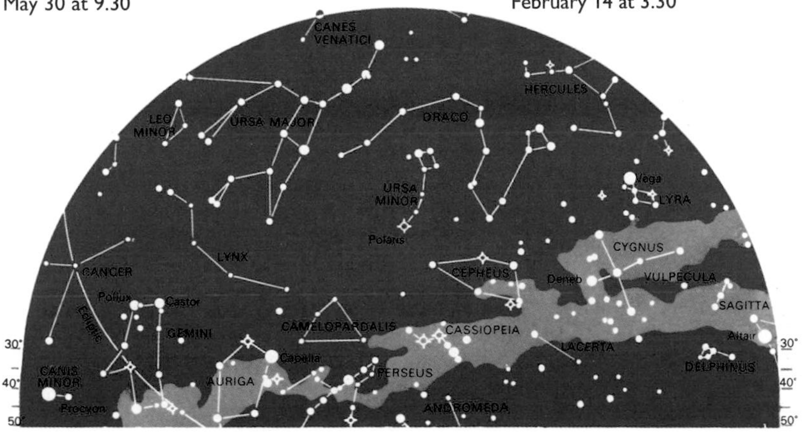

the celestial equator, while stars within 50° of the pole are circumpolar.

A star rises approximately four minutes earlier each night, so a pair of maps that represent the sky at a particular date and time serve equally well for a time approximately half an hour earlier on a date one week later, or two hours earlier one month later, and so on – as shown by the dates and times given. Of course, the positions of the planets and the Moon cannot be included on these maps, since they constantly vary in relation to the stellar background.

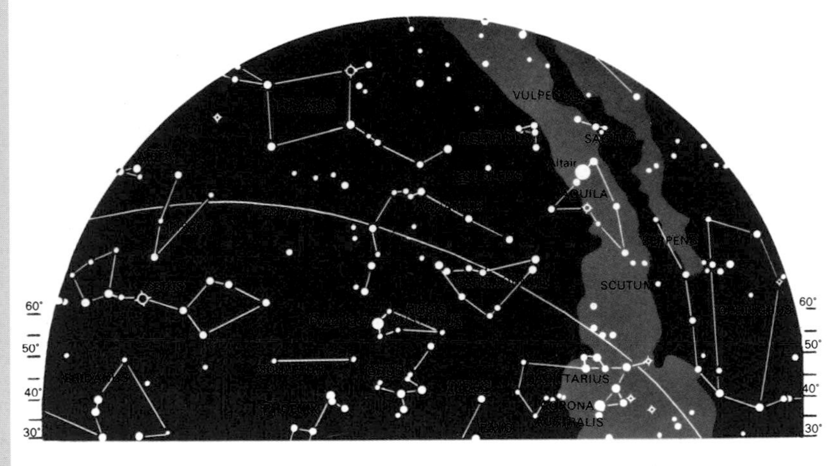

Evening
September 1 at 11.30
September 15 at 10.30
September 30 at 9.30

Morning
June 15 at 4.30
June 30 at 3.30
July 15 at 2.30

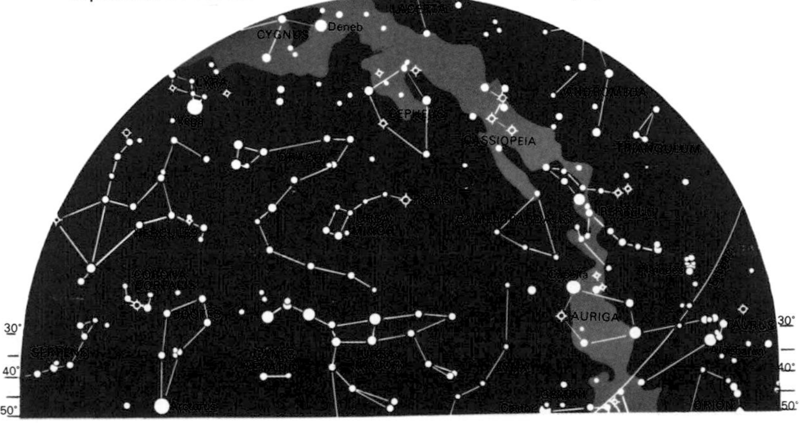

Guide to the stars (South)

The charts on this page and the two following show the changing appearance throughout the year of the constellations as viewed from the southern hemisphere. The upper chart in each pair shows the skies looking northward, toward the equator; the lower one shows the skies looking toward the pole. The scales at the sides of each map show

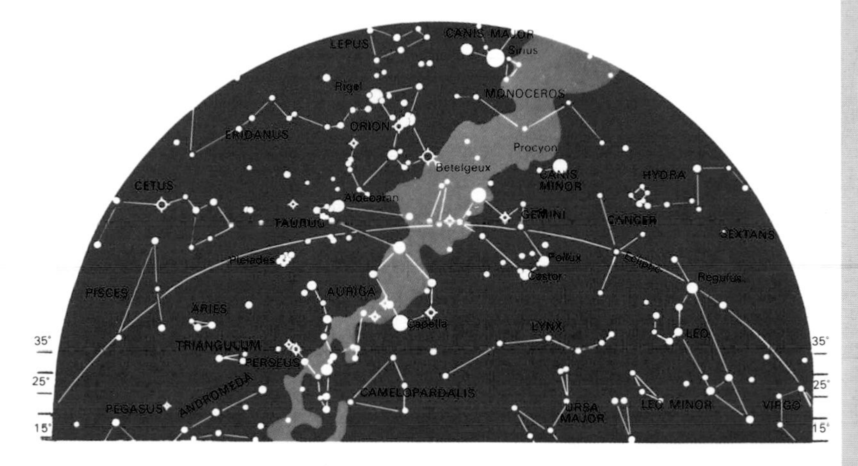

Evening
January 1 at 11.30
January 15 at 10.30
January 30 at 9.30

Morning
October 1 at 5.30
October 15 at 4.30
October 30 at 3.30

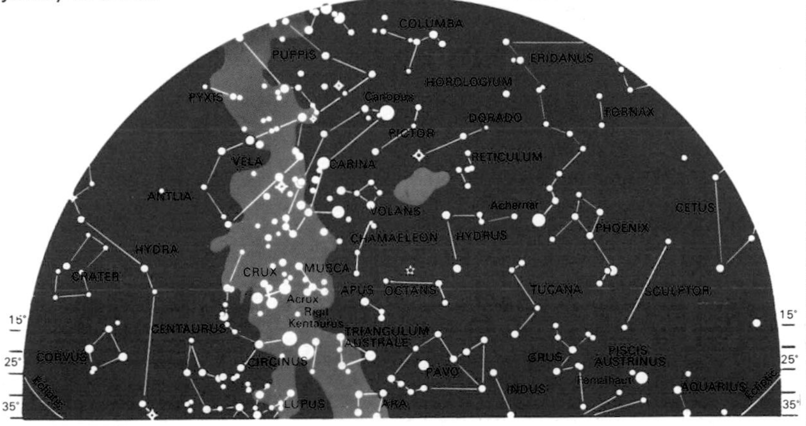

the horizon for the latitude indicated. Thus an observer at 15° south can see further north of the celestial equator than one at 35°, and the celestial pole appears closer to the horizon. He or she can in fact see 90° − 15° = 75° north of the celestial equator; the pole lies 15° above the horizon, though it is not marked by any star. Stars that are circumpolar for this observer – meaning that they never set – are those that lie less than 15° from the pole; and correspondingly for any other latitude. An observer at 35° S can see only 90° − 35° = 55° north of

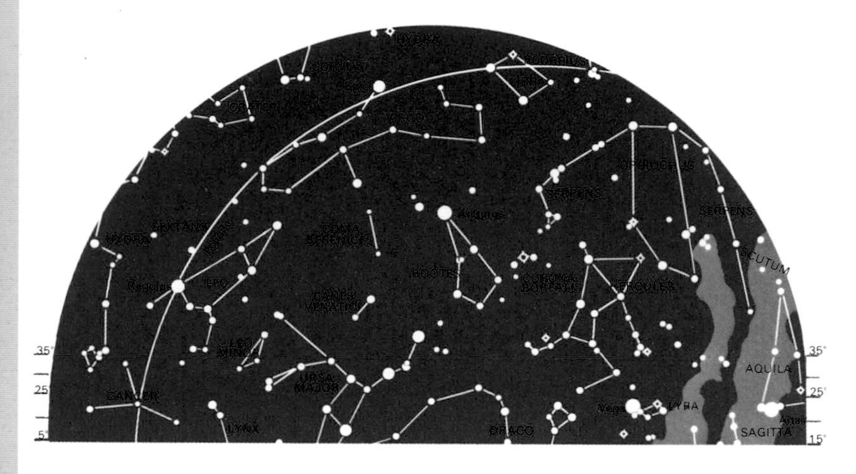

Evening
May 1 at 11.30
May 15 at 10.30
May 30 at 9.30

Morning
February 14 at 4.30
February 28 at 3.30
March 15 at 2.30

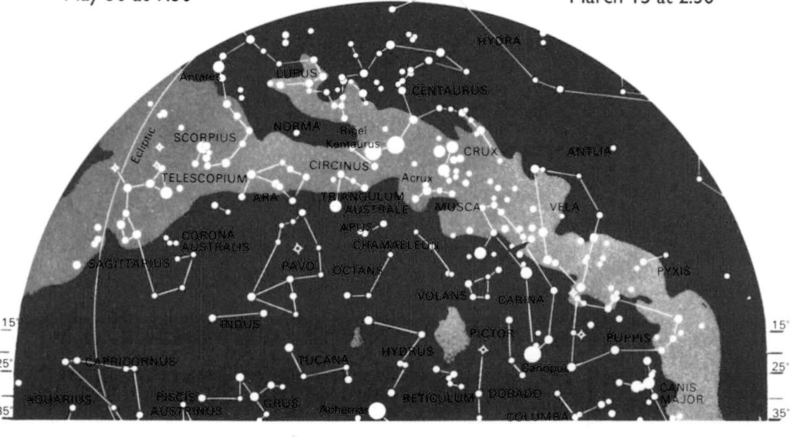

the celestial equator, while stars within 35° of the pole are circumpolar.

A star rises approximately four minutes earlier each night, so a pair of maps that represent the sky at a particular date and time serve equally well for a time that is approximately half an hour earlier on a date one week later, or two hours earlier one month later, and so on – as shown by the dates and times given. Of course, the positions of the planets and the Moon cannot be included on these maps, since they constantly vary in relation to the stellar background.

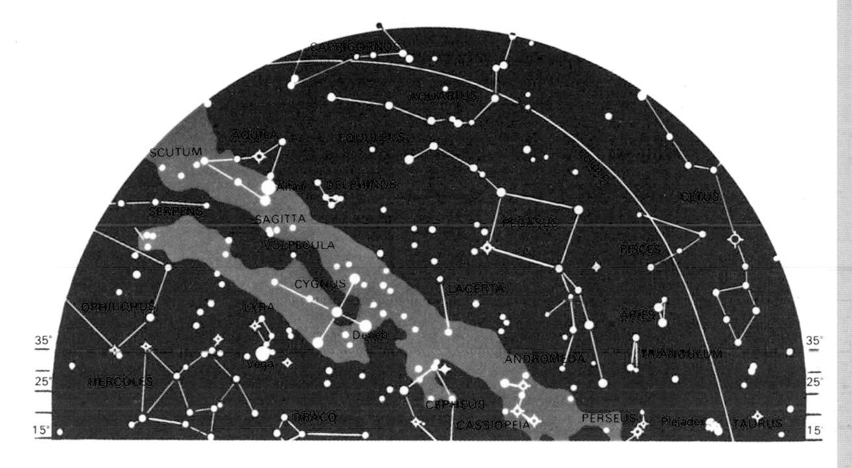

Evening
September 1 at 11.30
September 15 at 10.30
September 30 at 9.30

Morning
May 15 at 6.30
June 1 at 5.30
June 15 at 4.30

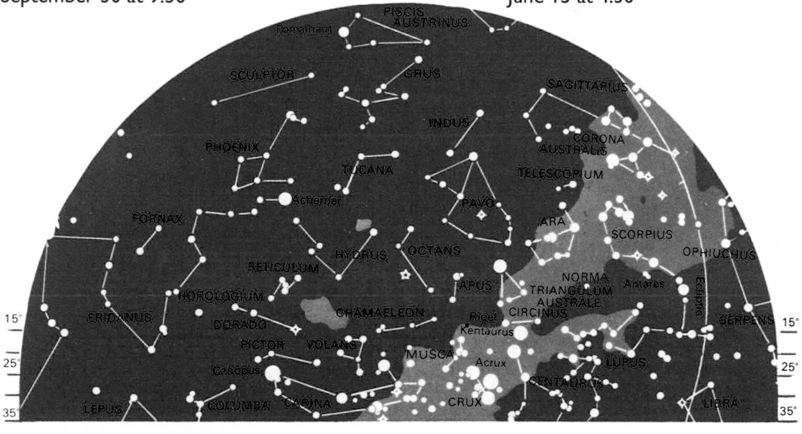

Andromeda

This is a large, important constellation, leading off from the Square of Pegasus; indeed one of the stars in the Square, Alpheratz, has been officially, if illogically, transferred to Andromeda. It used to be δ Pegasi, and is now α Andromedae. In mythology, Andromeda was the daughter of King Cepheus and Queen Cassiopeia; Cassiopeia was unwise enough to boast that her daughter was more beautiful than the sea nymphs, and thereby caused very grave offence to the Sea King, Neptune, who sent a monster to ravage Cepheus' kingdom. To placate the wrathful god, Andromeda was chained to a rock on the seashore to be devoured by the monster, but was rescued by the hero Perseus at the eleventh hour. This is one of the few legends to have a happy ending!

Andromeda is easy to locate, as its chief stars lie in a rough line. The Great Spiral, M31, is just visible with the naked eye on a clear night; binoculars show it easily. The open cluster NGC 752, not far from Almaak, is easy to find with binoculars, though it is loose and not at all conspicuous.

Telescopically, Almaak is seen to be a beautiful double, with an orange primary and a bluish companion, which is itself a very close double. The separation between the two main components is 9″.8.

The Mira variable R Andromedae lies close to the little triangle consisting of θ, σ and ϱ. Its position is RA 00h 24m, dec. +38° 35′. At maximum the magnitude is 5.8, within binocular range, but at minimum the star falls to 14.9; the period is 409 days.

Of course M31 is the most celebrated object in Andromeda, and a small telescope will show it well, together with its companion galaxies M32 and NGC 205. It is over 2 million light-years away, and is larger than our own Galaxy. Unfortunately it lies at a narrow angle to us, and the full beauty of the spiral form is lost. Its position is RA 00h 43m, dec. +41° 16′.5.

Leading stars	*RA* h m s	*Declination* ° ′ ″	*Magnitude*	*Spectrum*	*Distance* ly
β Mirach	01 09 43.8	+35 37 14	2.06	M0	88
α Alpheratz	00 08 23.2	+29 05 26	2.06	A0p	72
γ Almaak	02 03 53.9	+42 19 47	2.14	K2	121
δ	00 39 19.6	+30 51 40	3.27	K3	160

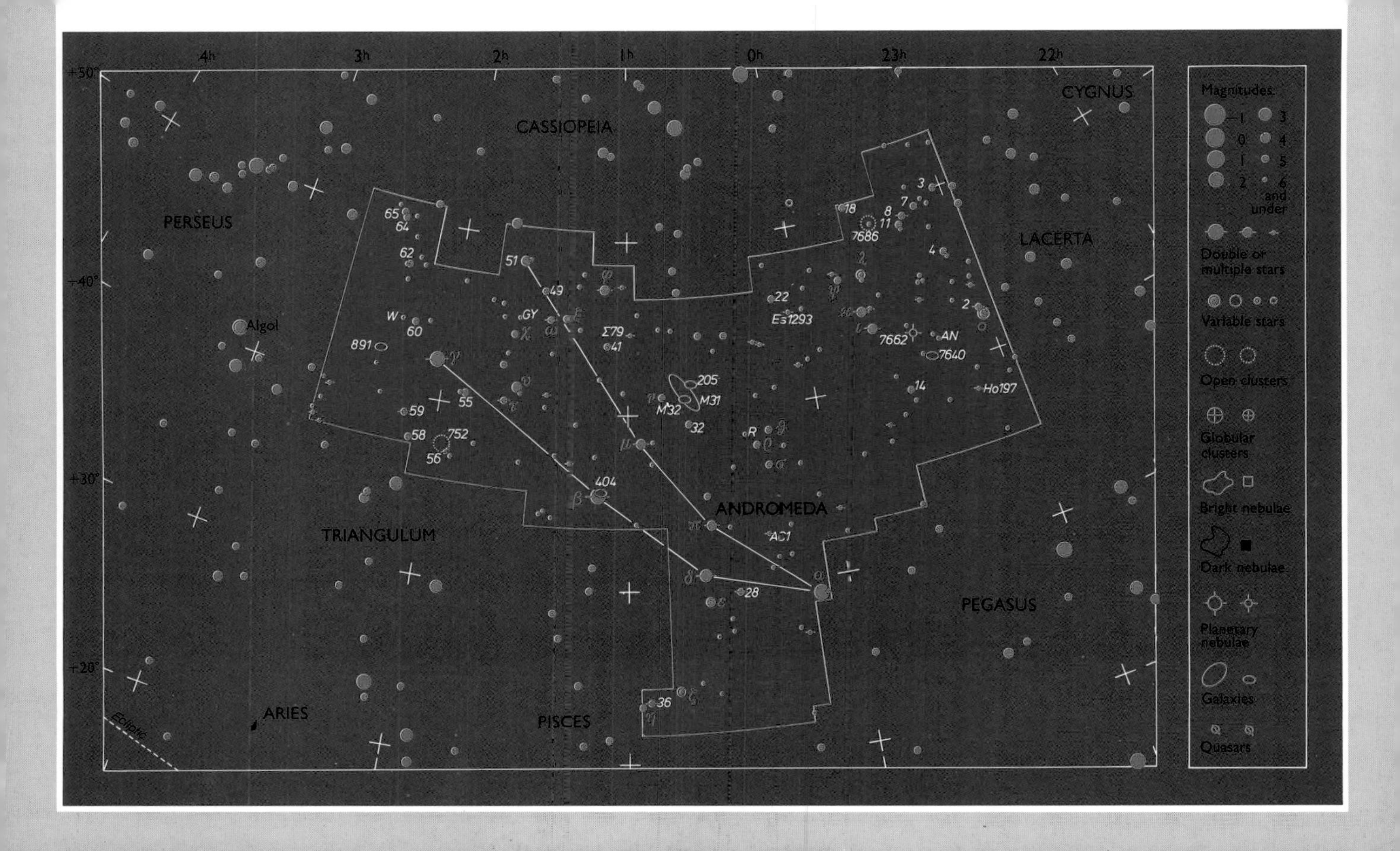

4h
3h
2h
1h
0h
23h
22h
+50°
+40°
+30°
+20°
CASSIOPEIA
CYGNUS
PERSEUS
LACERTA
ANDROMEDA
TRIANGULUM
PEGASUS
PISCES
ARIES
Ecliptic
Algol
65
64
62
51
49
GY
W
60
891
Σ79
41
205
M31
M32
32
59
55
58
752
56
404
R
22
Es1293
18
8
11
7686
3
7
4
2
7662
AN
7640
14
Ho197
AC1
28
36
Magnitudes:
−1
0
1
2
3
4
5
6
and under
Double or multiple stars
Variable stars
Open clusters
Globular clusters
Bright nebulae
Dark nebulae
Planetary nebulae
Galaxies
Quasars

Antlia/Pyxis

These are two very obscure constellations. Antlia, the Air Pump, was added to the sky in 1752 by the French astronomer Lacaille: its original name was Antlia Pneumatica. Pyxis, the Compass, was originally part of the huge constellation of Argo Navis, the Ship Argo, which carried Jason and his companions on their rather unprincipled quest for the Golden Fleece.

The brightest star in Antlia is α, magnitude 4.25; it is of type M0, and binoculars show it to be orange. Its position is RA 10h 27m 09s, dec. −31° 04′ 04″, so that from Britain it is always very low. There is no distinctive pattern to Antlia, and nothing of binocular interest. There are some galaxies within range of moderate telescopes, but even the brightest of them, the loose spiral NGC 2997, has an integrated magnitude of only 10.6: its position is RA 09h 45.6m, dec. −31° 11′.

Pyxis is almost equally obscure; the brightest stars are α (3.68), β (3.97) and γ (4.01), which lie more or less in a line. Again there is nothing of interest for the binocular user, and very little for the owner of a small telescope, though ζ is a fairly easy double: RA 08h 39.7m, dec. −29° 34′. The primary is of magnitude 4.9 and the secondary 9.1; the separation is 52″.4. It is worth watching the recurrent nova T Pyxidis, near ε; it is usually of magnitude 14, but flared up almost to naked-eye visibility in 1890, 1902, 1920, 1944 and 1966.

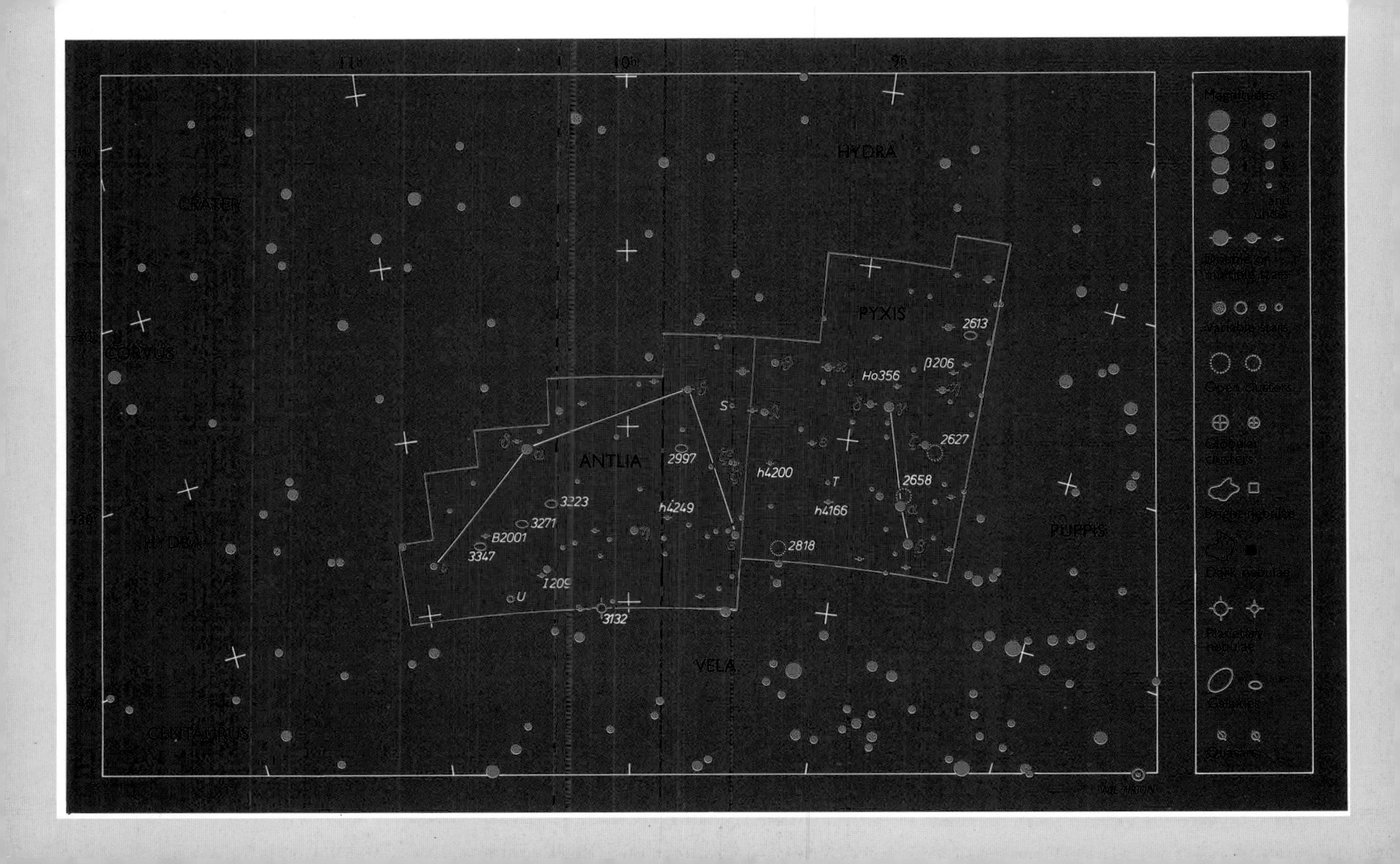

11h
10h
9h
HYDRA
CRATER
CORVUS
HYDRA
CENTAURUS
PYXIS
ANTLIA
VELA
PUPPIS
2613
β206
Ho356
2627
2658
h4200
T
h4166
2818
S
2997
h4249
3223
3271
B2001
3347
I209
U
3132
Magnitudes
1
0
1
2
3
4
5
6
and under
Double or multiple stars
Variable stars
Open clusters
Globular clusters
Bright nebulae
Dark nebulae
Planetary nebulae
Galaxies
Quasars

Apus/Triangulum Australe

Both these constellations are much too far south to be seen from anywhere in Europe. They were introduced by Bayer in his maps of 1603. Apus, the Bee, was originally Avis Indica, the Bird of Paradise. Triangulum Australe, the Southern Triangle, is one of the few constellations whose outline really recalls the object after which it is named.

Triangulum Australe is easy to find. It lies close to the brilliant Pointers to the Southern Cross, α and β Centauri. The orange color of α is very evident even with the naked eye, and is striking in binoculars. The most notable object is probably the open cluster NGC 6025 (RA 16h 3.7m, dec. −60° 30′), which is on the fringe of naked-eye visibility and is easy with binoculars.

The three brightest stars of Apus (α, γ and β) also make up a triangle. α is slightly yellowish, and is close to θ, which is decidedly red and is slightly variable, reaching the fringe of naked-eye visibility. δ is a very wide double; the components (magnitudes 4.7 and 5.1) are separated by 103 seconds of arc. The brighter member of the pair is obviously reddish.

TRIANGULUM AUSTRALE					
Leading stars	***RA*** h m s	***Declination*** ° ′ ″	***Magnitude***	***Spectrum***	***Distance*** ly
α	16 48 39.8	−69 01 39	1.92	K2	55
β	15 55 08.4	−63 25 50	2.85	F5	32
γ	15 18 54.5	−68 40 46	2.89	A0	91

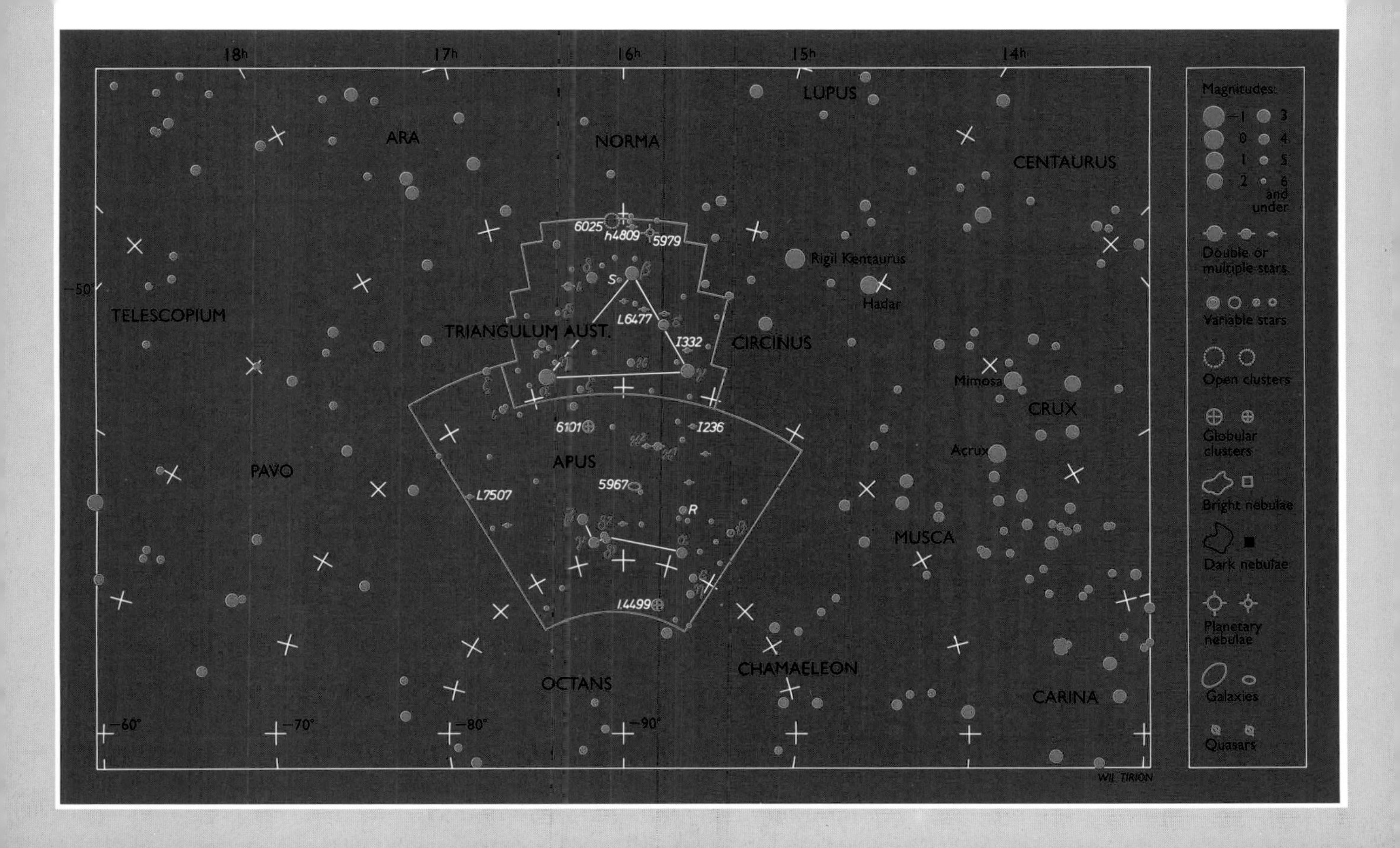
18h
17h
16h
15h
14h
LUPUS
ARA
NORMA
CENTAURUS
TELESCOPIUM
6025
h4809
5979
Rigil Kentaurus
Hadar
TRIANGULUM AUST.
L6477
I332
CIRCINUS
Mimosa
CRUX
6101
I236
Acrux
APUS
PAVO
5967
L7507
R
MUSCA
I.4499
CHAMAELEON
OCTANS
CARINA
−50°
−60°
−70°
−80°
−90°
Magnitudes:
−1
0
1
2
3
4
5
6 and under
Double or multiple stars
Variable stars
Open clusters
Globular clusters
Bright nebulae
Dark nebulae
Planetary nebulae
Galaxies
Quasars
WIL TIRION

Aquarius

Aquarius, the Water Bearer, is one of the Zodiacal constellations, but no well-defined legends seem to be attached to it, though it has been linked with Ganymede, cup bearer to the Olympian gods.

The constellation occupies much of the area between Pegasus and Fomalhaut. It is by no means distinctive, but the little group of stars lettered ψ – several of which are orange – can give the false impression of a loose cluster. ζ is a fine binary, with an orbital period of 856 years; the separation is 2 seconds of arc.

There are four notable non-stellar objects in Aquarius. M2 is one of the finest globular clusters in the sky. It is said to be a naked-eye object under good conditions; certainly it is very clear with binoculars, and is not hard to resolve. It is about 50,000 light-years away, with a diameter of about 150 light-years. M72 is much fainter, and is rather hard to resolve. Incidentally, number 73 in Messier's catalog was also in Aquarius, near M72, but is not a true cluster, and consists merely of four faint stars.

The two planetaries are of special interest. NGC 7009 is known as the Saturn Nebula, and NGC 7293 as the Helix. The Helix is usually regarded as the largest of all planetary nebulae, but it has a rather low surface brightness.

Note the variable R Aquarii (RA 23h 43.8m, dec. −15° 17′), which can rise to magnitude 5.8 at times, though it is generally much fainter. It shows rapid, unpredictable fluctuations, and is worth following.

Leading stars	RA h m s	Declination ° ′ ″	Magnitude	Spectrum	Distance ly
β Sadalsuud	21 31 33.3	−05 34 16	2.91	G0	980
α Sadalmelik	22 05 46.8	−00 19 11	2.96	G2	950
δ Scheat	22 54 38.8	−15 49 15	3.27	A2	98
ζ	22 28 49.5	−00 01 13	3.6	F2 + F2	98

Clusters and nebulae	RA h m	Declination ° ′	Magnitude	Nature
M2	21 33.5	−00 49	6.5	Globular cluster
M72	20 53.5	−12 32	9.3	Globular cluster
NGC 7009	21 04.2	−11 22	8.3	Planetary nebula
NGC 7293	22 29.6	−20 48	6.5	Planetary nebula

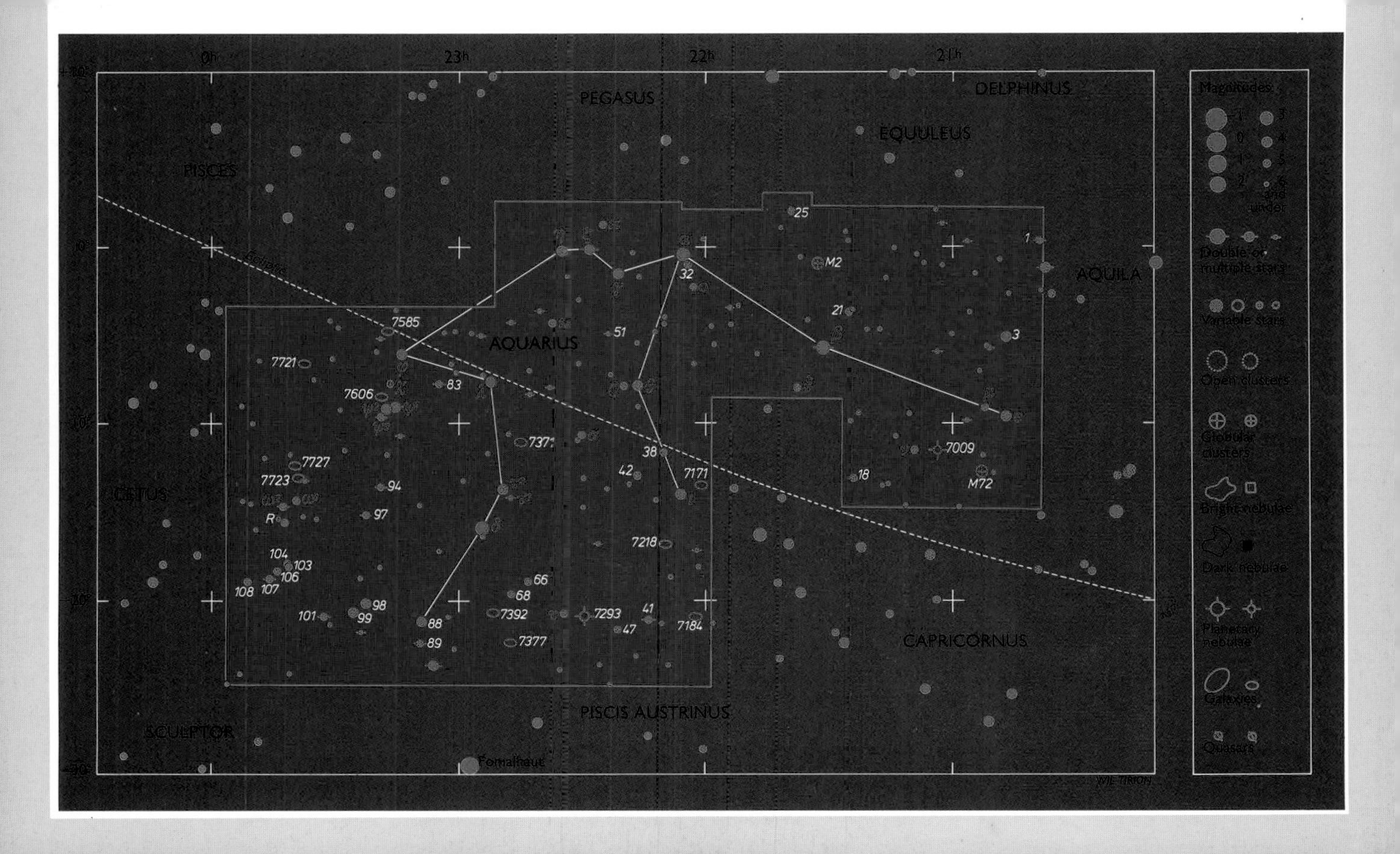
0h
23h
22h
21h
+10°
0°
-10°
-20°
-30°
PEGASUS
DELPHINUS
EQUULEUS
PISCES
AQUILA
AQUARIUS
CETUS
CAPRICORNUS
PISCIS AUSTRINUS
SCULPTOR
Fomalhaut
Ecliptic
M2
M72
7009
7585
7721
7606
7727
7723
7371
7171
7218
7184
7293
7392
7377
1
3
18
21
25
32
38
41
42
47
51
66
68
83
88
89
94
97
98
99
101
103
104
106
107
108
R
Magnitudes
Double or multiple stars
Variable stars
Open clusters
Globular clusters
Bright nebulae
Dark nebulae
Planetary nebulae
Galaxies
Quasars
WIL TIRION

Aquila

Aquila, the Eagle, is very prominent, and the Milky Way is very rich here. The constellation represents a mythological eagle sent by Jupiter to collect a shepherd boy, Ganymede, who was to become cup bearer to the gods.

Altair is one of the brightest stars in the sky, though its eminence is due mainly to its relative closeness; it is "only" 10 times more luminous than the Sun. It is flanked to either side by a fainter star, γ or Tarazed, which is obviously reddish, and β or Alshain (magnitude 3.71). Further south lies a line of three stars, θ (3.23), η (variable) and δ (3.36). η is a Cepheid with a period of 7.2 days and a magnitude range of 3.5 to 4.4; its fluctuations are easy to follow with the naked eye.

There are two easy open clusters in Aquila: NGC 6709 (magnitude 6.7; about 40 stars) and NGC 6755 (magnitude 7.5, 100 stars). The Mira variable R Aquilae (RA 19h 06.4m, dec. +08° 14′) has a period of 284 days and a magnitude range from 5.5 to 12.0.

Leading stars	RA h m s	Declination ° ′ ″	Magnitude	Spectrum	Distance ly
α Altair	19 50 46.8	+08 52 06	0.77	A7	16.6
γ Tarazed	19 46 15.4	+10 36 48	2.72	K3	280
ζ Dheneb	19 05 24.4	+13 51 48	2.99	B9	104

The Milky Way is a magnificent spectacle in Aquila.

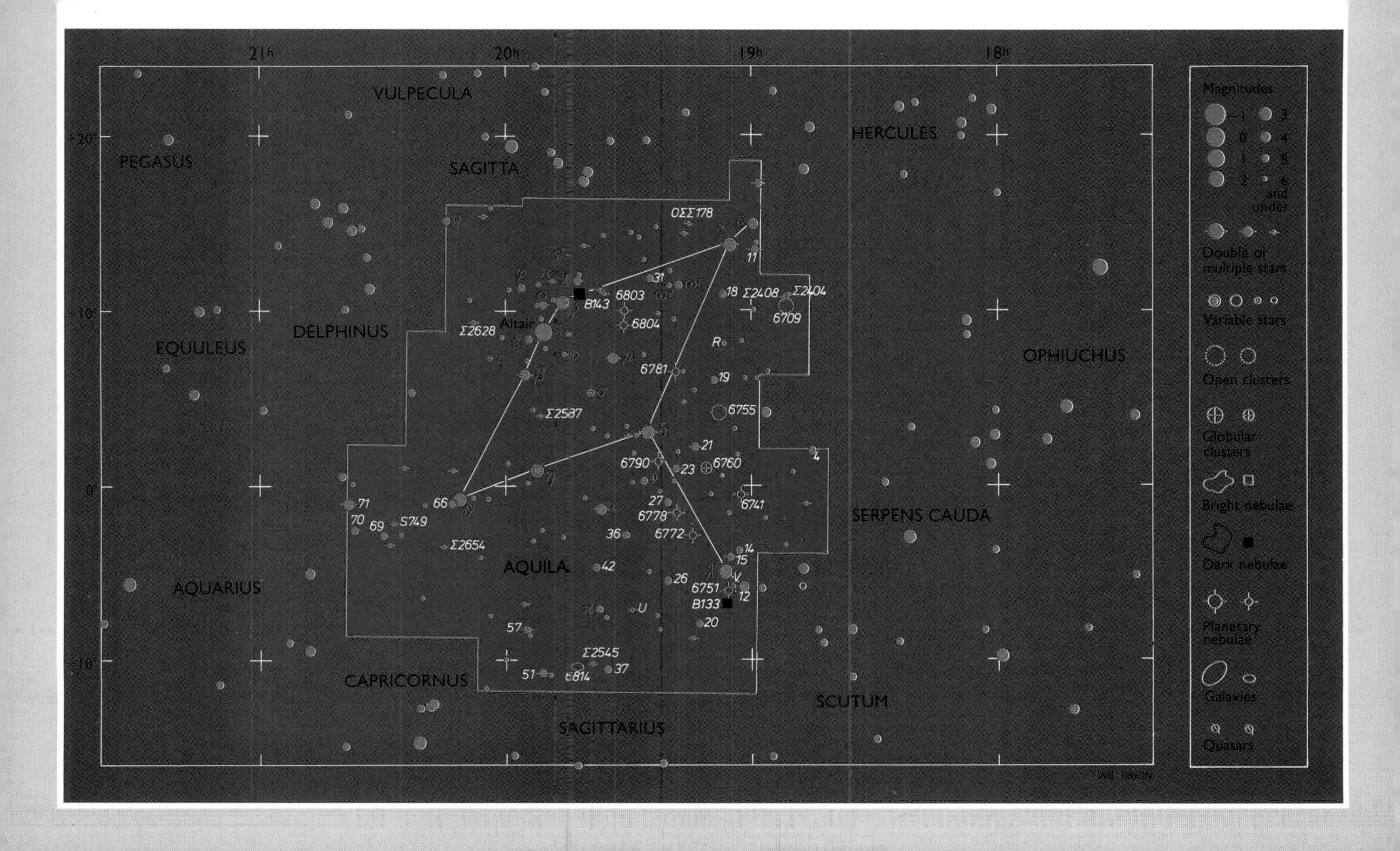

21h
20h
19h
18h
+20°
+10°
0°
−10°
VULPECULA
PEGASUS
SAGITTA
HERCULES
DELPHINUS
EQUULEUS
OPHIUCHUS
SERPENS CAUDA
AQUILA
AQUARIUS
CAPRICORNUS
SAGITTARIUS
SCUTUM
Altair
OΣΣ178
11
31
B143
6803
6804
18
Σ2408
Σ2404
6709
R
Σ2628
6781
19
6755
Σ2537
21
4
6790
23
6760
6741
66
71
70
69
S749
27
6778
6772
36
Σ2654
14
15
42
26
V
6751
12
U
B133
20
57
Σ2545
37
51
E814
Magnitudes:
−1
0
1
2
3
4
5
6
and under
Double or multiple stars
Variable stars
Open clusters
Globular clusters
Bright nebulae
Dark nebulae
Planetary nebulae
Galaxies
Quasars
WIL TIRION

Ara

Ara, the Altar, lies between θ Scorpii on one side and α Trianguli Australis on the other. The Milky Way flows through part of it. There seem to be no definite mythological legends attached to it, though it has been suggested that it may have been the altar on which Lupus, the Wolf, was to be sacrificed.

Ara has a fairly distinctive shape; other fairly bright stars, apart from those listed in the table, are γ (magnitude 3.34), δ (3.62), θ (3.66) and η (3.76). Three of the leading stars – β, ζ and η – are orange. R Arae, in the same binocular field with ζ and η, is an Algol-type eclipsing binary with a period of 4.4 days and a magnitude range of 6.0 to 6.9.

There are several globular clusters in Ara, of which the brightest is NGC 6397, close to β. It is an easy binocular object, and at 8,200 light-years is one of the closest of all globular clusters. Its position is RA 17h 40.7m, dec. −53° 40′.

Leading stars	*RA* h m s	*Declination* ° ′ ″	*Magnitude*	*Spectrum*	*Distance* ly
β	17 25 17.9	−55 31 47	2.85	K3	780
α	17 31 50.3	−49 52 34	2.95	B3	190
ζ	16 58 37.1	−55 59 24	3.13	K5	137

Dense star fields, with "lanes" of dark nebulosity, in Ara and Norma.

19h
18h
17h
16h
-40°
-50°
-60°
SAGITTARIUS
CORONA AUSTRALIS
SCORPIUS
TELESCOPIUM
INDUS
PAVO
APUS
ARA
NORMA
LUPUS
TRIANGULUM AUSTRALE
CIRCINUS
CENTAURUS
MUSCA
CRUX
Rigil Kentaurus
Hadar
Mimosa
h4949
6250
Brs13
6204
6352
I.4651
6193
6188
6253
6326
S397
h4978
6208
Rmk22
R
h4901
6221/15
6300
6362
WIL TIRION

Magnitudes
6 and fainter
Double or multiple stars
Variable stars
Open clusters
Globular clusters
Bright nebulae
Dark nebulae
Planetary nebulae
Galaxies
Quasars

Aries

Aries is officially the first constellation of the Zodiac, though in fact precession has now carried the "First Point of Aries," the vernal equinox, into the adjacent constellation of Pisces. In mythology, Aries represents the ram with the golden fleece, sent by the god Mercury (Hermes) to rescue the two children of the King of Thebes from assassination by their wicked stepmother. The ram could fly, and was carrying the two children to safety when the girl, Helle, overbalanced and fell to her death in the straits we now call the Hellespont; the boy, Phryxus, survived. After the ram's death, the golden fleece was hung in a sacred grove until removed forcibly by Jason and the Argonauts.

Aries is not hard to identify, near the line of stars marking Andromeda. α, β and γ make up a fairly conspicuous little trio. γ, or Mesartim, is a beautiful double, rather too close to be split with binoculars (the separation is 7.8 seconds of arc) but easy in a telescope; the components are equal at magnitude 4.8, so that with the naked eye Mesartim shows as a star of magnitude 3.9. The spiral galaxy NGC 772 is a fairly easy telescopic object, with an integrated magnitude just below 10: position RA 01h 59.3m, dec. +19° 01′. Three red variables, R, T and U, can exceed magnitude 8 when at maximum.

Leading stars	*RA* h m s	*Declination* ° ′ ″	*Magnitude*	*Spectrum*	*Distance* ly
α Hamal	02 07 10.3	+23 27 45	2.00	K2	85
β Sheratan	01 54 38.3	+20 48 29	2.64	A5	46

α, β and γ Arietis are noticeable near the center of the field here.

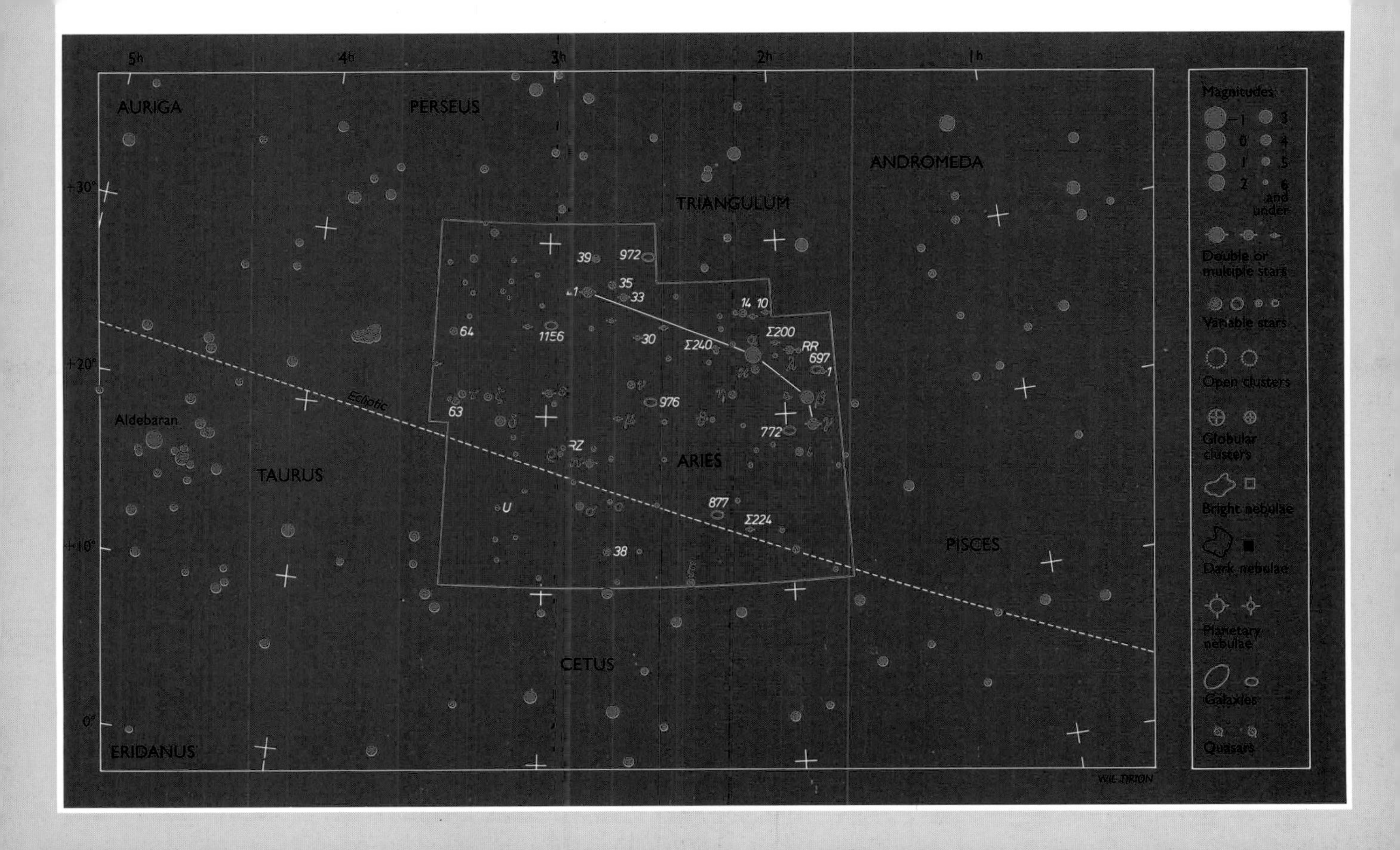

5h
4h
3h
2h
1h
+30°
+20°
+10°
0°
AURIGA
PERSEUS
ANDROMEDA
TRIANGULUM
ARIES
TAURUS
PISCES
CETUS
ERIDANUS
Aldebaran
Ecliptic
39
972
35
33
1
14
10
Σ200
RR
697
64
1156
30
Σ240
976
63
772
RZ
877
Σ224
U
38
Magnitudes
−1
0
1
2
3
4
5
6
and
under
Double or
multiple stars
Variable stars
Open clusters
Globular
clusters
Bright nebulae
Dark nebulae
Planetary
nebulae
Galaxies
Quasars
WIL TIRION

Auriga

Auriga, the Charioteer, is a very important northern constellation. It is named in honor of Erechthonius, son of Vulcan, the blacksmith of the gods; he became King of Athens, and is said to have invented the four-horse chariot. The star Al Nath used to be called γ Aurigae, but has now been transferred to Taurus as β Tauri.

Capella is the sixth brightest star in the sky, and this alone makes Auriga very easy to find; during winter evenings it is not far from the zenith as seen from Britain. Capella is yellow, and 70 times as luminous as the Sun; it is actually a very close binary. Close beside it is a little triangle of stars known as the Haedi or Kids: ε, η and ζ. Two of these, ε and ζ, are eclipsing binaries; ε is exceptionally luminous – over 200,000 times more powerful than the Sun – while the eclipsing companion has never been seen. The period is 27 years. ζ, or Sadatoni, is an eclipsing binary of less extreme type; the range is from magnitude 3.7 to 4.1, and the period is 972 days. The main component is red, and the secondary blue. At times the light of the blue star comes to us via the outer layers of the red supergiant, providing some interesting spectroscopic effects.

θ is an easy telescopic double (magnitudes 2.6 and 7.1, separation 3″.6) and there is a third star in the field, at a distance of 50″.

The Milky Way passes through Auriga, and there are three fine open clusters. The brightest, M37, is in the same binocular field as θ.

Leading stars	***RA*** h m s	***Declination*** ° ′ ″	***Magnitude***	***Spectrum***	***Distance*** ly
α Capella	05 16 41.3	+45 59 53	0.08	G8	42
β Menkarlina	05 50 31.7	+44 56 51	1.90	A2	72
θ	05 59 43.2	+37 12 45	2.62	A0p	82
ι Hassaleh	04 56 59.5	+33 09 58	2.69	K3	267
ε Almaaz	05 01 58.2	+43 49 24	2.99v	F0	4,600
η	05 06 30.8	+41 14 04	3.17	B3	200

Clusters	***RA*** h m	***Declination*** ° ′	***Magnitude***	***No. of stars***
M36	05 36.1	+34 08	6.0	60
M37	05 52.4	+32 33	5.6	150
M38	05 28.7	+35 50	6.4	100

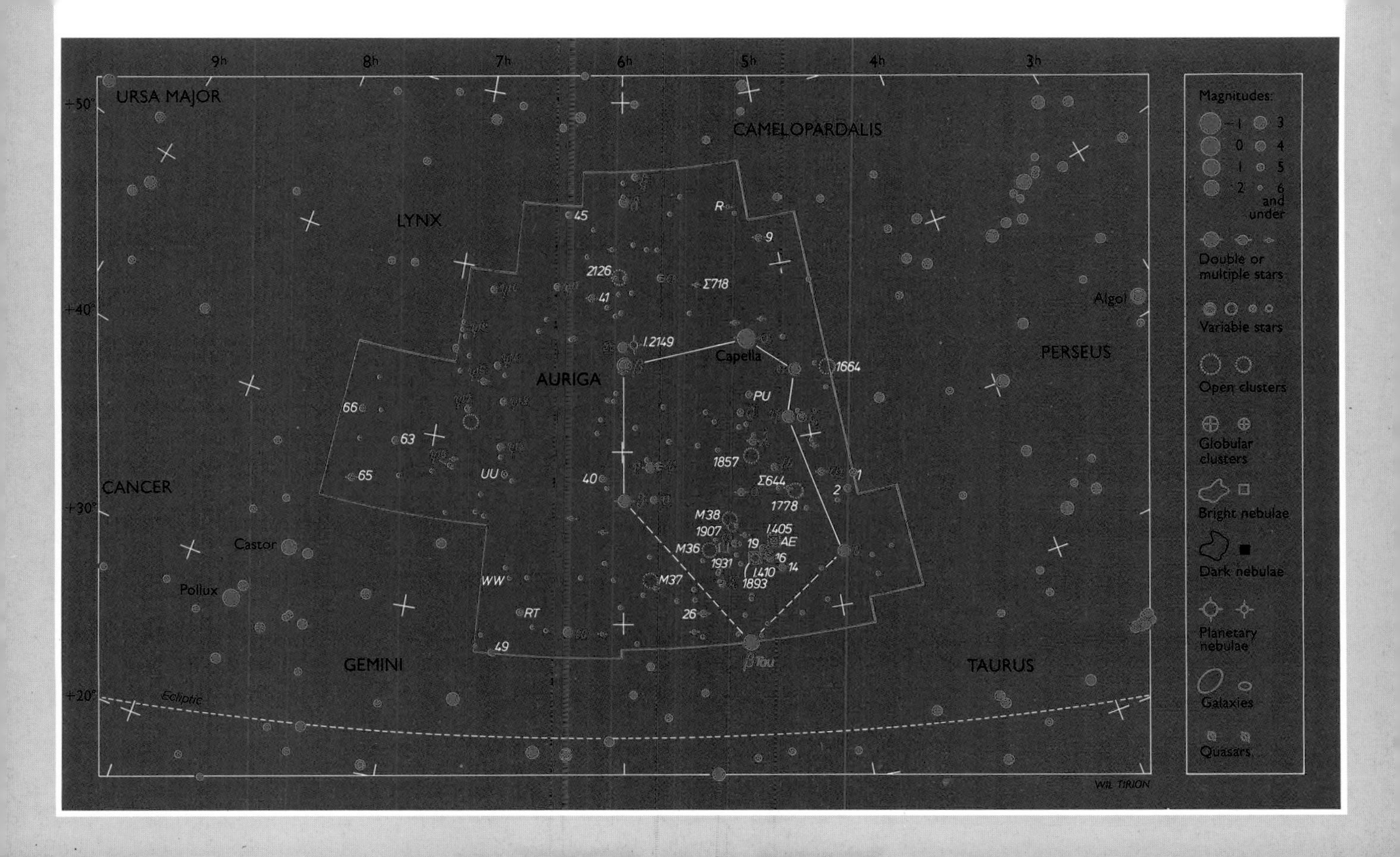
9h
8h
7h
6h
5h
4h
3h
+50°
+40°
+30°
+20°
URSA MAJOR
CAMELOPARDALIS
LYNX
AURIGA
PERSEUS
CANCER
GEMINI
TAURUS
Capella
Algol
Castor
Pollux
β Tau
Ecliptic
45
R
9
2126
Σ718
41
I.2149
1664
PU
66
63
UU
40
1857
Σ644
1
2
1778
M38
1907
I.405
AE
19
M36
16
1931
I.410
14
1893
M37
WW
RT
26
49
Magnitudes:
−1
0
1
2
3
4
5
6
and
under
Double or
multiple stars
Variable stars
Open clusters
Globular
clusters
Bright nebulae
Dark nebulae
Planetary
nebulae
Galaxies
Quasars
WIL TIRION

Boötes

Boötes, the Herdsman, is distinguished by the presence of Arcturus, the brightest star in the northern hemisphere of the sky; its three superiors, Sirius, Canopus and α Centauri, are well to the south of the celestial equator. There are no definite legends attached to Boötes, but according to one version he was placed in the sky as a reward for having invented the plow drawn by two oxen.

Quite apart from the lovely orange Arcturus – 115 times as luminous as the Sun – Boötes is easy to find, because of the characteristic Y-formation made up of Arcturus, η, ε, γ and α Coronae Borealis. ε is a good double, with magnitudes 2.5 and 4.9, separation 2″.8; the primary is orange, and the secondary looks bluish by contrast. ζ is a close binary (magnitudes 4.6 and 4.6, separation 1″, period 123 years) and there is a third star in the field, at a distance of 99″.

In April 1860 a star of magnitude 9.7 was observed telescopically in the same binocular field as Arcturus. It has been listed as T Boötis, but has never been seen again; it may have been a nova, but the position is worth checking (RA 14h 11m, dec. +19° 18′), as it may well reappear in the future if it is in fact a recurrent nova or an irregular variable.

β Boötis or Nekkar (magnitude 3.50) is the naked-eye star nearest to the radiant of the meteor shower of early January. This is known as the Quadrantid shower, because this part of Boötes once formed a now-discarded constellation, Quadrans (the Quadrant). Quadrans was added to the sky in 1775 by Johann Bode, but has since been deleted.

Leading stars	***RA*** h m s	***Declination*** ° ′ ″	***Magnitude***	***Spectrum***	***Distance*** ly
α Arcturus	14 15 39.6	+19 10 57	−0.04	K2	36
ε Izar	14 44 59.1	+27 04 27	2.37	K0	150
η	13 54 41.0	+18 23 51	2.68	G0	32
γ Seginus	14 32 4.6	+38 13 30	3.03	A7	104

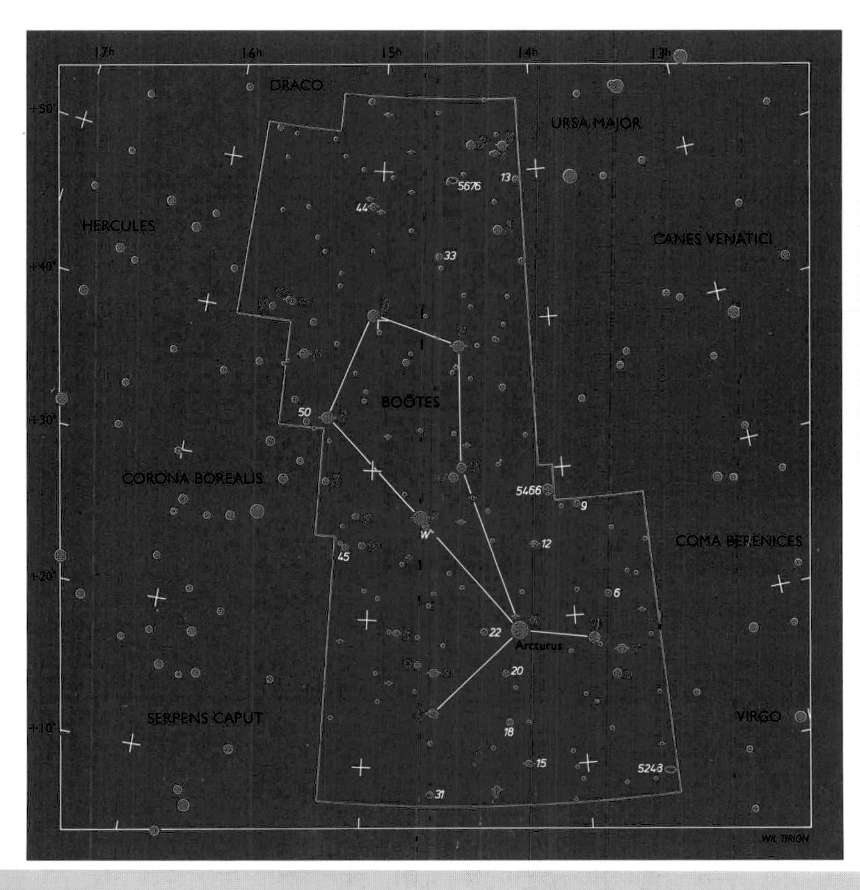
17h
16h
15h
14h
13h
+50°
+40°
+30°
+20°
+10°
DRACO
URSA MAJOR
HERCULES
CANES VENATICI
BOÖTES
CORONA BOREALIS
COMA BERENICES
SERPENS CAPUT
VIRGO
Arcturus
5676
13
44
33
50
5466
9
W
45
12
6
22
20
18
15
5243
31
WIL TIRION

Magnitudes:
−1
0
1
2
3
4
5
6
and under
Double or multiple stars
Variable stars
Open clusters
Globular clusters
Bright nebulae
Dark nebulae
Planetary nebulae
Galaxies
Quasars

Caelum/Columba

These are two modern constellations. Caelum (the Graving Tool) has no mythological associations, though Columba was originally Columba Noachi, Noah's Dove, presumably the bird which was released from the Ark.

Frankly, Caelum contains nothing of real interest; its brightest star is α, magnitude 4.45 (RA 04h 40m 33.6s, dec. −41° 51′ 50″). Columba does at least have two reasonably bright stars (see table). Both constellations are too low to be seen well from Britain, and η Columbae, declination −43 degrees, does not rise at all. There is little of interest here. The Mira variable T Columbae (RA 05h 19.3m, dec. −33° 42′) can reach magnitude 6.6 at maximum; at minimum it drops to 12.7, with a period of 226 days. The brightest galaxy is NGC 1808 (RA 05h 07.7m, dec. −37° 32′), which has an integrated magnitude of about 10, but is inconspicuous. Columba can be identified by the curved line of stars, including α and β, south of Lepus.

COLUMBA

Leading stars	*RA* h m s	*Declination* ° ′ ″	*Magnitude*	*Spectrum*	*Distance* ly
α Phakt	05 39 38.9	−34 04 27	2.64	B8	121
β Wazn	05 50 57.5	−35 46 06	3.12	K2	143

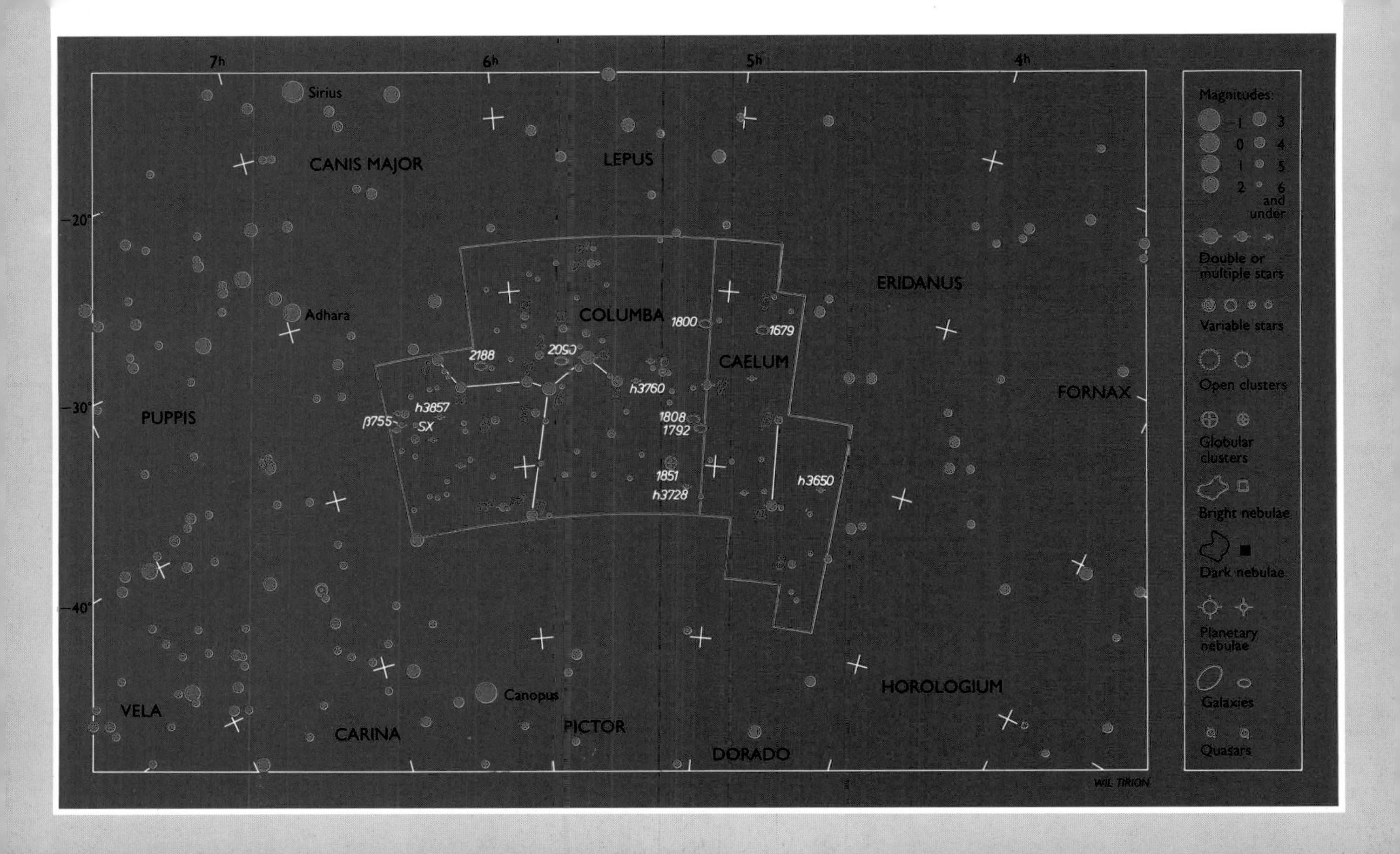
7h
6h
5h
4h
-20°
-30°
-40°
Sirius
CANIS MAJOR
LEPUS
ERIDANUS
FORNAX
Adhara
COLUMBA
1800
1679
CAELUM
2188
2050
h3760
h3857
β755
SX
PUPPIS
1808
1792
1851
h3728
h3650
HOROLOGIUM
VELA
CARINA
Canopus
PICTOR
DORADO
WIL TIRION
Magnitudes:
-1
0
1
2
3
4
5
6
and
under
Double or
multiple stars
Variable stars
Open clusters
Globular
clusters
Bright nebulae
Dark nebulae
Planetary
nebulae
Galaxies
Quasars

Camelopardalis

The Giraffe is one of the most barren regions of the sky. It lies between Auriga and the Pole Star, and though it covers a wide area it has no star as bright as the fourth magnitude; the leader is β (magnitude 4.03; RA 05h 03m 25s.1, dec. +60° 26′ 32″). The M-type semiregular variable VZ Camelopardalis (RA 07h 31.1m, dec. +82° 25′) is noticeably red when seen in binoculars; it has a small range of magnitude, 4.8 to 5.2, and a rough period of about 24 days.

Interestingly, several of the dim stars in Camelopardalis are in fact very luminous and remote; α, which is of magnitude 4.3, is well over 20,000 times as bright as the Sun.

There are some faint clusters and nebulae in the constellation, none of them particularly easy to identify. The most notable is probably NGC 2403 (RA 07h 36.9m, dec. +65° 36′), which is a loose spiral, only about 8 million light-years away and therefore not too far beyond the Local Group.

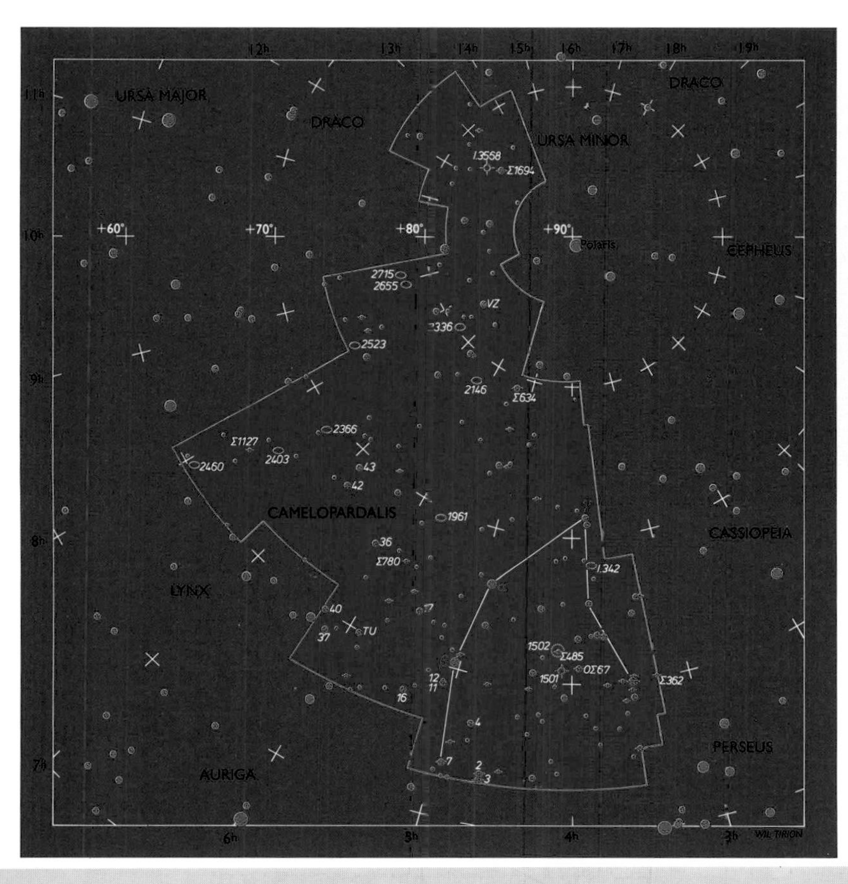

URSA MAJOR
DRACO
URSA MINOR
DRACO
CEPHEUS
Polaris
CAMELOPARDALIS
CASSIOPEIA
LYNX
AURIGA
PERSEUS
+60°
+70°
+80°
+90°
WIL TIRION

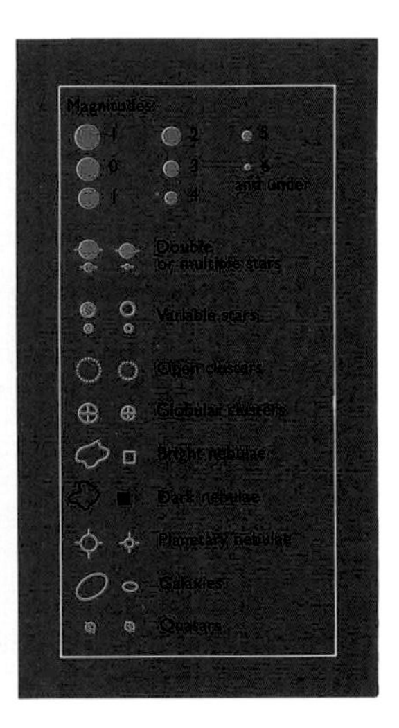

Magnitudes
-1
0
1
2
3
4
5
6 and under
Double or multiple stars
Variable stars
Open clusters
Globular clusters
Bright nebulae
Dark nebulae
Planetary nebulae
Galaxies
Quasars

Cancer

Cancer, the Crab, is one of the more obscure of the Zodiacal constellations. At least there is a legend attached to it. It represents a crab which Juno, the queen of Olympus, sent to the rescue of the multi-headed monster Hydra, which was doing battle with the hero Hercules. Not surprisingly, Hercules trod on the crab and squashed it – but as a reward for its efforts, Juno placed it in the sky.

In form Cancer resembles a very dim and ghostly Orion; the main pattern is formed by β or Altarf (magnitude 3.52), α or Acubens (4.25), δ or Asellus Australis (3.94), ι (4.02) and χ (5.1). Cancer lies between the Twins on one side and the Sickle of Leo on the other, so that it occupies the large triangle formed by Pollux, Procyon and Regulus.

There are two red variables within binocular range. X Cancri, near δ, has a range of 5.6 to 7.5; there is a rough period of 195 days, and the color is striking. R Cancri, near β, is an ordinary Mira variable with a range of 6.1 to 11.8, and a period of 362 days. The semiregular RS Cancri, not far from ι, has a period of 120 days and a range of 6.2 to 7.7.

ζ, or Tegmine, is a multiple star. The two most prominent components are of magnitude 5.0 and 6.2, with separation 5″.7. The brighter star is itself a close binary, and there is a third star in the same field; position RA 08h 12.2m, dec. +17° 39′.

Cancer is redeemed by the presence of two fine open clusters. M44 (Praesepe) is an easy naked-eye object; it is flanked by δ and γ (magnitude 4.7), which are known as the "Asses" because Praesepe itself is often nicknamed the Manger. (It is also called the Beehive, while for some unknown reason the ancient Chinese gave it the unromantic name of "the Exhalation of Piled-up Corpses.") Praesepe is probably best seen with binoculars, and is easy to resolve into stars.

The second open cluster is M67, near α: RA 08h 50.4m, dec. +11° 49′. It is on the fringe of naked-eye visibility and easy in binoculars; it is not hard to resolve. It lies well away from the galactic plane, and is one of the oldest known open clusters; it is well over 2,500 light-years away, and must include something like a thousand stars – very populous for a cluster of this type.

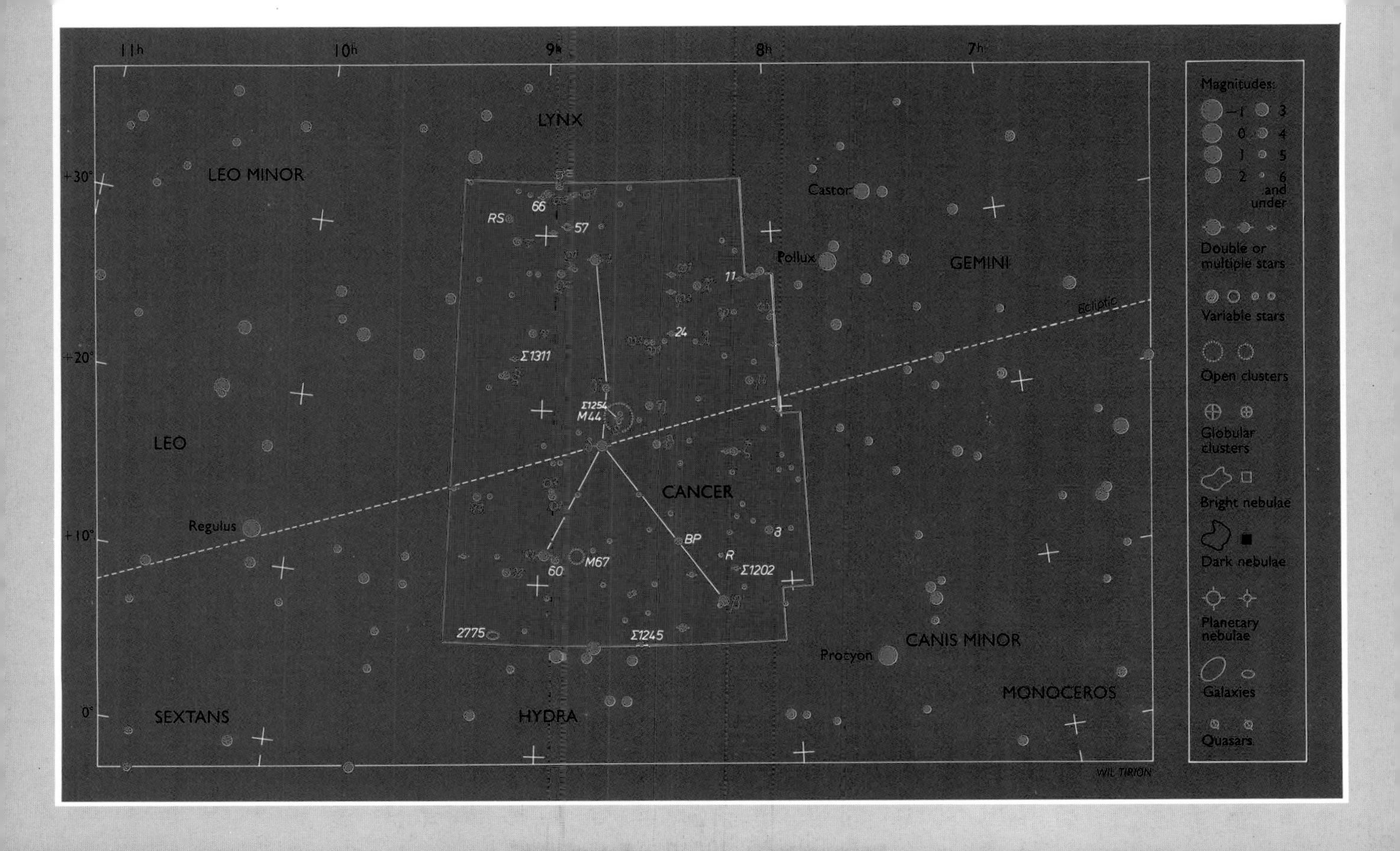
11h
10h
9h
8h
7h
+30°
+20°
+10°
0°
LYNX
LEO MINOR
LEO
SEXTANS
HYDRA
CANCER
GEMINI
CANIS MINOR
MONOCEROS
Castor
Pollux
Procyon
Regulus
Ecliptic
RS
66
57
11
24
Σ1311
Σ1254
M44
BP
R
Σ1202
60
M67
2775
Σ1245
WIL TIRION
Magnitudes:
−1
0
1
2
3
4
5
6 and under
Double or multiple stars
Variable stars
Open clusters
Globular clusters
Bright nebulae
Dark nebulae
Planetary nebulae
Galaxies
Quasars

Canes Venatici

This constellation was added to the sky by Hevelius in 1690; the two hunting dogs were originally called Asterion and Chara. They are being held by the herdsman Boötes, and it is suggested that this may be to prevent them from chasing the two Bears round the celestial pole. The only bright star is α, or Cor Caroli, named by Edmond Halley in honor of King Charles I of England: RA 12h 56.2m, dec. +38° 19′ 06″. It is of type A0p, and lies 65 light-years from us. It is the brightest member of the class of magnetic variables, but optically it is to all intents and purposes constant in light. Its nearness to Ursa Major makes it easy to find. It is an easy double: magnitudes 2.9 and 5.5, separation 19″.7.

There is one splendid globular cluster in the constellation: M3 (RA 13h 42.4m, dec. +28° 23′). It lies near the edge of Canes Venatici, not far from the star β Comae (4.26); it is not far below naked-eye visibility, and is a fine sight in binoculars.

Canes Venatici is rich in galaxies, and there are several in Messier's list. Of these the most famous is M51, the Whirlpool, which was actually the first spiral to be identified as such (by Lord Rosse, using his great Birr reflector in 1845). A 30-centimeter telescope is adequate to show the spiral form; the galaxy is linked with a companion galaxy, NGC 5195. Its distance is about 37 million light-years. M63 is also a spiral, but the arms are much less obvious. M94 is face-on to us, but the arms are rather tightly wound, with a bright, condensed nucleus. M106 – added later to the original Messier catalog – has one arm which is within the range of a 25-centimeter reflector. It lies about midway between β Canum Venaticorum (magnitude 4.26) and Phad, or γ Ursae Majoris, in the Great Bear.

Galaxies	RA h m	Declination ° ′	Magnitude	Type
M51	13 29.9	+47 12	8.4	Sc
M63	13 15.8	+42 02	8.6	Sc
M94	12 50.9	+41 07	8.2	Sb
M106	12 19.0	+47 18	8.3	Sb

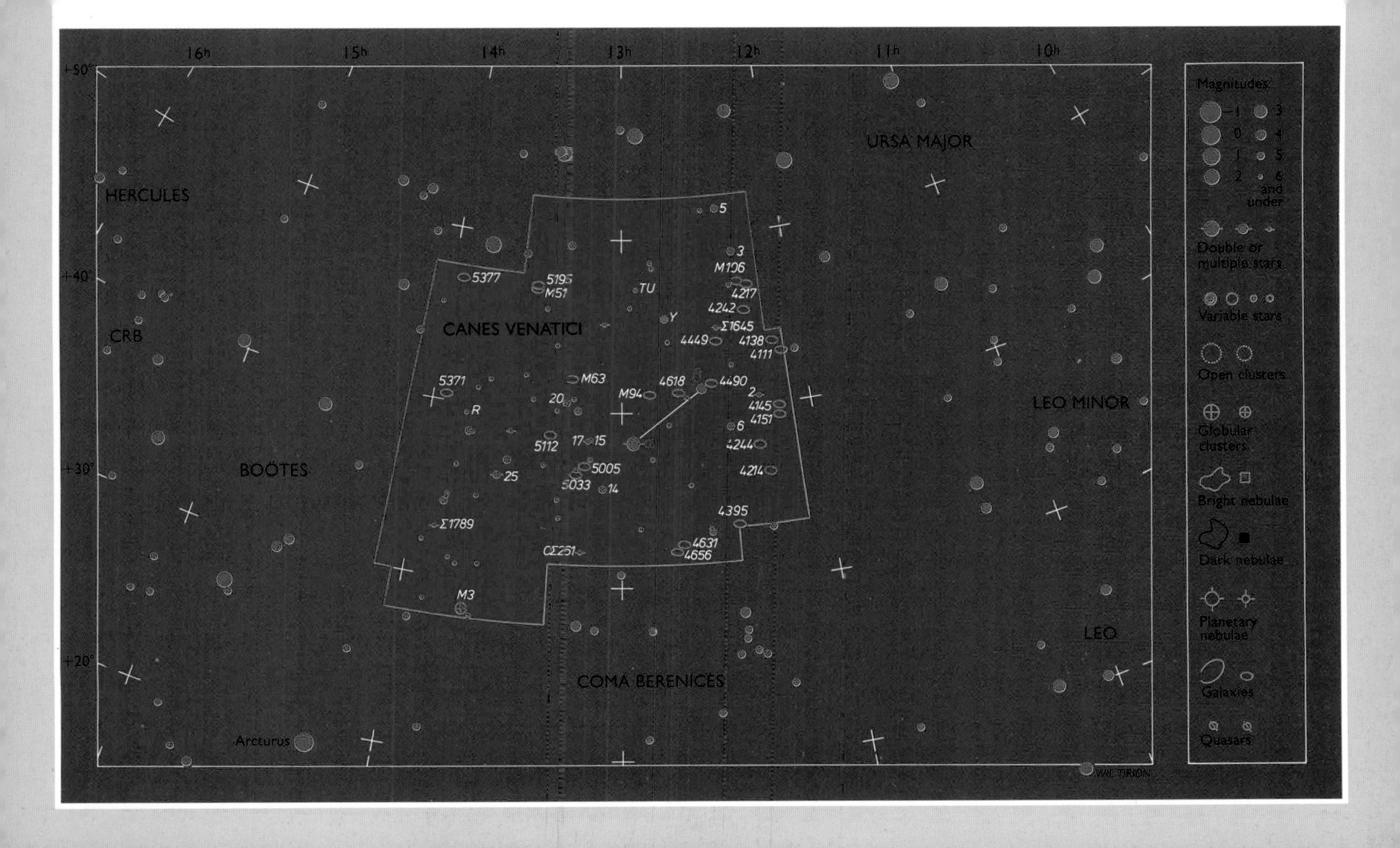
16h
15h
14h
13h
12h
11h
10h
+50°
+40°
+30°
+20°
HERCULES
CRB
BOÖTES
Arcturus
URSA MAJOR
LEO MINOR
LEO
COMA BERENICES
CANES VENATICI
5377
5195
M51
TU
Y
5
3
M106
4217
4242
Σ1645
4449
4138
4111
5371
M63
M94
4618
4490
2
4145
4151
6
R
20
17
15
5112
4244
5005
5033
14
4214
25
Σ1789
4395
CΣ251
4631
4656
M3
Magnitudes:
-1
0
1
2
3
4
5
6 and under
Double or multiple stars
Variable stars
Open clusters
Globular clusters
Bright nebulae
Dark nebulae
Planetary nebulae
Galaxies
Quasars
WIL TIRION

Canis Major/Lepus

Canis Major (the Great Dog) and Lepus (the Hare) belong to Orion's retinue. Canis Major is distinguished by the presence of Sirius, the brightest star in the sky. Sirius has a white dwarf companion with a period of 50 years. Its magnitude is 8.5, but in the mid-1990s the apparent separation is at its minimum, so that a powerful telescope will be needed to show it.

ε has a magnitude-7.4 companion at a separation of 7″.5. R Canis Majoris (RA 07h 19.5m, dec. −16° 24′) is an Algol-type eclipsing binary, with range 5.7 to 6.3 and period 1.14 days. UW Canis Majoris (RA 07h 18.4m, dec. −24° 34′) is a β Lyrae-type eclipsing binary, with magnitude range 4.0 to 5.3 and period 4.39 days.

The open cluster M41, near Sirius, is visible with the naked eye, and is resolvable with powerful binoculars. The brightest star in it is orange, with a K-type spectrum, at a distance of 2,400 light-years.

Lepus has two stars above the third magnitude. R Leporis, near μ (magnitude 3.31) is an intensely red variable of type N: RA 0h 59.6m, dec. −14° 48′. The range is 5.5 to 11.7, so at maximum it is an easy binocular object; period 432 days. Also visible with binoculars is the fine globular cluster M79 (RA 05h 24.5m, dec. −24° 33′).

CANIS MAJOR					
Leading stars	***RA*** h m s	***Declination*** ° ′ ″	***Magnitude***	***Spectrum***	***Distance*** ly
α Sirius	06 45 08.9	−16 42 58	−1.46	A1	8.6
ε Adhara	06 58 37.5	−28 58 20	1.50	B2	490
δ Wezea	07 08 23.4	−26 23 36	1.86	F8	3,060
β Mirzam	06 22 41.9	−17 57 22	1.98	B1	720
η Aludra	07 24 05.6	−29 18 11	2.44	B5	2,500
ζ Phurad	06 20 18.7	−30 03 48	3.02	B3	290
o^2	07 03 01.4	−23 50 00	3.03	B3	2,800

LEPUS					
Leading stars	***RA*** h m s	***Declination*** ° ′ ″	***Magnitude***	***Spectrum***	***Distance*** ly
α Arneb	05 32 43.7	−17 49 20	2.58	F0	950
β Nihal	05 28 14.7	−20 45 35	2.84	G2	316

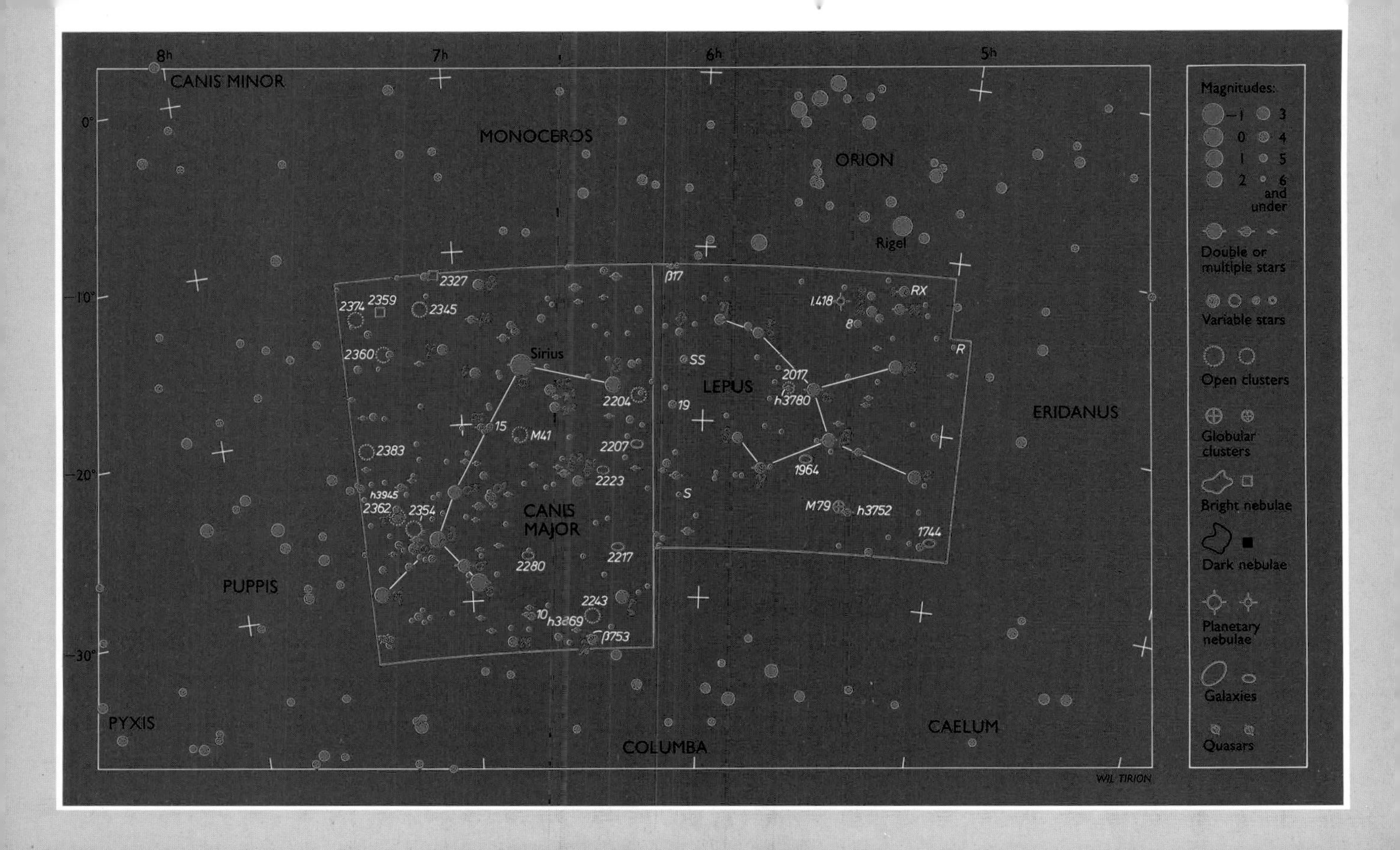
8h
7h
6h
5h
0°
−10°
−20°
−30°
CANIS MINOR
MONOCEROS
ORION
Rigel
ERIDANUS
PUPPIS
PYXIS
COLUMBA
CAELUM
CANIS MAJOR
LEPUS
Sirius
2327
2374
2359
2345
2360
2383
h3945
2362
2354
15
M41
2204
2207
2223
2280
2217
2243
10
h3869
β753
β17
SS
19
S
1.418
8
RX
R
2017
h3780
1964
M79
h3752
1744
WIL TIRION
Magnitudes:
−1
0
1
2
3
4
5
6 and under
Double or multiple stars
Variable stars
Open clusters
Globular clusters
Bright nebulae
Dark nebulae
Planetary nebulae
Galaxies
Quasars

Canis Minor/Monoceros

Canis Minor, Orion's lesser dog, has two bright stars. Procyon is very brilliant and, like Sirius, has a white dwarf companion; this is, however, a difficult object, as the companion is much fainter (magnitude 13). The period is 40.7 years. There is nothing else of real interest in the constellation.

Monoceros represents the fabled unicorn, but it is not an ancient constellation, and no particular legends are attached to it. It occupies the area enclosed by lines joining Procyon, Alhena in Gemini and Saiph in Orion, but there is no distinctive pattern. The brightest star is β, magnitude 3.7 (RA 06h 28m 48″.9, dec. −07° 01′ 58″).

Monoceros is crossed by the Milky Way, and there are many faint clusters and nebulae. β is a triple star; the main components are of magnitudes 4.5 and 6.5, separated by 7″.3, and there is a third star of magnitude 6.1 at a separation of 10″.

The open cluster M50 is an easy binocular object (RA 07h 03.2m, dec. −08° 20′) and is not hard to resolve telescopically. Of more interest is the open cluster NGC 2244 (RA 06h 32.4m, dec. +04° 52′) round the star 12 Monocerotis; it is easy to find with binoculars. Surrounding the cluster is the Rosette Nebula, which when photographed with large telescopes is truly glorious; it is around 4,500 light-years away, and its stars are probably no more than half a million years old. Visually it is decidedly elusive, but it is not difficult to photograph.

Two more nebulae worth locating are Hubble's variable nebula round R Monocerotis (RA 06h 39.2m, dec. +08° 44′) and the Cone Nebula, round S Monocerotis (RA 06h 40.9m, dec. +09° 54′).

CANIS MINOR					
Leading stars	***RA*** h m s	***Declination*** ° ′ ″	***Magnitude***	***Spectrum***	***Distance*** ly
α Procyon	07 39 18.1	+05 13 30	0.38	F5	11.4
β Gomeisa	07 27 09.0	+08 17 21	2.90	B8	137

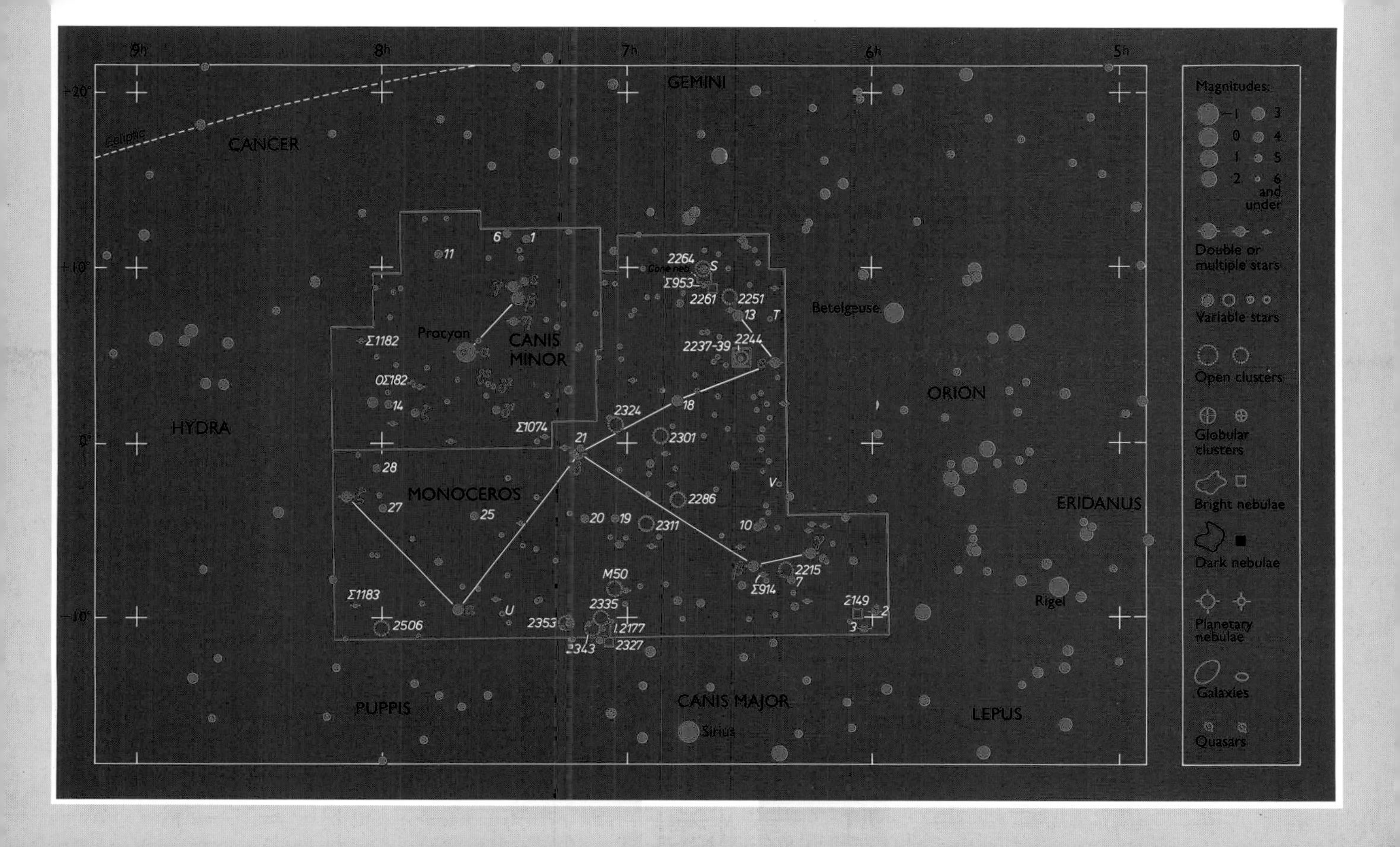
9h
8h
7h
6h
5h
+20°
+10°
0°
−10°
Ecliptic
CANCER
GEMINI
HYDRA
CANIS MINOR
MONOCEROS
ORION
ERIDANUS
PUPPIS
CANIS MAJOR
LEPUS
Procyon
Betelgeuse
Rigel
Sirius
Cone neb.
2264
S
Σ953
2261
2251
13
T
2237-39
2244
18
2324
2301
21
2286
V
10
2311
20
19
2215
7
Σ914
2149
2
3
M50
2335
2353
L2177
2343
2327
U
Σ1183
2506
27
25
28
Σ1182
ΟΣ182
14
Σ1074
11
6
1
Magnitudes:
−1
0
1
2
3
4
5
6
and under
Double or multiple stars
Variable stars
Open clusters
Globular clusters
Bright nebulae
Dark nebulae
Planetary nebulae
Galaxies
Quasars

Capricornus

Capricornus, the Sea Goat, has been vaguely identified with the demigod Pan, but there seem to be no definite legends attached to it. It lies between Fomalhaut and Aquila and is not hard to identify, despite its lack of bright stars. Capricornus is in the Zodiac, which is probably its best claim to fame! The brightest star is δ (Deneb al Giedi), RA 21h 47m 02.3s, dec. −16° 07′ 38″, magnitude 2.87. It is of type A5, and is 49 light-years away.

α^1 and α^2 Capricorni make up a wide naked-eye pair. The separation is 378 seconds of arc: the magnitudes are 4.2 and 3.57 respectively. They are not really associated: α^2 is only 117 light years away, α^1 as much as 1,600 light-years. Each star is double, and the fainter component of α^2 is again double. β, or Dabih, is also a very wide double – magnitudes 3.1 and 6.0, separation 205 seconds of arc – and here too the secondary is a close binary.

The globular cluster M30 lies not far from ζ, at RA 21h 40.4m, dec. −23° 11′. The outer edges are fairly easy to resolve telescopically, but the center of the cluster is densely packed. The integrated magnitude is 7.5, so binoculars will show it as a dim patch.

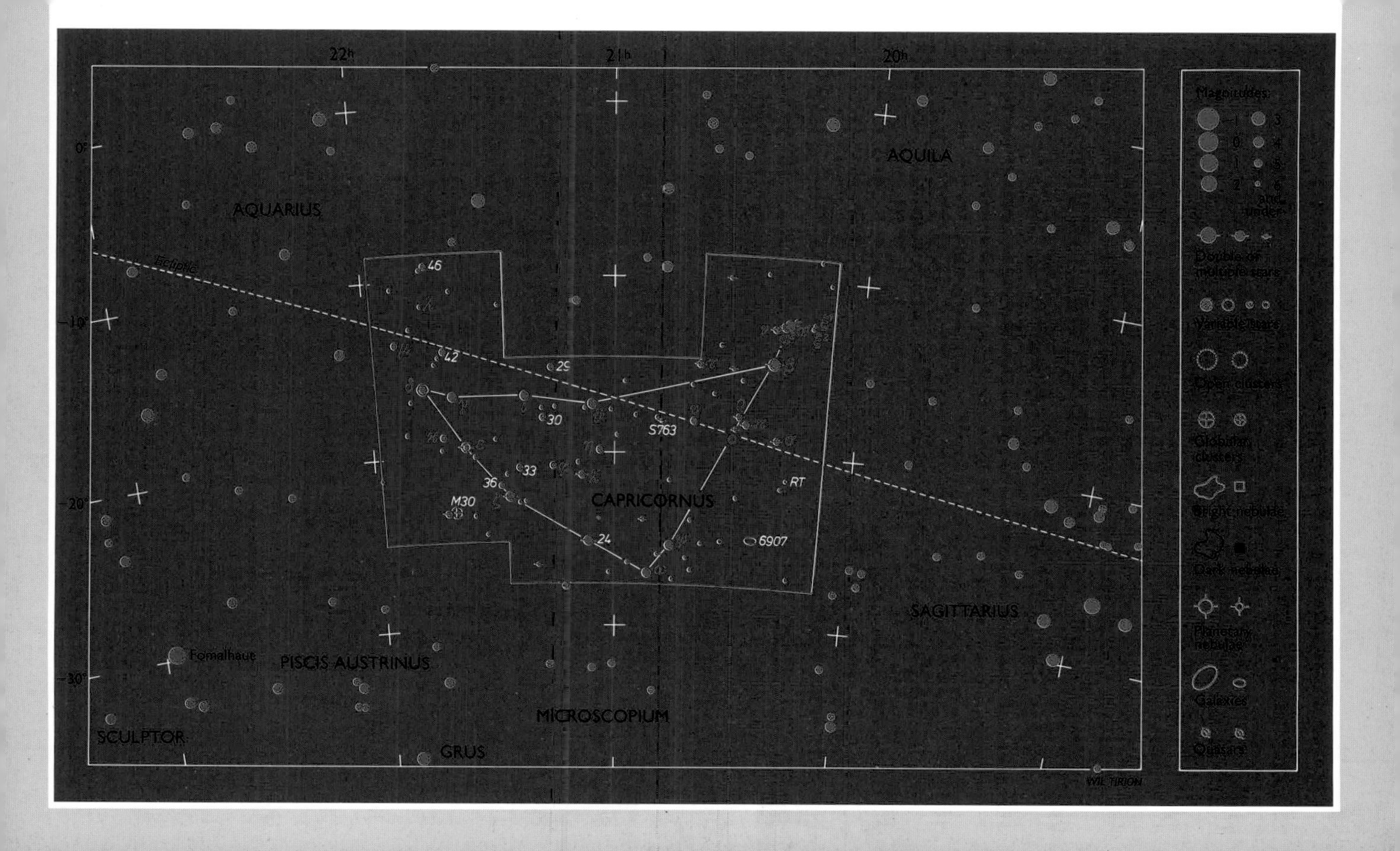
22h
21h
20h
0°
-10°
-20°
-30°
AQUARIUS
AQUILA
Ecliptic
46
42
29
30
S763
33
36
M30
CAPRICORNUS
RT
24
6907
SAGITTARIUS
Fomalhaut
PISCIS AUSTRINUS
MICROSCOPIUM
SCULPTOR
GRUS
WIL TIRION
Magnitudes:
-1
0
1
2
3
4
5
6 and under
Double or multiple stars
Variable stars
Open clusters
Globular clusters
Bright nebulae
Dark nebulae
Planetary nebulae
Galaxies
Quasars

Carina/Volans

Carina, graced by the presence of the brilliant Canopus, is the main part of the now-dismembered Ship Argo; Canopus, which used to be α Argûs, has therefore become α Carinae. Carina is crossed by the Milky Way, and is very rich in interesting objects of all kinds. Canopus stands out: it is a supergiant, and its spectral class indicates that it should be yellowish, though most people will certainly call it white. According to the authoritative Cambridge catalog, it is 200,000 times more luminous than the Sun – and it is interesting to compare it with Sirius, which looks almost a magnitude brighter but is very feeble when set aside Canopus! Look for the "False Cross," made up of ε and ι Carinae and δ and ϰ Velorum; its form is like that of the Southern Cross, but it is larger, more symmetrical and less brilliant.

R Carinae (RA 09h 32.3m, dec. −62° 47′) is a Mira variable, range 3.9 to 10.5, period 309 days. However, pride of place must go to η Carinae (RA 10h 45.1m, dec. −59° 41′), the most remarkable variable in the sky; for a while a century and a half ago it surpassed even Canopus, but is now just below naked-eye visibility. It is very massive and unstable; at its peak it may have equaled 6 million Suns, and is still almost as powerful, though at present much of its emission is in the infra-red. It is surrounded by a magnificent nebula, and gives the impression of being an "orange blob" rather than an ordinary star.

The cluster IC 2602, round θ, is an easy naked-eye object, and is very fine. Many other clusters are to be found in the constellation; and the whole region will repay sweeping with binoculars or a telescope on a low power.

Volans (originally Piscis Volans, the Flying Fish) intrudes into Carina; its brightest star is γ, RA 07h 08m 42.3s, dec. −70° 29′ 50″. It is double; the components are of magnitudes 3.8 and 5.7 and are separated by 13″.6. There is nothing else of note in Volans.

CARINA					
Leading stars	***RA*** h m s	***Declination*** ° ′ ″	***Magnitude***	***Spectrum***	***Distance*** ly
α Canopus	06 23 57.1	−52 41 44	−0.72	F0	1,200
β Miaplacidus	09 13 12.2	−69 43 02	1.68	A0	85
ε Avior	08 22 30.8	−59 30 34	1.86	K0	200
ι Tureis	09 17 05.4	−59 16 31	2.25	F0	900
θ	10 42 57.4	−61 23 39	2.76	B0	750
υ	09 47 06.1	−65 04 18	2.97	A7	320

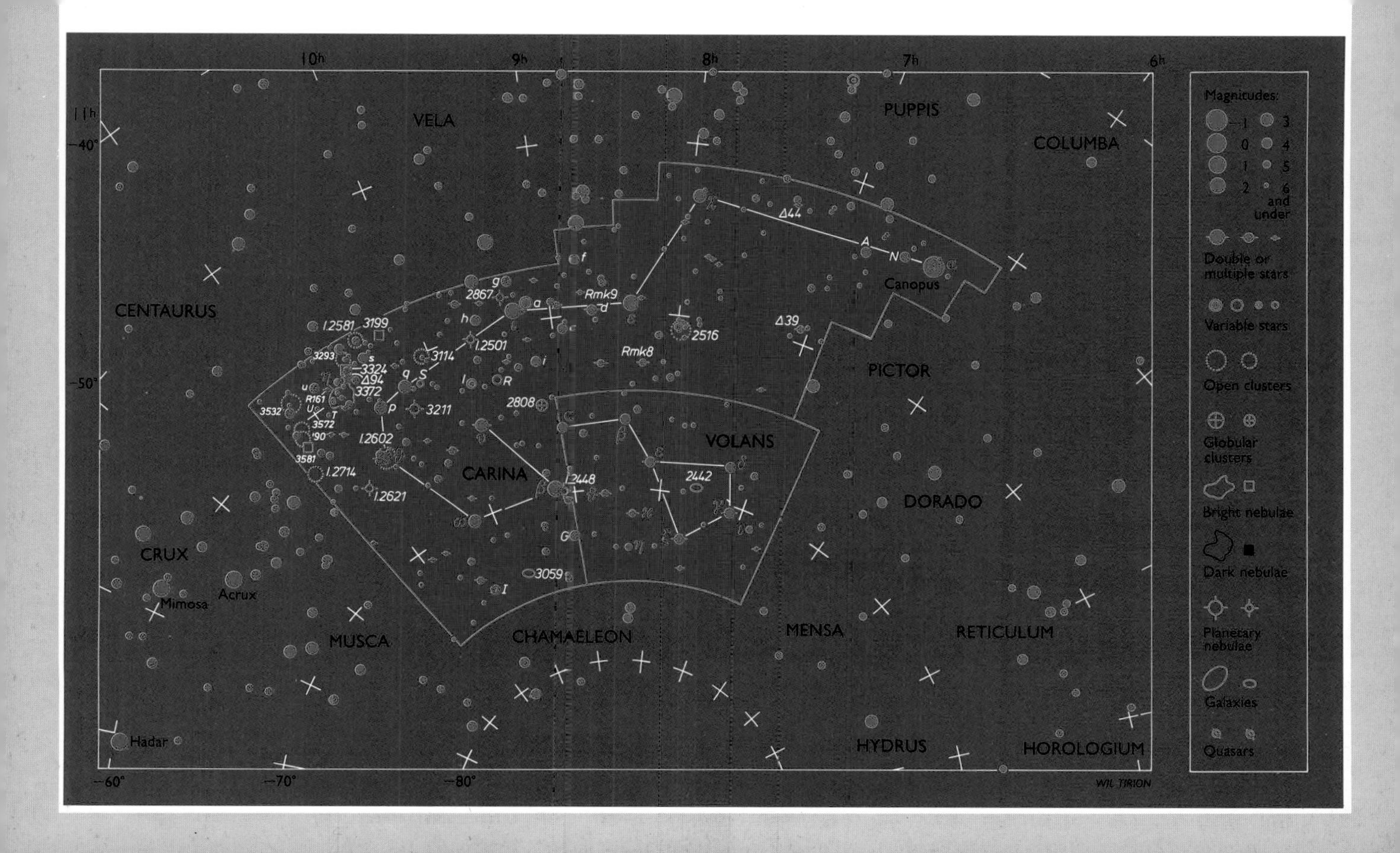

10h
9h
8h
7h
6h
11h
−40°
−50°
−60°
−70°
−80°
VELA
PUPPIS
COLUMBA
CENTAURUS
PICTOR
VOLANS
CARINA
DORADO
CRUX
MUSCA
CHAMAELEON
MENSA
RETICULUM
HYDRUS
HOROLOGIUM
Canopus
Mimosa
Acrux
Hadar
Δ44
Δ39
Δ94
Rmk9
Rmk8
2516
2867
I.2501
3114
3211
2808
I.2581
3199
3293
3324
3372
R161
3532
3572
3581
I.2602
I.2714
I.2621
I.2448
2442
3059
WIL TIRION
Magnitudes:
−1
0
1
2
3
4
5
6 and under
Double or multiple stars
Variable stars
Open clusters
Globular clusters
Bright nebulae
Dark nebulae
Planetary nebulae
Galaxies
Quasars

Cassiopeia

The striking northern constellation Cassiopeia is named after the proud queen of the Perseus legend. Both its brightest stars are variable, γ particularly so: it is unstable, and occasionally throws off "shells." It has been known to rise to magnitude 1.6 and fade to 3.3, and is quite unpredictable. α seems to fluctuate by a few tenths of a magnitude. Binoculars bring out its orange hue well.

ϱ Cassiopeiae, near β, is a most unusual variable. Generally it is around magnitude 4.9, but has been known to fade to below 6. Its type is unknown; it is an exceptionally luminous and remote supergiant with an F8 spectrum. Suitable comparison stars are σ (4.88) and τ (4.87). Of the several Mira variables, the brightest is R (RA 23h 58.4m, dec. +51° 24′): magnitude range 4.7–13.5, period 431 days.

There are two easy doubles in Cassiopeia and some interesting clusters (see table). NGC 457 lies around ψ Cassiopeiae (magnitude 4.74), though whether ψ is a foreground star or a genuine cluster member is uncertain.

Leading stars	*RA* h m s	*Declination* ° ′ ″	*Magnitude*	*Spectrum*	*Distance* ly
γ	00 56 42.4	+60 43 00	2.2v	B0p	780
α Shedir	00 40 30.4	+56 32 15	2.2v	K0	120
β Chaph	00 09 10.6	+59 08 59	2.27	F2	42
δ Ruchbah	01 25 48.9	+60 14 07	2.68	A5	62
ε Segin	01 54 23.6	+63 40 13	3.38	B3	520

Multiple stars	*RA* h m	*Declination* ° ′	*Magnitude* (separation ″)	*Distance* ly
η	00 49.1	+57 49	3.4, 7.5 (12.2)	480
ι	02 29.1	+67 24	4.6, 6.8 (12.4)	840

Clusters of stars	*RA* h m	*Declination* ° ′	*Magnitude*	*No. of stars*
M52	23 24.2	+61 35	6.9	100
M103	01 33.2	+60 42	7.4	25
NGC 663	01 46.0	+61 15	7.1	80
NGC 457	01 19.1	+58 20	6.4	80

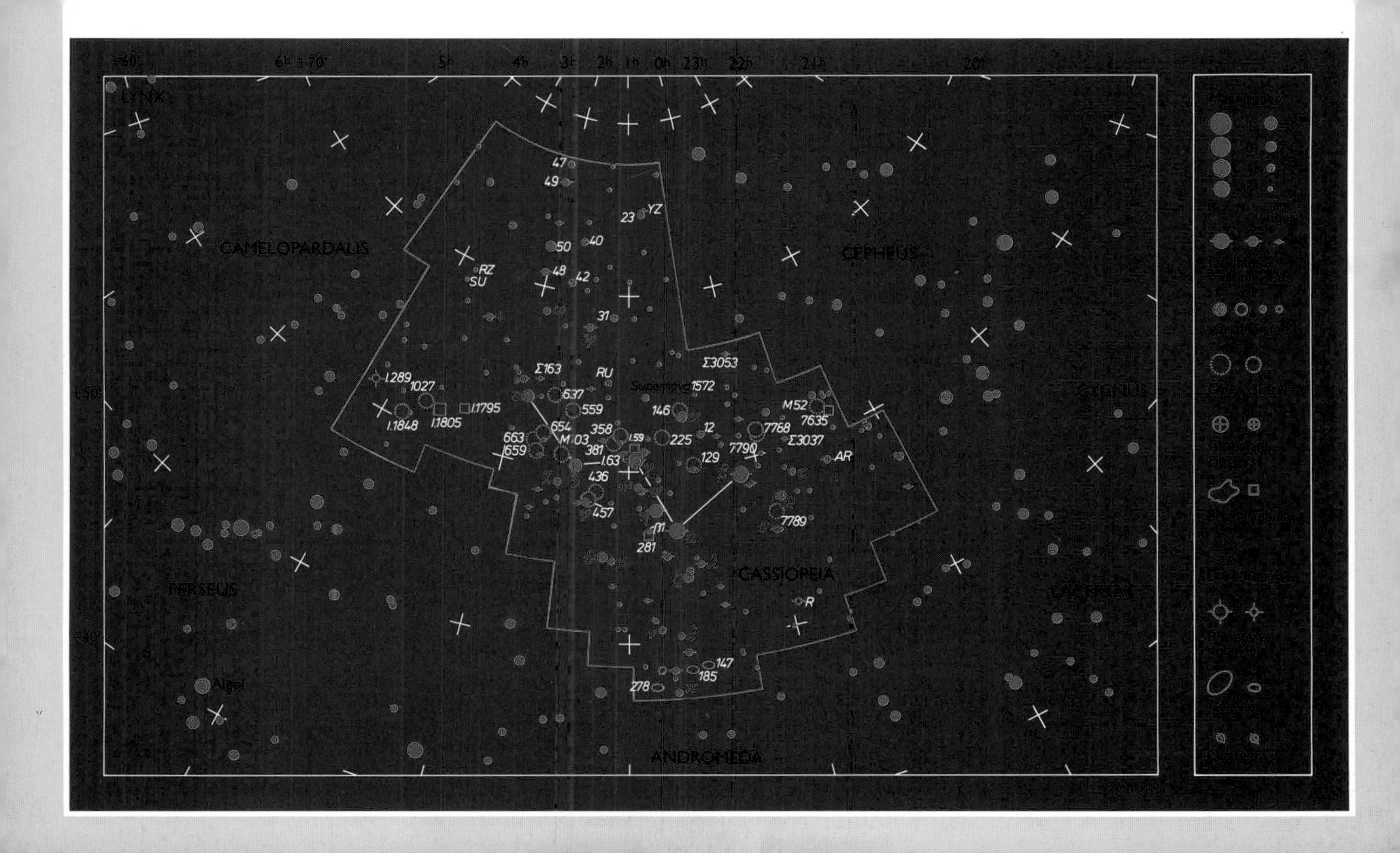
LYNX
CAMELOPARDALIS
CEPHEUS
CYGNUS
CASSIOPEIA
PERSEUS
ANDROMEDA
Algol
6h +70°
5h
4h
3h
2h
1h
0h
23h
22h
21h
20h
+50°
+40°
47
49
23
YZ
50
40
48
42
RZ
SU
31
Σ163
RU
Σ3053
Supernova 1572
I.289
1027
637
559
146
M52
7635
I.1795
I.1805
I.1848
654
358
12
7768
663
M 103
I.59
225
7790
Σ3037
659
381
I.63
129
AR
436
457
I11
281
7789
R
147
185
278

Centaurus

The Centaur is one of the most imposing of all constellations; British observers always regret that it lies so far south. Its mythological associations are uncertain, but it seems reasonable to identify it with Chiron, the wise and kindly centaur who taught Jason and many other heroes.

α and β are the Pointers to the Southern Cross. α has never had an official proper name, though it has been referred to as Rigel Kent and as Toliman; Agena has also been called Hadar.

Centaurus abounds in interesting objects. α is at present a wide double – magnitudes 0.0 and 1.2, separation 19″.7 – though this alters quickly since the binary period is only 80 years. γ is a closer binary: magnitudes 2.9 and 2.9, separation 1″.4, period 84.5 years. The Mira variable R Centauri lies between the Pointers: range 5.3 to 11.8, period 546 days (RA 14h 16.6m, dec. −59° 55′). The cluster IC 2944 lies round λ Centauri (magnitude 3.13) and is easy to resolve. However, the most spectacular object is the globular cluster ω Centauri (RA 13h 26.8m, dec. −47° 29′). Its diameter is over 36 minutes of arc, so that it is best seen with a very wide-field eyepiece. It is around 17,000 light-years away, and may well contain over a million stars; it is a very easy naked-eye object, and telescopically it is superb. The average separation between the stars at its center is no more than a tenth of a light-year, so that an astronomer living on a planet in this region would indeed have a spectacular view of the night sky!

Leading stars	***RA*** h m s	***Declination*** ° ′ ″	***Magnitude***	***Spectrum***	***Distance*** ly
α	14 39 36.7	−60 50 02	−0.27	G2+K1	4.3
β Agena	14 03 49.4	−60 22 22	0.61	B1	460
θ Haratan	14 06 40.9	−36 22 12	2.06	K0	46
γ Menkent	12 41 30.9	−48 57 34	2.17	A0	110
ε	13 39 53.2	−53 27 58	2.30	B1	490
η	14 35 30.3	−42 09 28	2.31v	B3	360
ζ Al Nair al Kentaurus	13 55 32.3	−47 17 17	2.55	B2	360
δ	12 08 21.5	−50 43 20	2.60v	B2	325
ι	13 20 35.8	−36 42 44	2.75	A2	52

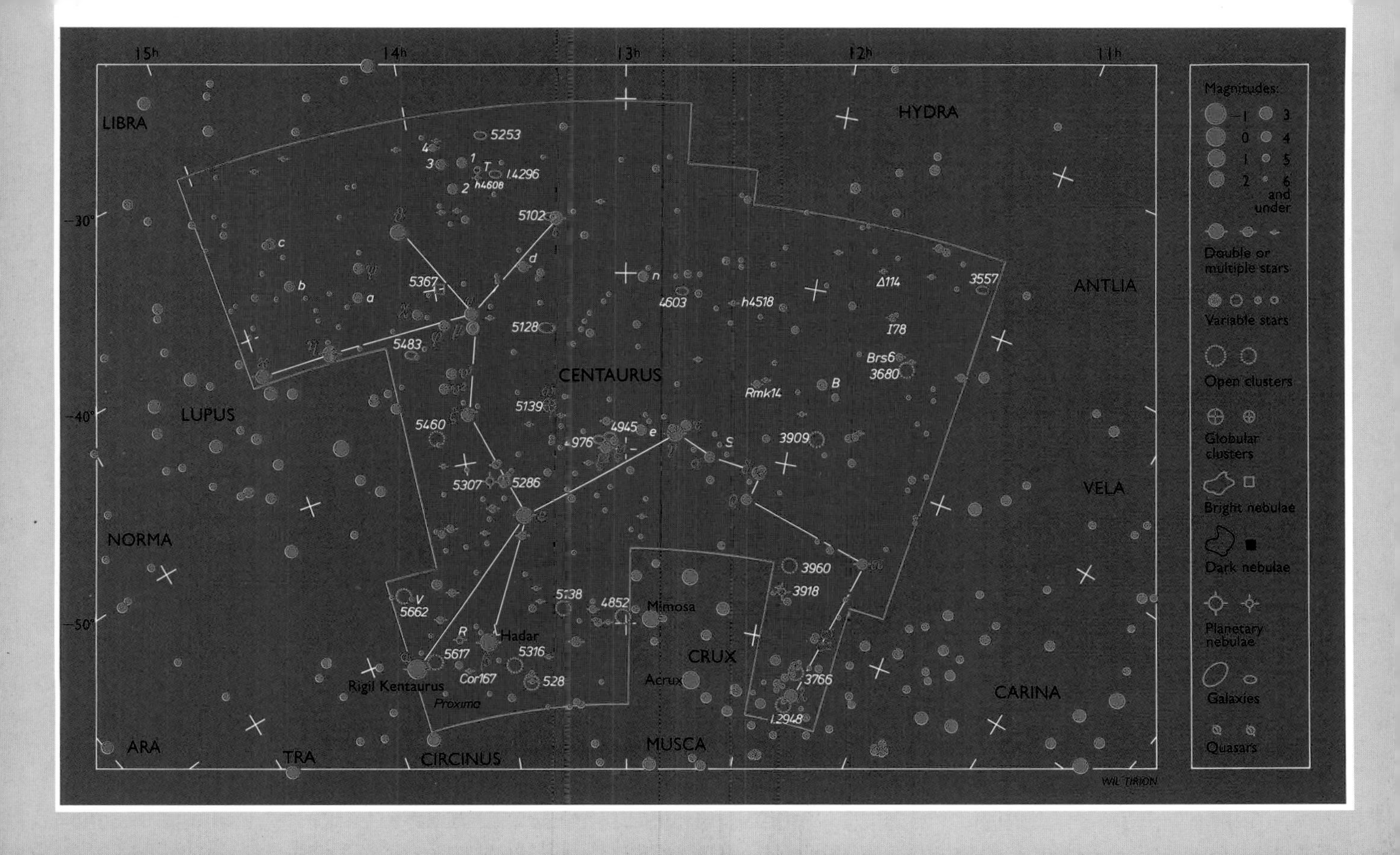
15h
14h
13h
12h
11h
−30°
−40°
−50°
LIBRA
HYDRA
ANTLIA
VELA
CARINA
LUPUS
NORMA
CENTAURUS
CRUX
MUSCA
CIRCINUS
TRA
ARA
5253
1.4296
h4608
5102
5367
5483
5128
5139
5460
5307
5286
4945
4976
4603
h4518
Δ114
3557
I78
Brs6
3680
Rmk14
3909
3960
3918
3766
1.2948
5138
4852
Mimosa
Acrux
Hadar
5316
5617
Cor167
528
5662
Rigil Kentaurus
Proxima
WIL TIRION
Magnitudes:
−1
0
1
2
3
4
5
6 and under
Double or multiple stars
Variable stars
Open clusters
Globular clusters
Bright nebulae
Dark nebulae
Planetary nebulae
Galaxies
Quasars

Cepheus

Cepheus, the King of the Perseus legend, is not nearly so prominent as his wife Cassiopeia. The constellation extends between Cassiopeia and the Pole Star, and has no really distinctive shape. The brightest star is α or Alderamin: magnitude 2.44, RA 21h 18m 34.6s, dec. +62° 35′ 08″. It is a white A7 star, 46 light-years away, forming a rough square with γ (magnitude 3.21), ζ (3.35) and ι (3.52). The reddish star γ or Alrai is closer to the pole, and will actually succeed Polaris as the pole star in the remote future.

There are two important variables in the constellation. One is δ, which forms a small triangle with ζ (magnitude 3.35) and ε (4.19). This is the prototype Cepheid variable, and has given its name to the entire class; its range is from magnitude 3.5 to 4.4 in a period of 5.37 days. Its fluctuations are easy to follow, as ζ and ε act as good comparison stars.

Not far away is μ Cephei (RA 21h 43m 30.2s, dec. +56° 46′ 48″). The range is from 3.4 to 5.1, but usually the star is around magnitude 5; there seems to be no definite period. It is so red that it has been nicknamed the Garnet Star; seen with binoculars or a telescope it is really striking. It has an M-type spectrum, and is actually much more luminous than Betelgeux in Orion, but it is also much further away – around 1,560 light-years. It could match well over 50,000 Suns.

δ has a 7.5-magnitude companion, at a separation of 41″. β is also a double, with magnitudes of 3.2 and 7.9, and separation 13″.3. Incidentally, β is also a variable with a very small magnitude range; stars of this type are known either as β Cephei or β Canis Majoris variables.

The nebula IC 1396 (RA 21h 39.1m, dec. +57° 30′) is easy to photograph, though rather elusive visually. The open cluster NGC 188, near the pole (RA 00h 44.4m, dec. +85° 20′), is old by open-cluster standards, because it is well away from the Galactic plane and is therefore less subject to disruption by passing stars.

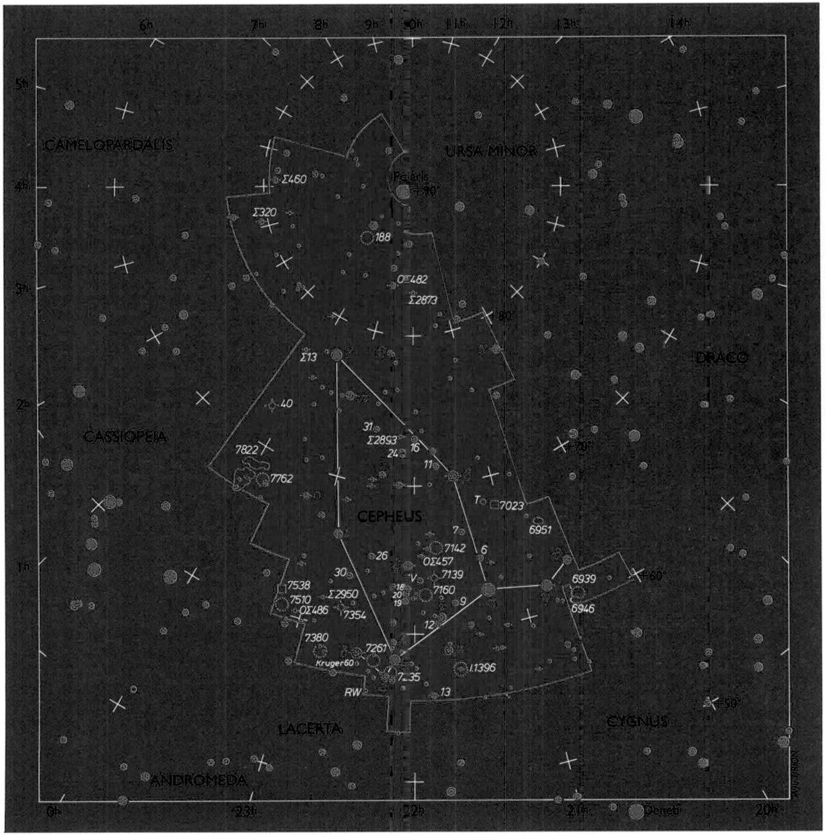

6h
7h
8h
9h
10h
11h
12h
13h
14h
5h
4h
3h
2h
1h
0h
23h
22h
21h
20h
CAMELOPARDALIS
URSA MINOR
Polaris
+90°
DRACO
+80°
+70°
+60°
+50°
CASSIOPEIA
CEPHEUS
LACERTA
CYGNUS
ANDROMEDA
Deneb
Σ460
Σ320
188
OΣ482
Σ2873
Σ13
40
31
Σ2853
24
16
11
7822
7762
T
7023
6951
7
7142
6
26
OΣ457
30
V
7139
18
20
19
7160
9
7538
Σ2950
7510
OΣ486
7354
6939
6946
12
7380
7261
Kruger 60
I.1396
7235
RW
13
WIL TIRION

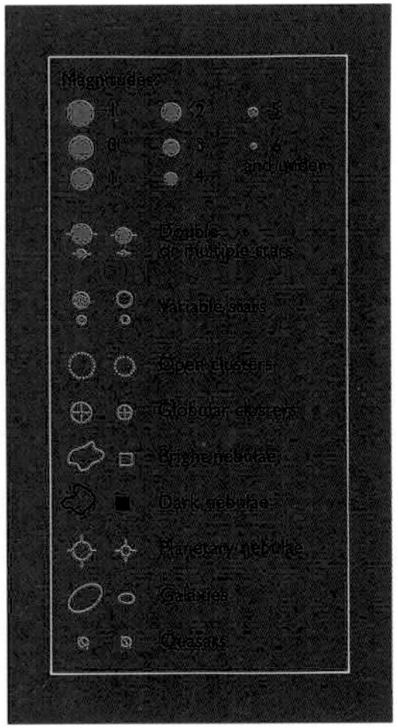

Magnitudes:
-1
0
1
2
3
4
5
6
and under
Double or multiple stars
Variable stars
Open clusters
Globular clusters
Bright nebulae
Dark nebulae
Planetary nebulae
Galaxies
Quasars

Cetus

Cetus is sometimes linked with the sea monster of the Perseus legend, though it has also been relegated to the status of a harmless whale. It is very large, but not particularly rich. The head of Cetus adjoins Taurus, and apart from Menkar (see table) contains γ (magnitude 3.47), χ (4.28) and μ (4.27). Diphda lies south of the Square of Pegasus, and two of the stars in the Square, Alpheratz and γ Pegasi, point more or less toward it; it has been confused with Fomalhaut, but is a magnitude fainter.

The most famous object in Cetus is Mira, which has given its name to a whole class of variable stars. It is a pulsating red giant, and at some maxima has been known to reach magnitude 1.7, though on average it is visible with the naked eye for only a few weeks per year. At minimum it drops to magnitude 10. UV Ceti (RA 01h 38.8m, dec. −17° 59′) is the prototype flare star, usually of magnitude 13, though on one occasion it flared abruptly to 6.8. There are several other Mira variables in the constellation (W, S, U Ceti) which can exceed magnitude 8 at maximum.

γ Ceti (Alkaffaljidhina), of magnitude 3.5, has a 7.5-magnitude companion, at a separation of 2″.8. Mira itself has a 12th-magnitude companion at a separation of 0″.3. τ Ceti (magnitude 3.50) is one of the nearest stars which is not too unlike the Sun; it is of spectral type G8 and is 11.7 light-years away. With ε Eridani it was selected for the initial SETI search as a possible center of a planetary system (SETI stands for Search for Extra-Terrestrial Intelligence).

There are various galaxies in Cetus. The most important is the giant elliptical M77 (RA 02h 42.7m, dec. −00° 01′), not far from δ (magnitude 4.07). The nucleus is condensed and active: M77 is a Seyfert galaxy, possibly with a central black hole. It is also a radio source (labeled Virgo A). The distance is given as 52 million light-years; it is the most massive system in the populous Virgo Cluster of galaxies.

Leading stars	***RA*** h m s	***Declination*** ° ′ ″	***Magnitude***	***Spectrum***	***Distance*** ly
β Diphda	00 43 35.3	−17 59 12	2.04	K0	68
α Menkar	03 02 16.7	+04 05 23	2.53	M2	130
o Mira	02 19 20.6	−02 58 39	var.	Md	95

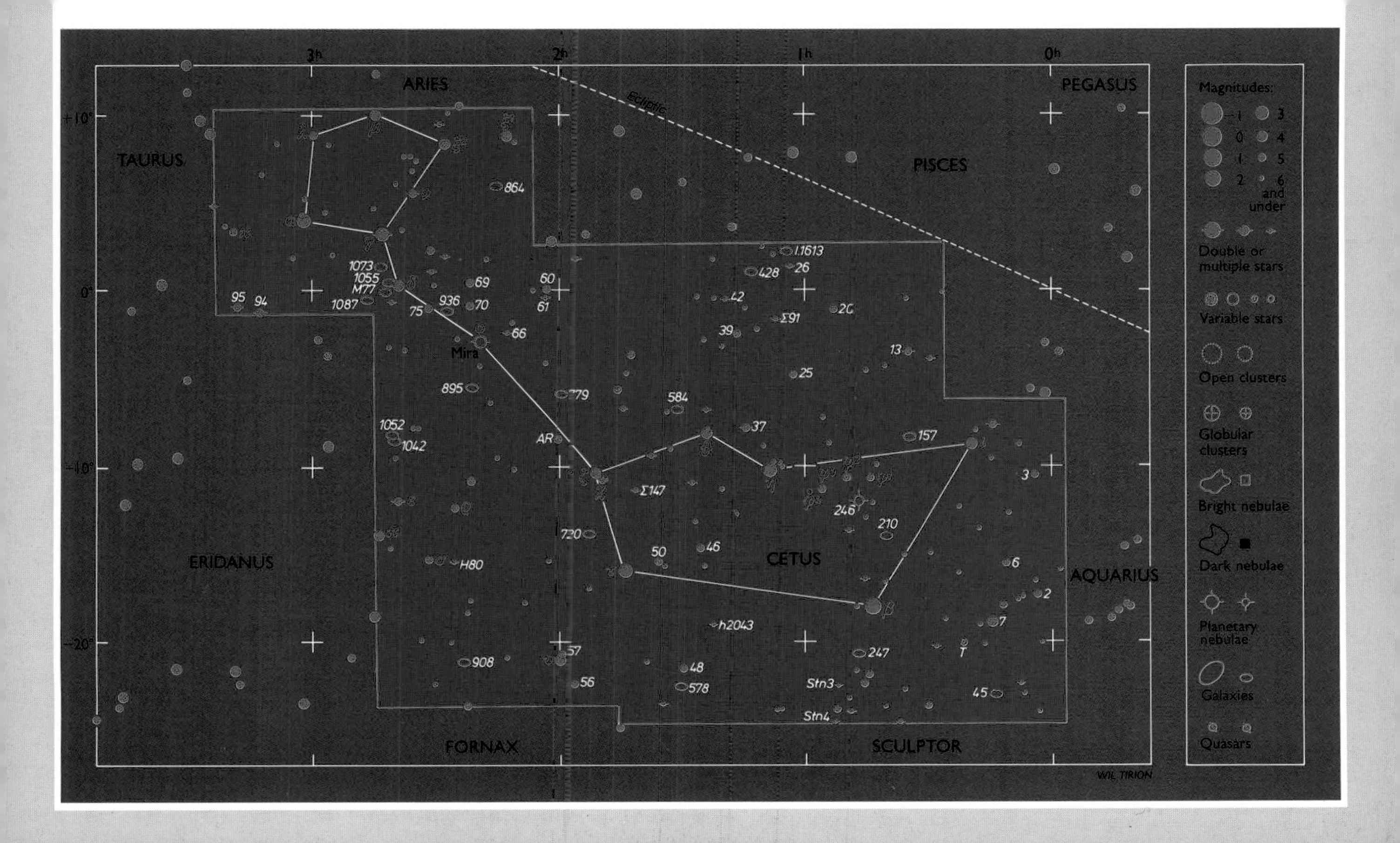

3h
2h
1h
0h
+10°
0°
−10°
−20°
ARIES
PEGASUS
TAURUS
PISCES
Ecliptic
ERIDANUS
CETUS
AQUARIUS
FORNAX
SCULPTOR
Mira
AR
864
1073
1055
M77
1087
95
94
75
936
69
70
60
61
66
895
779
584
1052
1042
Σ147
720
50
46
H80
908
57
56
48
578
h2043
Stn3
Stn4
247
1.1613
428
26
42
2C
Σ91
39
13
25
37
157
3
246
210
6
2
7
T
45
WIL TIRION
Magnitudes:
−1
0
1
2
3
4
5
6
and under
Double or multiple stars
Variable stars
Open clusters
Globular clusters
Bright nebulae
Dark nebulae
Planetary nebulae
Galaxies
Quasars

Chamaeleon/Octans

These are two obscure, far-south constellations added to the sky in "modern" times, and with no mythological associations. Chamaeleon adjoins Musca and is near Carina; its four leaders (α, β, γ, and δ) are all between magnitudes 4 and 4.5, and make a rough diamond outline. θ (magnitude 4.35) is in the same binocular field with α, making up what looks like a wide pair even though the two are not genuinely associated; δ is made up of two components 6′ apart – the brighter member is white and the secondary is orange, as is well shown with binoculars.

Octans is notable only because it contains the south celestial pole. The only star above the fourth magnitude is ν (magnitude 3.76). The star nearest to the south pole, σ Octantis, is only of magnitude 5.5; it is none too easy to see with the naked eye, and is almost a degree from the actual pole, so that it makes a very poor equivalent of the northern Polaris. Octans is a very barren group, and contains very little of interest. The Mira variable R Octantis (RA 05h 26.1m, dec. −86° 23′) can rise to 6.4 at maximum; it drops to 13.2 at minimum. The period is 406 days. Other Mira variables which can exceed magnitude 8 at maximum are U (303 days) and S (259 days); in each case the minimum magnitude is 14.

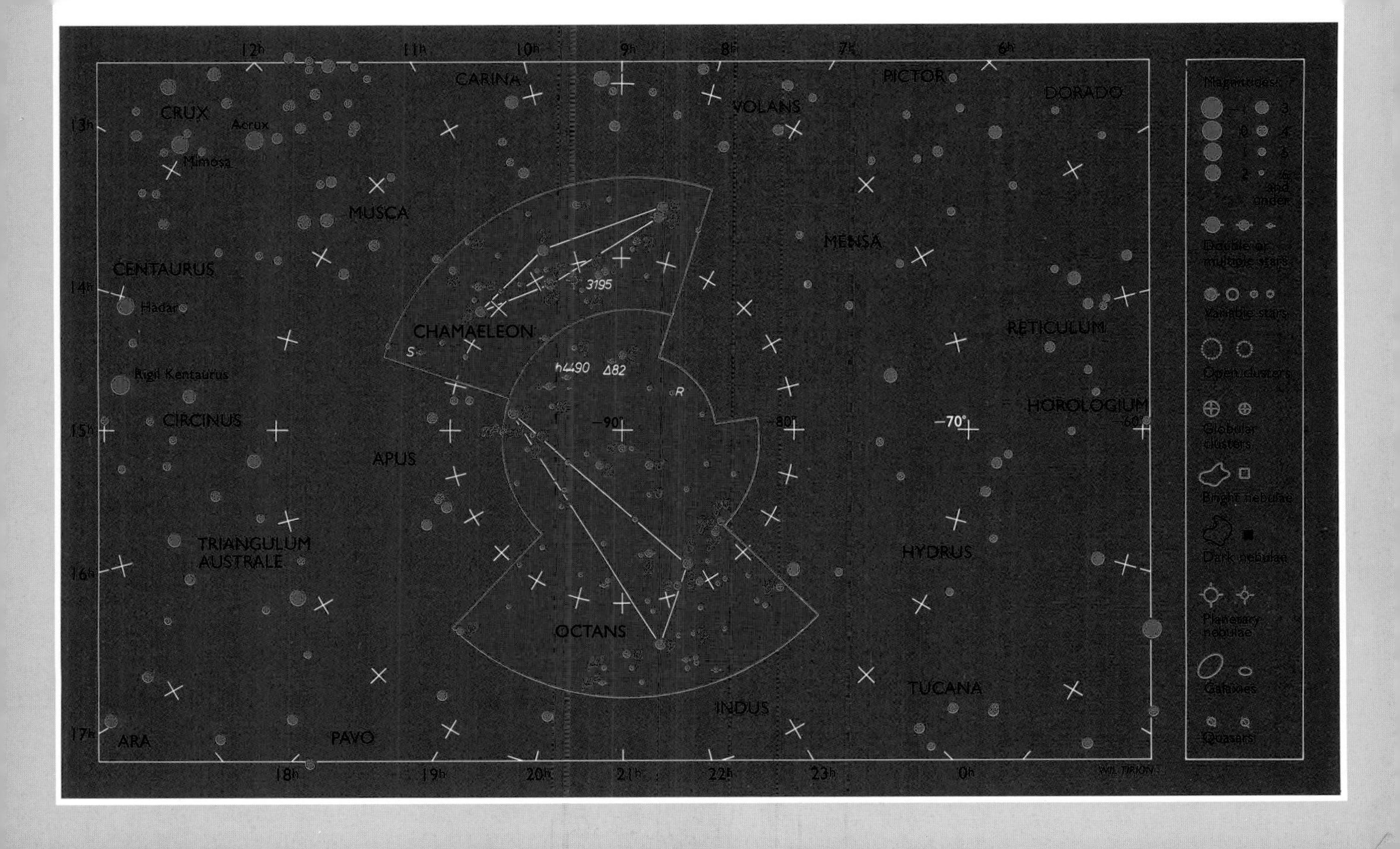
CRUX
Acrux
Mimosa
CARINA
VOLANS
PICTOR
DORADO
MUSCA
MENSA
CENTAURUS
Hadar
CHAMAELEON
RETICULUM
Rigil Kentaurus
HOROLOGIUM
CIRCINUS
APUS
HYDRUS
TRIANGULUM AUSTRALE
OCTANS
TUCANA
INDUS
ARA
PAVO
3195
h4490
Δ82
S
R
−90°
−80°
−70°
−60°
12h
11h
10h
9h
8h
7h
6h
13h
14h
15h
16h
17h
18h
19h
20h
21h
22h
23h
0h
Magnitudes:
−1
0
1
2
3
4
5
6 and fainter
Double or multiple stars
Variable stars
Open clusters
Globular clusters
Bright nebulae
Dark nebulae
Planetary nebulae
Galaxies
Quasars
WIL TIRION

Circinus

This is another very small and obscure southern constellation, unworthy of a separate identity. It lies close to α and β Centauri; its main star is α, of magnitude 3.19, which has a companion of magnitude 8.6 at 15″.7 separation. γ is a binary with a period of 180 years; the components are of magnitudes 5.1 and 5.5, and the current separation is only 0″.6. Circinus contains nothing of real interest. Note, however, that α Circini is in the same low-power binocular field with α Centauri.

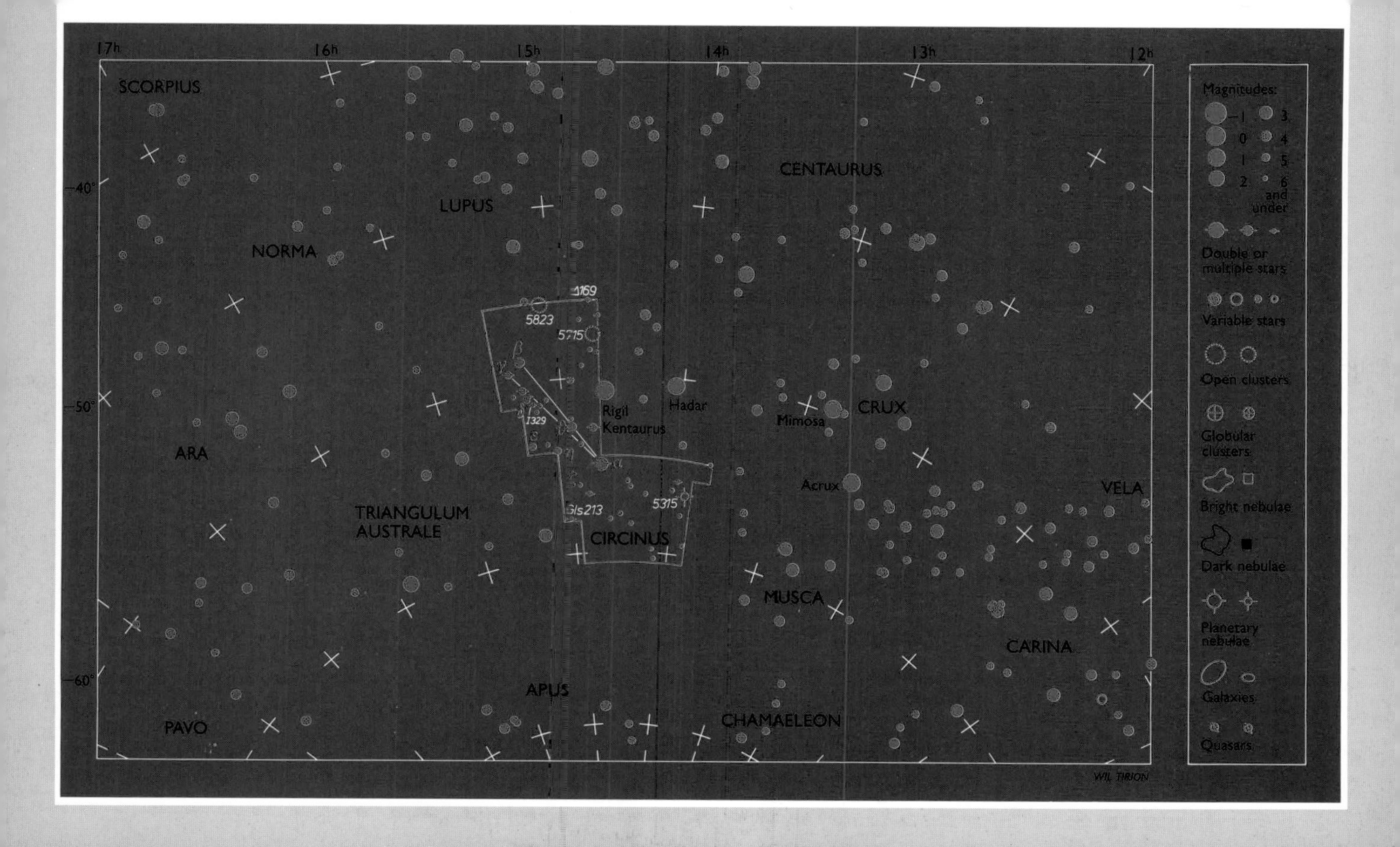
17h
16h
15h
14h
13h
12h
-40°
-50°
-60°
SCORPIUS
LUPUS
NORMA
CENTAURUS
ARA
TRIANGULUM AUSTRALE
CIRCINUS
CRUX
VELA
MUSCA
CARINA
APUS
PAVO
CHAMAELEON
I169
5823
5715
I329
Gls213
5315
Rigil Kentaurus
Hadar
Mimosa
Acrux
WIL TIRION
Magnitudes:
-1
0
1
2
3
4
5
6 and under
Double or multiple stars
Variable stars
Open clusters
Globular clusters
Bright nebulae
Dark nebulae
Planetary nebulae
Galaxies
Quasars

Coma Berenices

Berenice's Hair is not an ancient constellation – it was added to the sky by Tycho Brahe around 1590 – but there is a legend attached to it. When Ptolemy Euergetes, King of Egypt, set out on an expedition against Assyria, his wife Berenice vowed that when he returned safely she would cut off her lovely hair and place it in the temple of Venus. Ptolemy returned; Berenice kept her vow, and subsequently the golden tresses were placed in the sky.

Coma adjoins Boötes and Canes Venatici. Its brightest stars, α (Diadem) and β are only of magnitude 4.3, but the whole area gives the impression of being a large star cluster, and there are many galaxies as well as one globular cluster.

M64 is nicknamed the Black-eye Galaxy; some estimates give its integrated magnitude as above 7. The nickname comes from a dark region north of the center of the galaxy, but this cannot be seen with a telescope much below 25 centimeters in aperture. M64 is a very massive system at a distance of around 44 million light-years.

R Comae (RA 12h 4.0m, dec. +18° 49′) is a Mira variable, with magnitude range 7.1 to 14.6, period 363 days and spectral type M.

Non-stellar objects	**RA** h m	**Declination** ° ′	**Magnitude**	**Nature**
M53	13 12.9	+18 10	7.7	Globular cluster
M64	12 56.7	+21 41	8.5	Sb galaxy
M88	12 32.0	+14 25	9.5	SBb galaxy
M98	12 13.8	+14 54	10.1	Sb galaxy
M99	12 18.8	+14 25	9.8	Sc galaxy
M100	12 22.9	+15 49	9.4	Sc galaxy

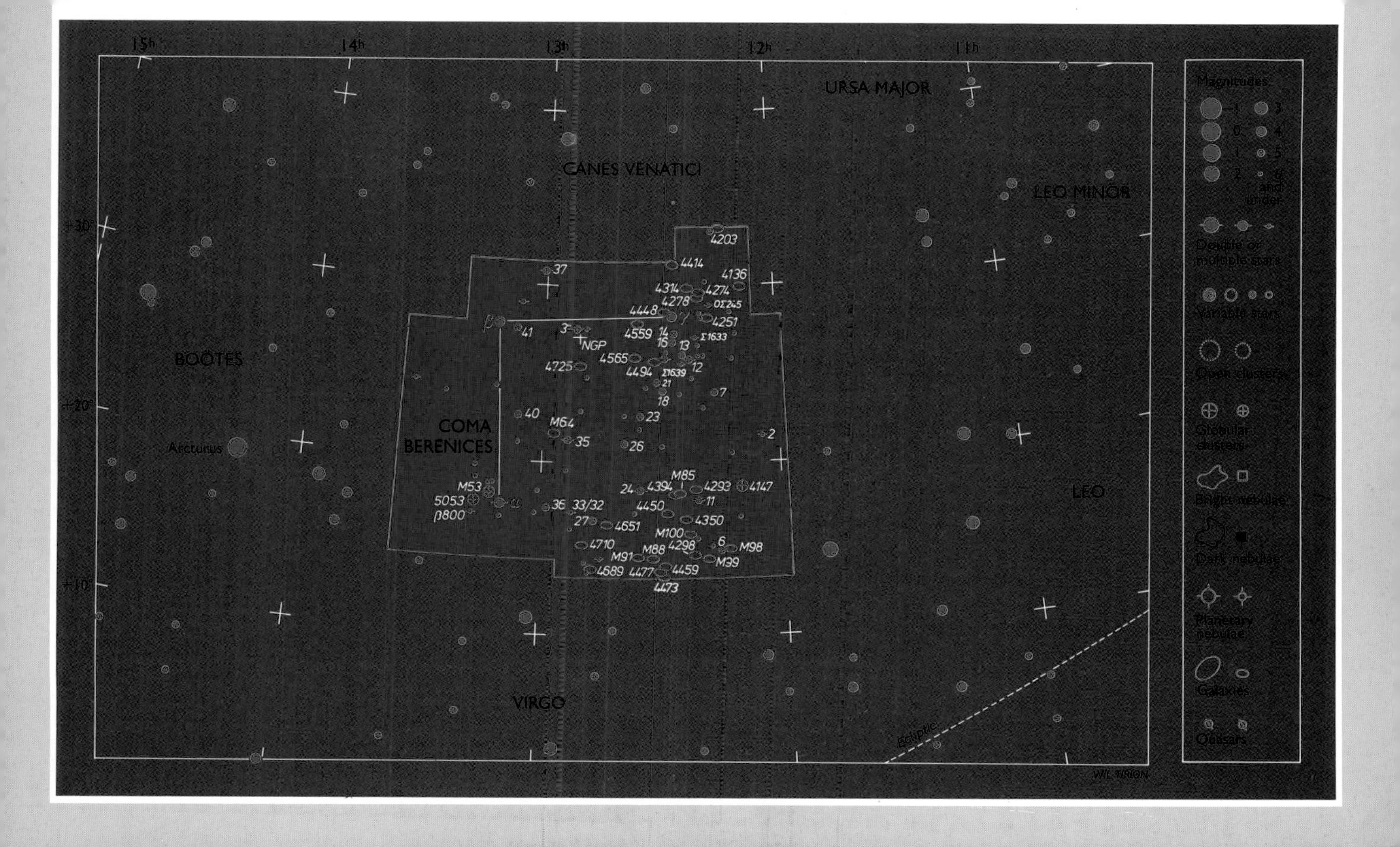

15h
14h
13h
12h
11h
+30°
+20°
+10°
URSA MAJOR
CANES VENATICI
LEO MINOR
BOÖTES
Arcturus
COMA BERENICES
LEO
VIRGO
Ecliptic
4203
4414
4136
4314
4274
4278
0Σ245
4448
4251
4559
14
16
13
Σ1633
4565
12
4494
Σ1639
21
4725
7
18
23
40
M64
35
26
2
M85
24
4394
4293
4147
M53
5053
β800
36
33/32
4450
11
27
4651
4350
M100
4710
4298
6
M98
M91
M88
M39
4689
4477
4459
4473
37
41
3
NGP
WIL TIRION
Magnitudes:
−1
0
1
2
3
4
5
6 and under
Double or multiple stars
Variable stars
Open clusters
Globular clusters
Bright nebulae
Dark nebulae
Planetary nebulae
Galaxies
Quasars

Corona Australis

The Southern Crown (sometimes called Corona Austrinus) is an ancient constellation, probably because, although it has no star brighter than magnitude 4, the short curve made by γ, α, β, δ and θ is distinctive; it lies near α Sagittarii, too far south to be seen from Britain. κ (RA 18h 33.4m, dec. −38° 44′) is an easy double; the components are equal at magnitude 5.9 and the separation is 21″.6.

Much the most interesting object in the constellation is the comet-like nebula NGC 6729, which is associated with the irregular variable R Coronae Australis at RA 19h 01.9m, dec. −36° 57′. The star is of type F5. Variations in the light of the nebula are usually linked with the unpredictable variations of the star, which may amount to as much as three magnitudes (9 to 12).

Corona Australis contains one globular cluster, NGC 6541 (RA 18h 08.0m, dec. −43° 42′), which is just detectable with binoculars; it lies between θ Coronae and θ Scorpii.

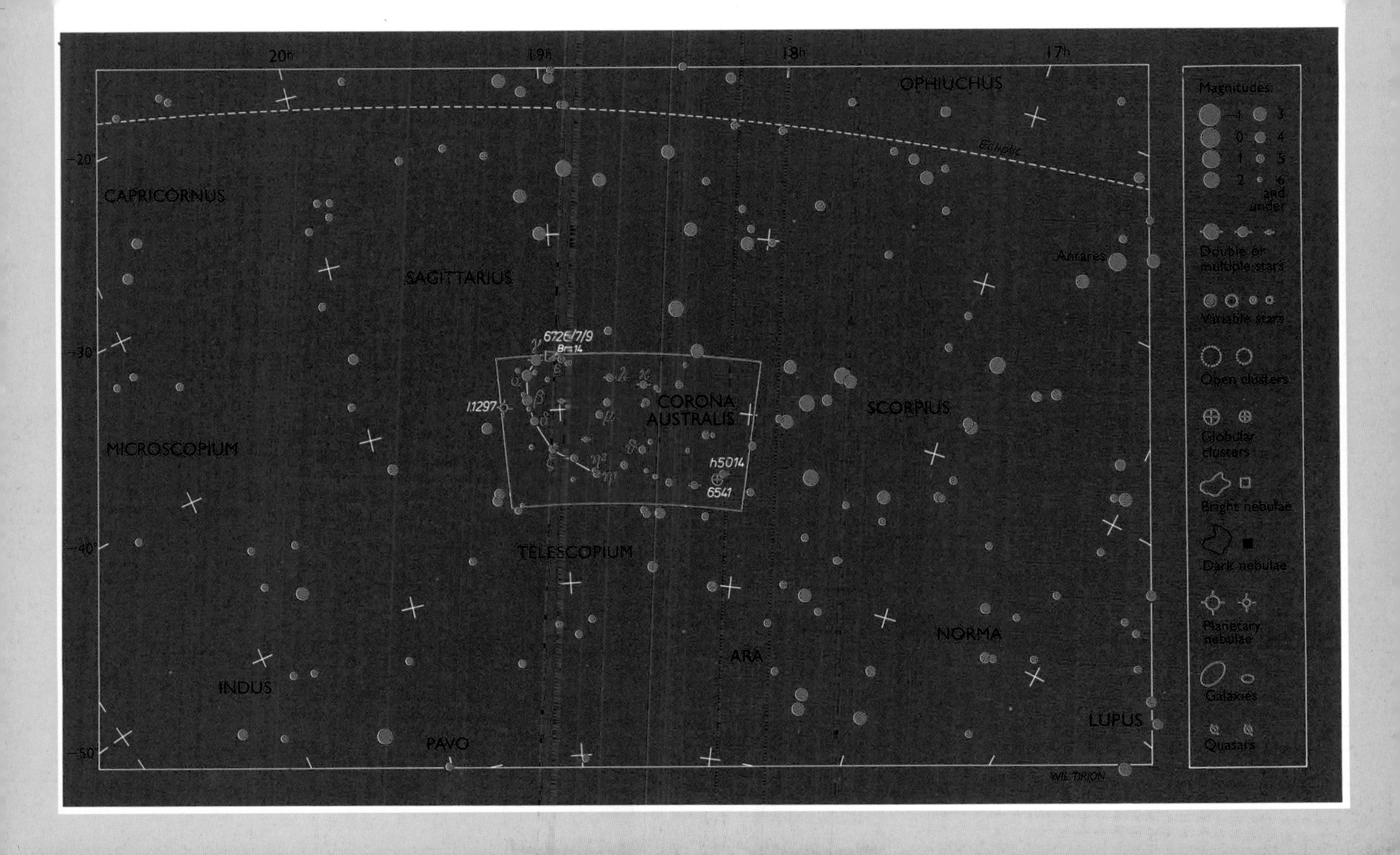

20h
19h
18h
17h
OPHIUCHUS
Ecliptic
-20°
-30°
-40°
-50°
CAPRICORNUS
SAGITTARIUS
Antares
6726/7/9
Br 14
I.1297
CORONA
AUSTRALIS
SCORPIUS
h5014
6541
MICROSCOPIUM
TELESCOPIUM
NORMA
ARA
INDUS
LUPUS
PAVO
WIL TIRION
Magnitudes
−1
0
1
2
3
4
5
6
and under
Double or multiple stars
Variable stars
Open clusters
Globular clusters
Bright nebulae
Dark nebulae
Planetary nebulae
Galaxies
Quasars

Corona Borealis/Serpens Caput

Corona Borealis is said to be the crown given by the wine-god Bacchus to Ariadne, the daughter of King Minos of Crete. Its brightest star, α or Alphekka, forms part of the prominent Y pattern with Arcturus, ε Boötis and γ Boötis. Alphekka's principal data are: RA 15h 34m 41.2s, dec. +26° 42′ 53″, magnitude 2.23, type A0, distance 78 light-years.

η is double, with magnitudes 5.6 and 12.5, and separation 57″.7; the brighter member is again double, with a separation of 1″. ζ is also double, with magnitudes 5.1 and 6.0, and separation 6″.3.

There is a rich quota of variables. T is the "Blaze Star," which flared up to naked-eye visibility in 1866 and again in 1946. R is the prototype of its class; usually it is on the fringe of naked-eye visibility, but at unpredictable intervals fades, because of soot accumulating in its atmosphere. Generally binoculars show two stars in the bowl of the Crown: R, and a star of magnitude 6.6. If you find only one star, you may be sure that R is experiencing one of its minima.

Serpens is the serpent with which Ophiuchus is struggling, and seems to have been pulled into two pieces, Caput (the head) and Cauda (the body). The leading star of Caput is α or Unukalhai: RA 15h 44m 16.0s, dec. +06° 25′ 32″, magnitude 2.65, type K2, distance 88 light-years. The triangle made up of Unukalhai, γ (magnitude 3.7) and β (2.8) is easy to find. Between β and γ is the Mira variable R Serpentis (RA 15h 50.7m, dec. +15° 08; range 5.1 – 14.4; period 356 days). δ is a binary with a long period, with magnitudes 4.1 and 5.2 and separation 4″.4.

The main object of interest is the globular cluster M5 (RA 15h 18.6m, dec. +02° 05′) which is on the fringe of naked-eye visibility, and is a splendid sight in a telescope. After M13 Herculis it is probably the most impressive globular visible from Britain and the USA.

CORONA BOREALIS					
Variable stars	***RA*** h m	***Declination*** ° ′	***Magnitude***	***Period***	***Type***
U	15 18.2	+31 39	7.7–8.8	3.45	Algol
S	15 21.4	+31 22	5.8–14.1	360	Mira
R	15 48.6	+27 08	5.7–15	–	R Coronae
V	15 49.5	+39 34	6.9–12.6	358	Mira
T	15 59.5	+25 55	2.0–10.8	–	Recurrent nova
W	16 15.4	+37 48	7.8–14.3	238	Mira

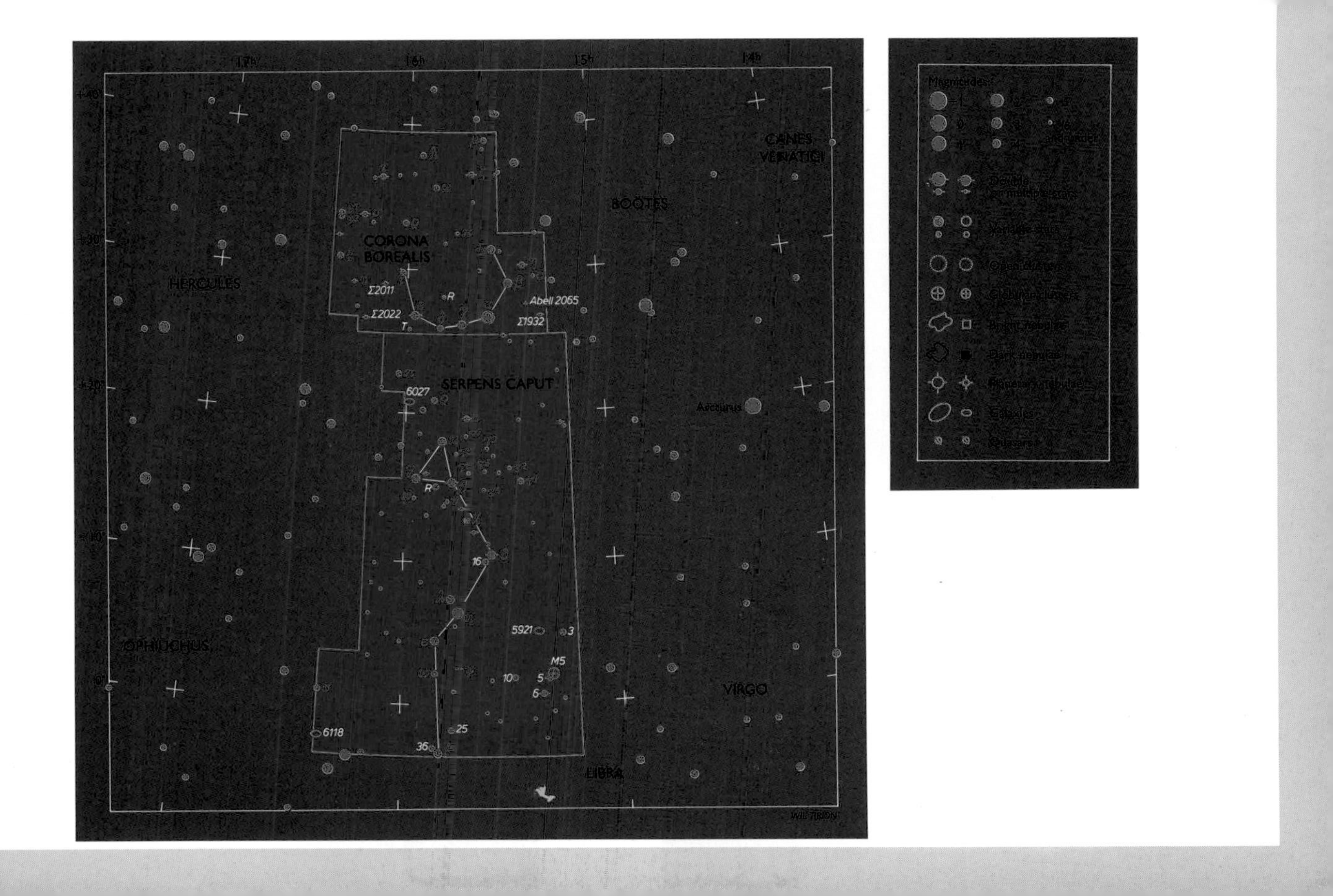

Magnitudes
Double or multiple stars
Variable stars
Open clusters
Globular clusters
Bright nebulae
Dark nebulae
Planetary nebulae
Galaxies
Quasars
CANES VENATICI
BOÖTES
Arcturus
VIRGO
LIBRA
CORONA BOREALIS
SERPENS CAPUT
HERCULES
OPHIUCHUS
Abell 2065
Σ1932
Σ2011
Σ2022
R
T
6027
5921
M5
3
5
6
10
16
25
36
6118
14h
15h
16h
17h
+40°
+30°
+20°
+10°
0°
WIL TIRION

Corvus/Crater

These are two small but ancient constellations, adjoining Hydra. Corvus represents a crow sent by the god Apollo to report on the behavior of a lady named Coronis, with whom he had become enamored. The crow's report was decidedly adverse, but the bird was rewarded with a place in the sky. Crater is said to represent the goblet of the wine-god Bacchus.

The four main stars of Corvus (γ, δ, β and ε) are all between magnitudes 2.5 and 3, and are distinctive because they lie in a barren area. (Curiously, the star lettered α is a magnitude fainter.) δ is double – magnitudes 3.0 and 9.2, separation 24″.2, so that this is an easy telescopic pair – but otherwise Corvus is remarkably devoid of interesting objects.

Crater is even less distinctive; its brightest star, α, is only of magnitude 3.56, forming a small triangle with α (4.08) and γ (also 4.08). γ has a 9.6-magnitude companion at a separation of 5″.2. There are several spiral galaxies in Crater, but none of them is as bright as magnitude 11.

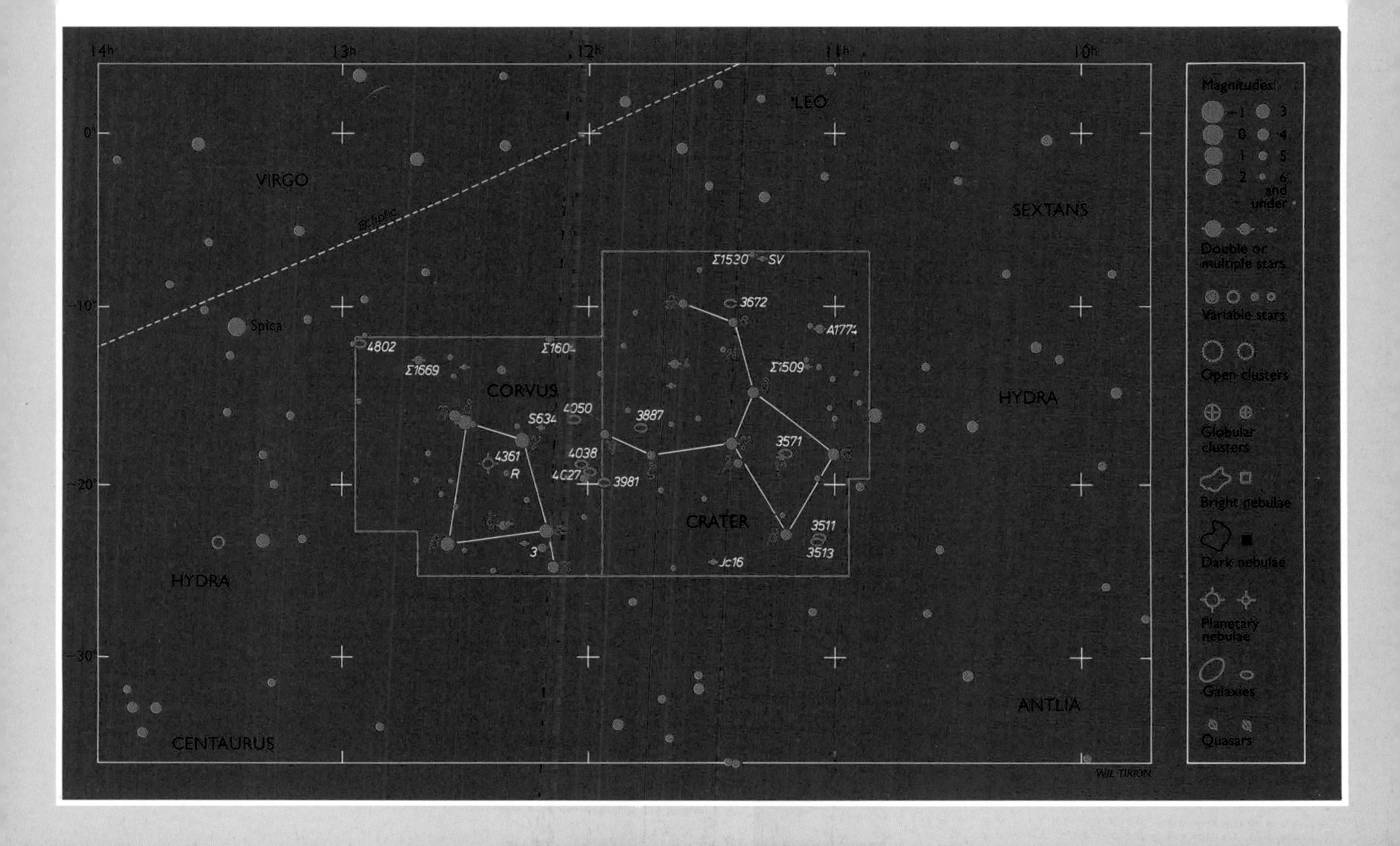
14h
13h
12h
11h
10h
0°
−10°
−20°
−30°
VIRGO
ecliptic
Spica
LEO
SEXTANS
HYDRA
CORVUS
CRATER
HYDRA
ANTLIA
CENTAURUS
Σ1530
SV
3672
A1774
Σ1509
3887
3571
3511
3513
Jc16
4802
Σ1669
Σ1604
4050
S634
4361
R
4038
4027
3981
3
WIL TIRION
Magnitudes
−1
0
1
2
3
4
5
6
and under
Double or multiple stars
Variable stars
Open clusters
Globular clusters
Bright nebulae
Dark nebulae
Planetary nebulae
Galaxies
Quasars

Crux/Musca

Crux Australis, the Southern Cross, is the most famous of all southern constellations, though it was not accepted as a separate group until 1679; before that it had been included in Centaurus, which almost surrounds it. The fifth-brightest star, ε (magnitude 3.59), rather spoils the symmetry of the constellation. Moreover, δ is fainter than the other three principal stars, and in any case Crux is more like a kite than a cross – there is no central star to make an X, as there is with Cygnus in the northern hemisphere.

Though Crux is the smallest of all recognized constellations, it is remarkably rich and is easy to find, if only because the brilliant α and β Centauri point to it. α Crucis is a wide, easy double: magnitudes 1.4 and 1.9, separation 4″.4, with a third star in the same field. γ is also double: magnitudes 1.6 and 6.7, again with a third star in the same field. Even casual observation will show that γ differs from the other stars of the Cross: it is a red giant, whereas the other three are hot and bluish-white.

The lovely cluster NGC 475 lies close to β around ϰ Crucis and is known as the Jewel Box. Most of its stars are bluish-white, but there is one red supergiant which stands out. The cluster is around 7,700 light-years away; three of its leading stars form a triangle, inside which is the red supergiant. By coincidence the Jewel Box lies at the edge of the Coal Sack, a dark nebula over 500 light-years away, which appears as an almost blank area, blocking the light of stars beyond.

Musca Australis (the Southern Fly) adjoins Crux, and is fairly distinctive; its brightest star is α, magnitude 2.69. In the same binocular field with α Crucis are λ (magnitude 3.64) and μ (4.72); λ is white, μ very red, so that they make up a striking pair. Musca also contains two fairly prominent globular clusters, NGC 4372 and 4833, which are on the fringe of binocular visibility, though not particularly easy to resolve into stars.

CRUX					
Leading stars	***RA*** h m s	***Declination*** ° ′ ″	***Magnitude***	***Spectrum***	***Distance*** ly
α Acrux	12 26 35.9	−63 05 56	0.83	B1+B3	360
β	12 47 43.2	−59 41 19	1.25	B0	125
γ	12 31 09.9	−57 06 47	1.63	M3	88
δ	12 15 08.6	−58 44 55	2.80	B2	257

15h
14h
13h
12h
11h
-40°
-50°
-60°
-70°
-80°
CENTAURUS
LUPUS
NORMA
CIRCINUS
TRIANGULUM AUSTRALE
ARA
PAVO
APUS
OCTANS
CHAMAELEON
MUSCA
CRUX
VELA
CARINA
VOLANS
Hadar
Rigil Kentaurus
Coal Sack
Mimosa
Acrux
4337
4439
4755
Tr20
4349
4103
4609
4052
4463
h4432
L4920
I.4191
5189
H6
R
4833
4372
Magnitudes
Double or multiple stars
Variable stars
Open clusters
Globular clusters
Bright nebulae
Dark nebulae
Planetary nebulae
Galaxies
Quasars

Cygnus

Cygnus is one of the most splendid of all constellations. It is said to represent the swan in the form of which Jupiter once visited the wife of the King of Sparta – for reasons which need not concern us here!

The leading stars form an X, with γ at the center; β is fainter than the rest and further away from the center, but is a superb double, with magnitudes 3.1 and 5.1 at a separation of 34″.4. The primary is golden yellow, the companion vivid blue – the most spectacular colored pair in the sky. δ is also double: magnitudes 2.9 and 6.3, separation 2″.4. γ has a 9.9-magnitude companion at 41″.2, itself a close double.

Cygnus abounds in variables, and several bright novae have appeared in it. χ is the brightest Mira-type star apart from Mira itself, and can become prominent; a good comparison star is η (magnitude 3.89). P Cygni hovers around magnitude 5: it is a very massive, unstable star. U is exceptionally red. SS is the prototype of the dwarf class, known either as SS Cygni or U Geminorum stars.

The Milky Way is rich in Cygnus, and there are several dark rifts due to obscuring matter. M29 and M39 are prominent open clusters. NGC 6960 and NGC 6992 – the Cygnus Loop – are parts of a supernova remnant; NGC 7000 is known as the North American Nebula because of its shape. It is visually dim, but easy to photograph; it lies at RA 20h 58.8m, dec. +44° 20′.

Leading stars	***RA*** h m s	***Declination*** ° ′ ″	***Magnitude***	***Spectrum***	***Distance*** ly
α Deneb	20 41 25.8	+45 16 49	1.25	A2	1,800
γ Sadr	20 22 13.5	+40 15 24	2.20	F8	750
ε Gienah	20 46 12.5	+33 58 13	2.46	K0	82
δ	19 44 58.4	+45 07 51	2.87	A0	124
β Albireo	19 30 43.1	+27 57 35	3.08	K5	390

Variable stars	***RA*** h m	***Declination*** ° ′	***Range***	***Period*** days	***Type***
CH	19 24.5	+50 14	6.4–8.7	97	Z Andromedae
R	19 36.8	+50 12	5.1–14.2	26	Mira
χ	19 50.6	+32 55	3.3–14.2	07	Mira
Z	20 01.4	+50 03	7.4–4.7	26	Mira
P	10 17.8	+38 02	3–6	–	
U	20 19.6	+47 54	5.9–12.1	462	Mira
W	21 36.0	+45 22	5.0–7.6	126	Semiregular
SS	21 42.7	+43 35	8.4–12.4	47	SS Cygni

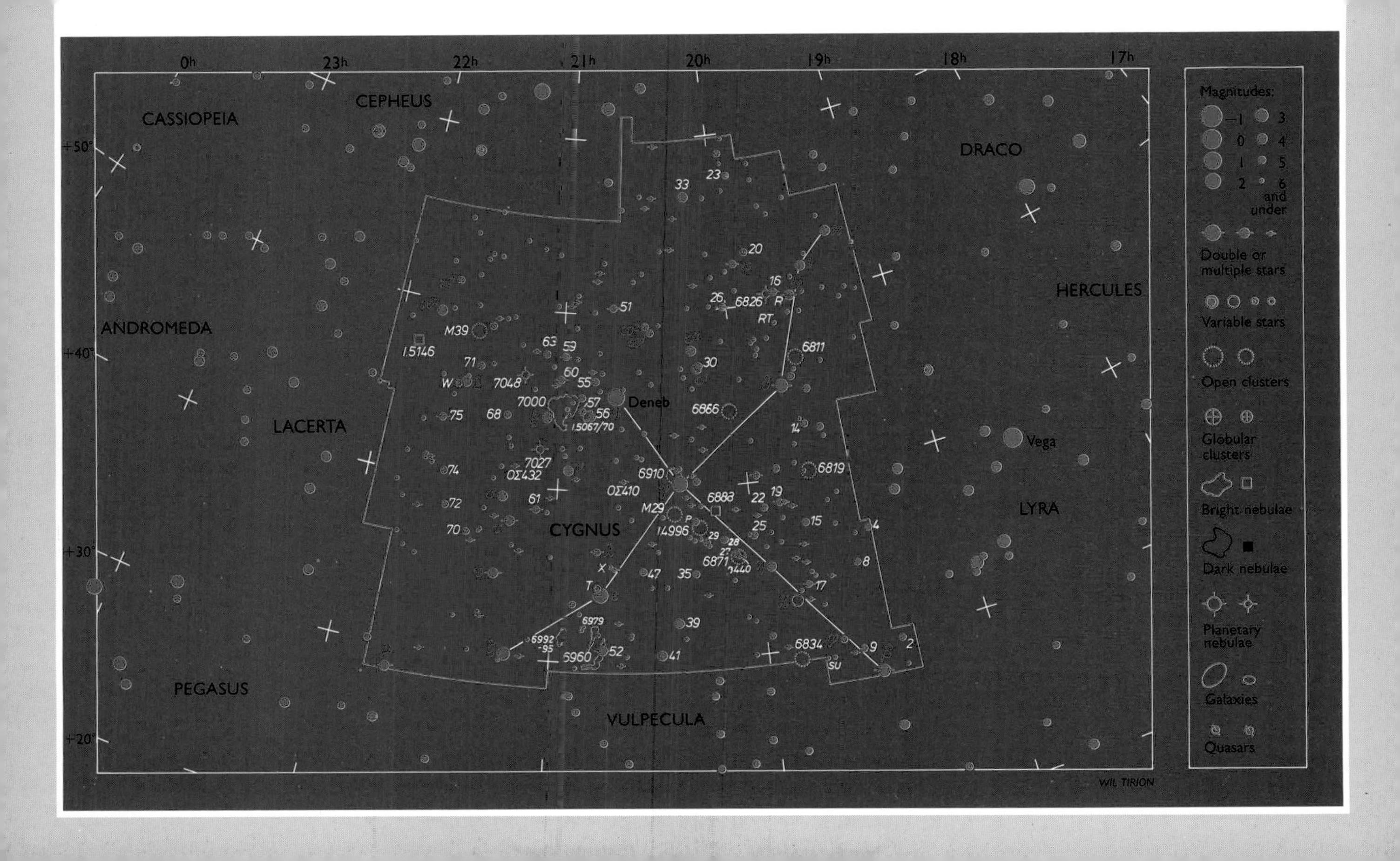
0h
23h
22h
21h
20h
19h
18h
17h
+50°
+40°
+30°
+20°
CASSIOPEIA
CEPHEUS
DRACO
ANDROMEDA
HERCULES
LACERTA
LYRA
CYGNUS
PEGASUS
VULPECULA
Deneb
Vega
M39
M29
I.5146
I.5067/70
I.4996
7000
7048
7027
6910
6888
6866
6826
6811
6819
6871
6834
6979
6960
6992
-95
OΣ432
OΣ410
W
P
R
RT
T
X
SU
Magnitudes:
–1
0
1
2
3
4
5
6
and under
Double or multiple stars
Variable stars
Open clusters
Globular clusters
Bright nebulae
Dark nebulae
Planetary nebulae
Galaxies
Quasars
WIL TIRION

Delphinus/Equuleus

These are two small constellations in the Cygnus–Aquila area. Delphinus is easy to identify; it looks almost like a cluster, and unwary observers have been known to mistake it for the Pleiades. It honors the dolphin which carried the great singer Arion to safety after he had been thrown overboard from the ship carrying him home from a competition in which he had taken part. When the dolphin died, at a great age, it was placed in the sky. Equuleus represents a foal which was given by Mercury to Castor, one of the Heavenly Twins.

The four stars α, β, γ and δ Delphini, all between magnitude 3.5 and 4.5, make up a small square, with ε (magnitude 4.03) close by. α is named Sualocin and β Rotanev – they were christened by one Nicolaus Venator (it should not take you long to see how the names were derived!). γ is a fine double, with magnitudes 4.5 and 5.5 and separation 9″.6; the primary is yellowish, and the companion looks slightly bluish, mainly by contrast.

There are two interesting red semiregular variables in the constellation: U (RA 20h 45.5m, dec. +18° 05′, range 7.6–8.9) and close to it the rather brighter EU (RA 20h 37.9m, dec. +18° 16′, range 5.9–6.9). These are good binocular objects. Near them is HR Delphini, a nova which flared up to naked-eye visibility in 1967. It was the "slowest" nova on record; its magnitude is now about its pre-outburst value (12.7) and it is unlikely to fade much further. NGC 6394 (RA 20h 34.2m, dec. +07° 24′) is a globular cluster which is an easy telescopic object.

Equuleus has only one star above the fourth magnitude: α (3.92), sometimes called Kitalpha. ε (RA 20h 59.1m, dec. +04° 18′) is a binary with magnitudes 6.0 and 6.3, period 101.4 years and separation 1″; there is a 4.7-magnitude star in the same field. δ is a binary with a period of only 5.7 years, but the separation is never more than 0″.3.

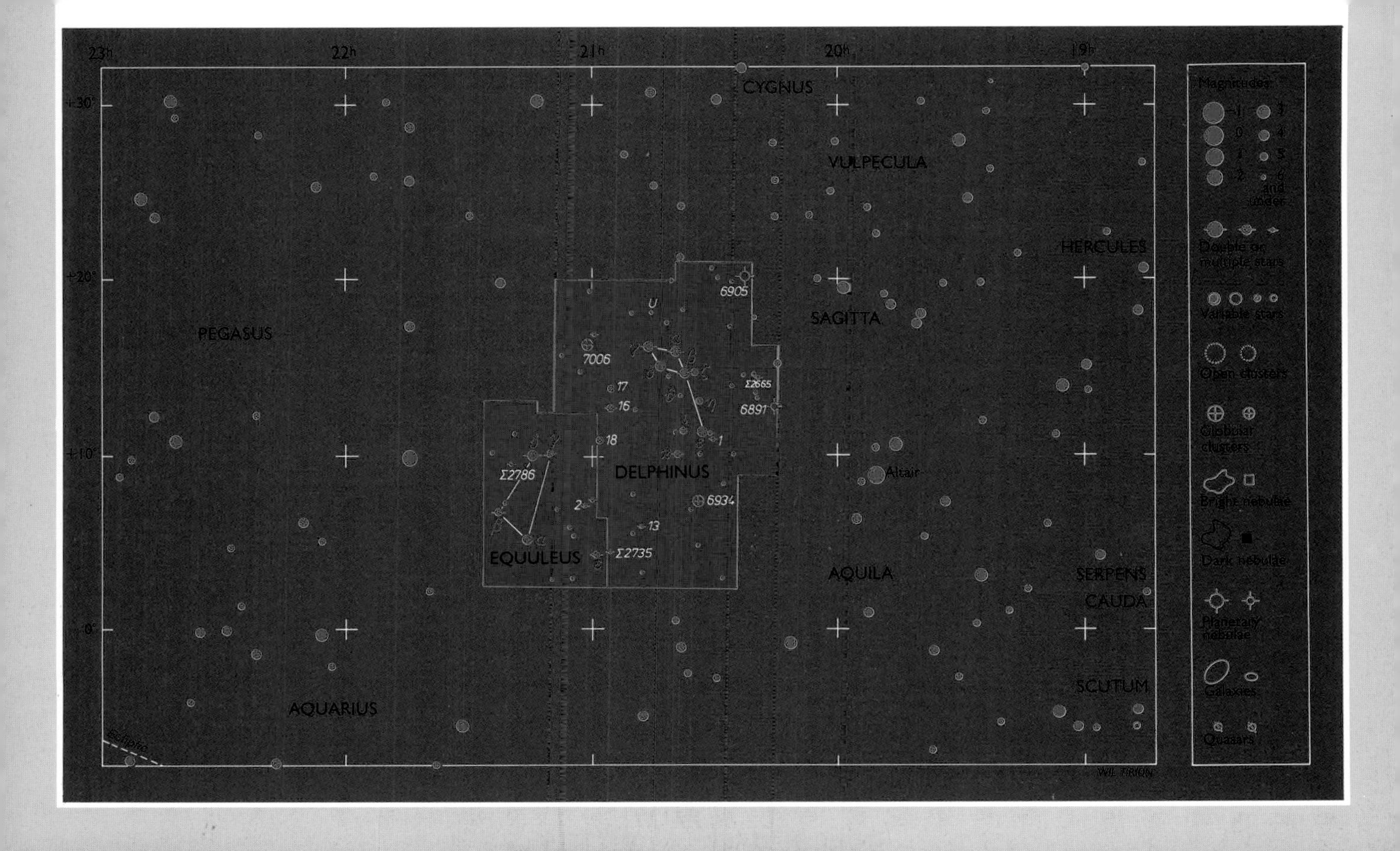
23h
22h
21h
20h
19h
+30°
+20°
+10°
0°
CYGNUS
VULPECULA
HERCULES
SAGITTA
PEGASUS
6905
U
7006
Σ2665
17
16
6891
18
1
2
Σ2786
DELPHINUS
Altair
6934
13
EQUULEUS
Σ2735
AQUILA
SERPENS
CAUDA
SCUTUM
AQUARIUS
Ecliptic
WIL TIRION
Magnitudes
-1
0
1
2
3
4
5
6
and
under
Double or
multiple stars
Variable stars
Open clusters
Globular
clusters
Bright nebulae
Dark nebulae
Planetary
nebulae
Galaxies
Quasars

Dorado/Mensa

These two far-southern constellations are notable only because they include the Large Magellanic Cloud. Dorado, the Swordfish, has only one star above the fourth magnitude: α, magnitude 3.27. It is a very close binary, and there is a 9.8-magnitude star at a separation of 77″.7. β (RA 05h 33.6m, dec. −62° 29′) is a Cepheid variable with a magnitude range of 3.7 to 4.1, and a period of 9.84 days. A convenient comparison star is δ (magnitude 4.35). Mensa is even less distinguished: its brightest star, α, is only of magnitude 5.1. It was originally Mons Mensae, Table Mountain, but the Mons has been tacitly dropped. In any case it is difficult to see why it is ranked as a separate constellation.

The Large Magellanic Cloud is 170,000 light-years away, and is one of the most important objects in the sky to astronomers. It contains objects of all kinds, notably the superb Tarantula Nebula, 30 Doradus, which is the most brilliant part of the whole Cloud and can be identified even with the naked eye. It is the largest known diffuse nebula – if it were as close to us as the Great Nebula in Orion, it would cast shadows! Many novae have been seen in the Cloud, and in 1987 there was even a supernova, 1987a, which became brighter than the third magnitude. It has now become very faint. Probably no object has been more intensely studied than this. It may be expected to produce a pulsar, but so far none has been found. Unexpectedly, the star that exploded was a blue giant rather than a red one.

The Cloud is very prominent with the naked eye, and even moonlight will not hide it. It shows indications of barred spiral form, though previously it was regarded as an irregular system. It is by far the brightest extragalactic object visible without optical aid, and European astronomers never cease to regret that it lies so far south.

The Large Cloud also contains the variable star S Doradus, which is at least a million times more luminous than the Sun and is therefore one of the most powerful stars known – and yet at this distance it is too faint to be seen with the naked eye!

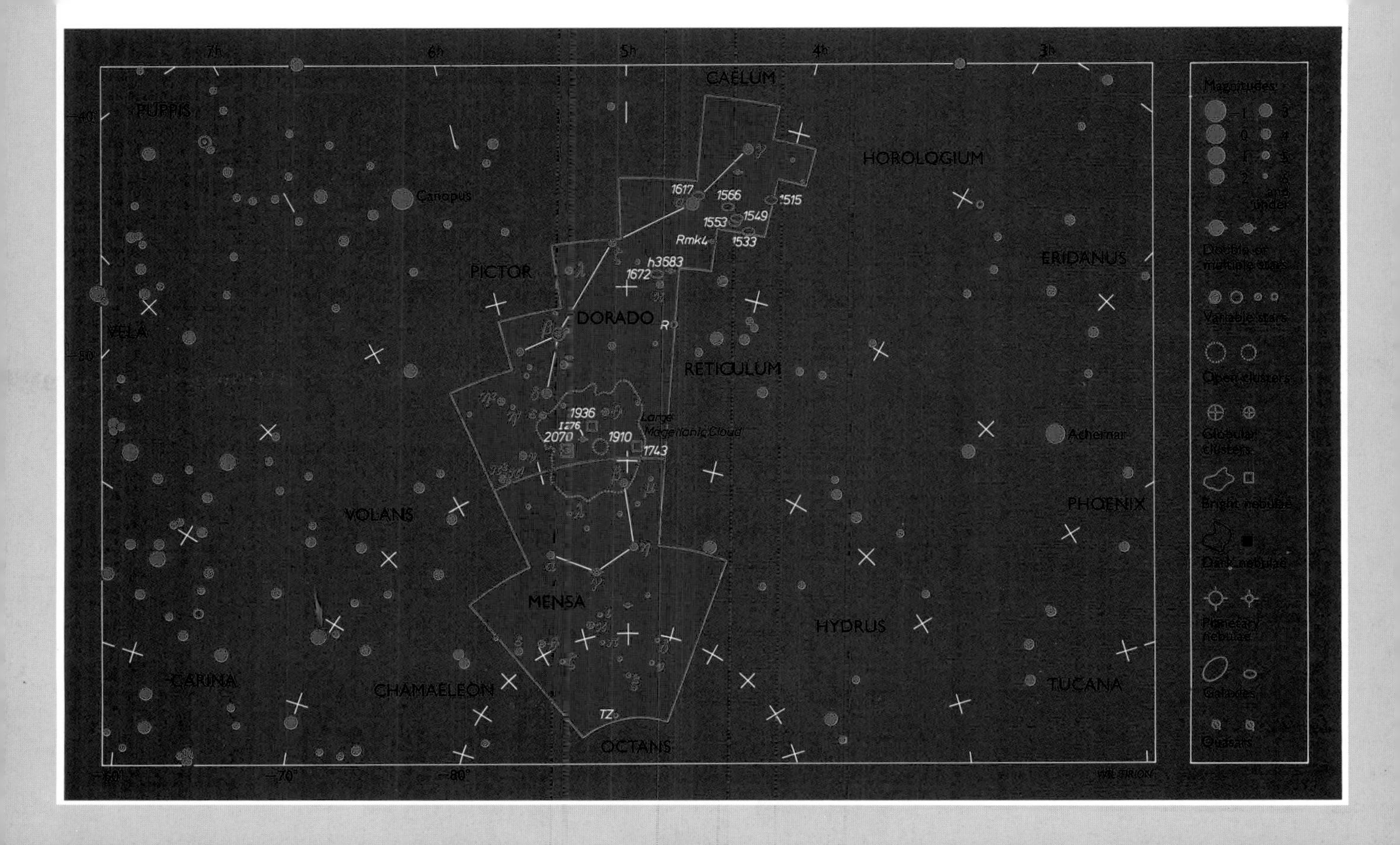
7h
6h
5h
4h
3h
−40°
−50°
−60°
−70°
−80°
CAELUM
HOROLOGIUM
ERIDANUS
PUPPIS
VELA
PICTOR
DORADO
RETICULUM
VOLANS
MENSA
HYDRUS
PHOENIX
TUCANA
CARINA
CHAMAELEON
OCTANS
Canopus
Achernar
Large Magellanic Cloud
1617
1566
1515
1553
1549
1533
Rmk4
h3583
1672
R
1936
1910
2070
1743
TZ
Magnitudes
Double or multiple stars
Variable stars
Open clusters
Globular clusters
Bright nebulae
Dark nebulae
Planetary nebulae
Galaxies
Quasars
WIL TIRION

Draco/Ursa Minor

These are the north polar constellations: Polaris (α Ursae Minoris) lies within one degree of the pole, while at the time when the Pyramids were built the pole star was Thuban or α Draconis.

Mythologically, Ursa Minor represents Arcas, the son of Callisto (see Ursa Major). The outline is like a very faint and distorted Ursa Major. β and γ are known as the Guardians of the Pole; β is very obviously orange.

Polaris has a 9th-magnitude companion at a separation of 18″.4. A telescope of 7.5 centimeters aperture will show it, and it has been seen with a 5-centimeter refractor, though with some difficulty. Polaris itself is actually a Cepheid variable with a very small range.

Draco, the Dragon, stretches from the area between the Bears almost as far as Vega. In mythology Draco is linked with the dragon killed by Cadmus before the founding of the city of Boeotia, though it may also represent the dragon which guarded the Golden Fleece and was disposed of by Jason. The Dragon's head is made up of γ, β (see table), ξ (magnitude 3.75) and ν (4.8). ν is a very wide, easy double; the components are equally bright at magnitude 4.9, and the separation is 61″.9. ε is double: magnitudes 3.8 and 7.4, separation 3″.1.

It is worth finding the planetary nebula NGC 6543 (RA 17h 58.6m, dec. +36° 38′), which is said to look like a blue disk; the central star is a difficult object. Draco contains numbers of galaxies, but all are faint.

Thuban (α Draconis) is of magnitude 3.65, and type A0. It lies midway between Mizar in Ursa Major and the Guardians of the Pole.

DRACO					
Leading stars	**RA** h m s	**Declination** ° ′ ″	**Magnitude**	**Spectrum**	**Distance** ly
γ Eltamin	17 56 36.2	+51 29 20	2.23	K5	101
η Aldhibain	16 23 59.3	+61 30 50	2.74	G8	82
β Alwaid	17 30 25.8	+52 18 05	2.79	G2	267

URSA MINOR					
Leading stars	**RA** h m s	**Declination** ° ′ ″	**Magnitude**	**Spectrum**	**Distance** ly
α Polaris	02 31 50.4	+89 15 51	1.99	F8	680
β Kocab	14 50 42.2	+74 09 10	2.08	K4	95
γ Pherkad Major	15 20 43.6	+71 50 02	3.05	A3	225

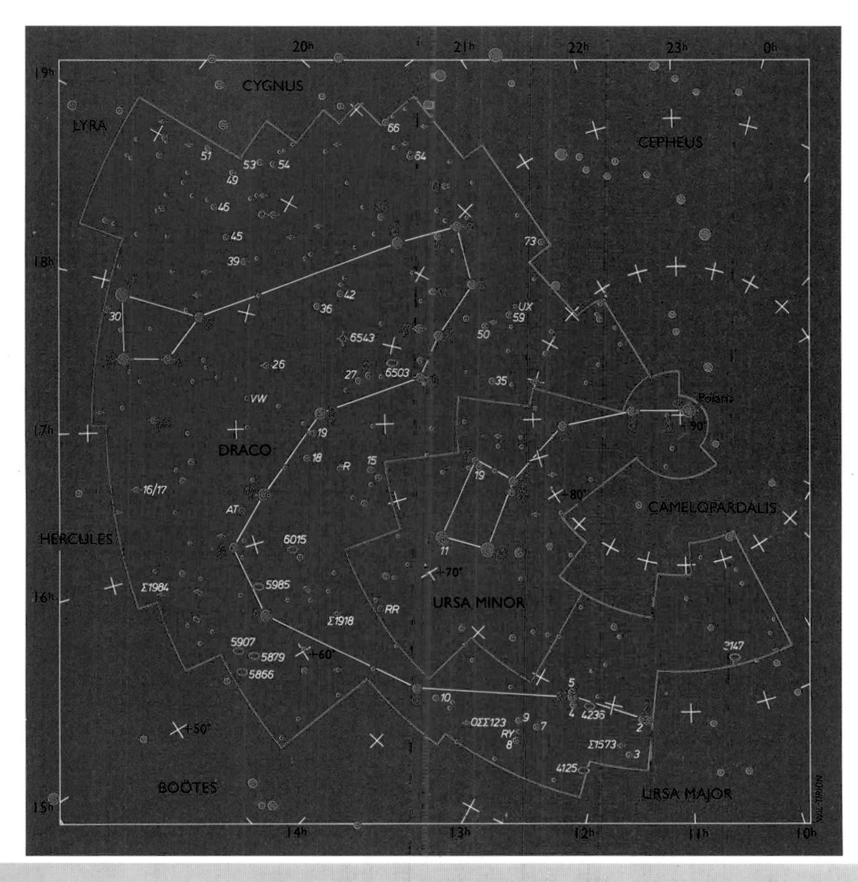
CYGNUS
LYRA
CEPHEUS
DRACO
HERCULES
URSA MINOR
CAMELOPARDALIS
URSA MAJOR
BOÖTES
Polaris
+90°
+80°
+70°
+60°
+50°
19h
18h
17h
16h
15h
20h
21h
22h
23h
0h
14h
13h
12h
11h
10h
W.L. TIRION

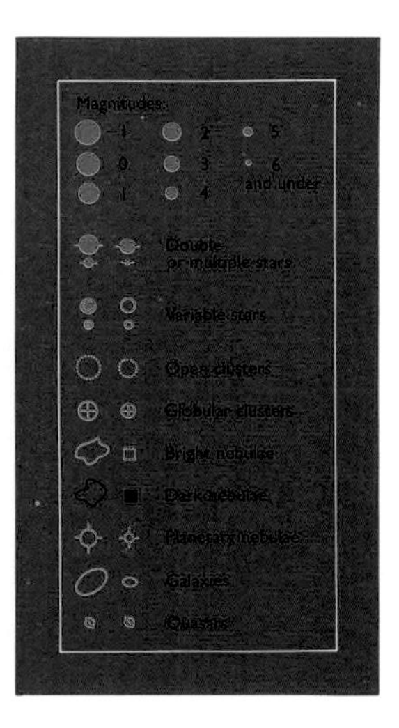
Magnitudes:
-1 0 1 2 3 4 5 6 and under
Double or multiple stars
Variable stars
Open clusters
Globular clusters
Bright nebulae
Dark nebulae
Planetary nebulae
Galaxies
Quasars

Eridanus

Eridanus is an immensely long constellation, stretching from near Rigel in Orion right to the south polar region. Mythologically it represents the Po – the river into which the reckless youth Phaethon fell when he was driving the Sun chariot and was struck down by a thunderbolt. There are not many bright stars apart from the far-southern Achernar. Achernar is one of the brightest stars in the sky, and much the most brilliant object anywhere near the south pole.

Acamar is known as "the Last in the River," and has been suspected of fading from magnitude 1 to magnitude 3 in historical times, though the evidence is decidedly slender; it is a fine double (magnitudes 3.4 and 4.5, separation 8″.2). ε (magnitude 3.73) is less than 11 light-years away; it and τ Ceti are the two nearest of the stars that are at all similar to the Sun, and may well be the centers of planetary systems. o^1 and o^2 (named Beid and Keid respectively) are close together in the sky, but not really associated; o^2 is made up of two stars, of which the fainter is a particularly feeble dwarf. Their distance from us is only 16 light-years. ζ, or Zibal, is now of magnitude 4.8, but used to be ranked as 3, and the evidence for secular fading is stronger than in most similar cases; the spectral type is A3 and the distance 52 light-years. Eridanus contains a good number of faint galaxies, but nothing of special interest. Achernar lies close to the third-magnitude star α Hydri.

Leading stars	***RA*** h m s	***Declination*** ° ′ ″	***Magnitude***	***Spectrum***	***Distance*** ly
α Achernar	01 37 42.9	−57 14 12	0.46	B5	85
β Kursa	05 07 50.9	−05 05 11	2.79	A3	91
θ Acamar	02 58 15.6	−40 18 17	2.92	A3+A2	55
γ Zaurak	03 58 01.7	−13 30 31	2.95	M0	143

TAURUS
ORION
Rigel
ERIDANUS
LEPUS
CAELUM
COLUMBA
HOROLOGIUM
FORNAX
CETUS
Mira
PHOENIX
WIL TIRION
Magnitudes:
−1 0 1 2 3 4 5 6 and under
Double or multiple stars
Variable stars
Open clusters
Globular clusters
Bright nebulae
Dark nebulae
Planetary nebulae
Galaxies
Quasars

Fornax/Sculptor/Phoenix

These are three southern-hemisphere constellations; only Phoenix is at all prominent.

Fornax, the Furnace – originally Fornax Chemica, the Chemical Furnace – adjoins Eridanus. The only star above the fourth magnitude is α (3.87), at RA 03h 12m 04.2s, dec. −28° 59′ 13″. Fornax contains a considerable number of faint galaxies. So too does Sculptor, the Sculptor (originally Apparatus Sculptoris, the Sculptor's Apparatus). Sculptor contains no star above the fourth magnitude.

Phoenix is one of the "Southern Birds" (the others are Grus, the Crane; Pavo, the Peacock; and Tucana, the Toucan). In mythology Phoenix was a bird which was periodically burned to ashes, but soon grew again to be as lively as before! The leading star is α or Ankaa (magnitude 2.39, RA 00h 26m 17.00s, dec. −42° 18′ 22″), of type K0 and therefore definitely orange; it is 78 light-years away. Next come β (magnitude 3.31) and γ (3.41). ζ is an Algol-type eclipsing binary at RA 01h 08m.4, dec. −55° 15′; its magnitude range is 3.9 to 4.4, and its period 1.67 days. Both the components are hot stars of type B. β is a fairly close double; the components are almost equal at magnitudes 4.0 and 4.2, and the separation is 1″.4.

ζ Phoenicis lies close to Achernar in Eridanus. The triangle made up of γ, β and δ (magnitude 3.95) is fairly distinctive.

The Fornax cluster of galaxies lies about 55 million light-years away.

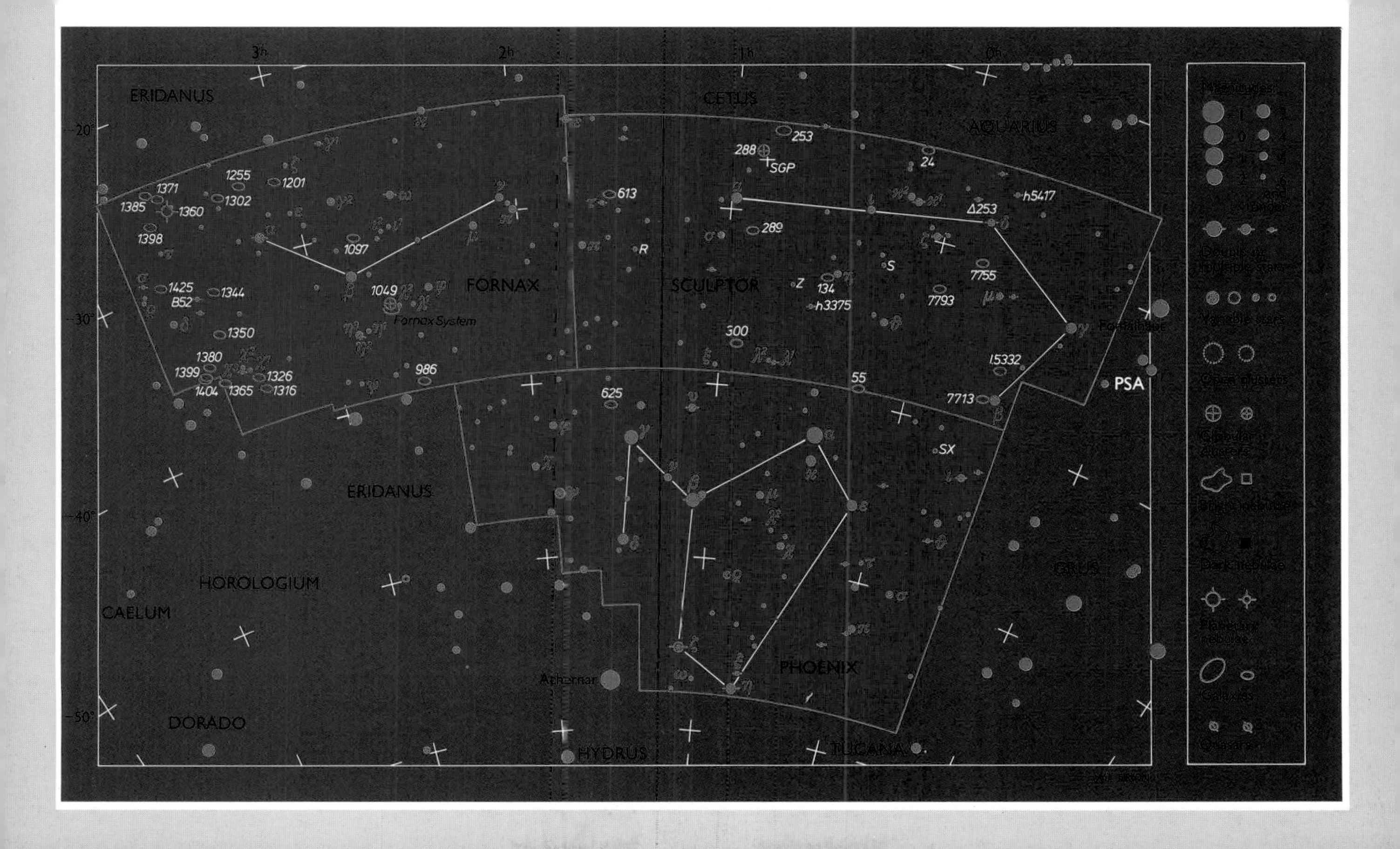
3h
2h
1h
0h
20°
30°
40°
50°
ERIDANUS
CETUS
AQUARIUS
FORNAX
SCULPTOR
PSA
Fomalhaut
ERIDANUS
HOROLOGIUM
CAELUM
DORADO
HYDRUS
Achernar
PHOENIX
TUCANA
GRUS
Fornax System
SGP
253
288
24
h5417
Δ253
613
289
R
S
Z
134
h3375
7755
7793
300
15332
55
7713
625
SX
986
1049
1097
1201
1255
1302
1371
1385
1360
1398
1425
B52
1344
1350
1380
1399
1404
1365
1326
1316
Magnitudes
Double or multiple stars
Variable stars
Open clusters
Globular clusters
Bright nebulae
Dark nebulae
Planetary nebulae
Galaxies
Quasars

Gemini

Gemini, the Twins, is a large and important northern Zodiacal constellation. Castor and Pollux were twin sons of the Queen of Sparta; Castor was mortal, but Pollux was not – his father was none other than Jupiter, who had paid a clandestine visit to the Queen for totally discreditable reasons! When Castor was killed in battle, Pollux pleaded to be allowed to share his immortality with his brother; this was granted, and both boys were transferred to the sky.

Today Pollux is the brighter of the two, though ancient astronomers gave precedence to Castor; it is unlikely that any real change has occurred. Castor is a fine double. The components are of magnitudes 1.9 and 2.9, and the present separation is 2″.5, so that Castor can still be split with a small telescope; the revolution period is 420 years. Each component is a spectroscopic binary, and there is a third component, Castor C or YY Geminorum, at 72″.5 (magnitude 8.8), which is also a binary – in fact, an eclipsing variable, so Castor consists of six suns. Pollux is a single star; its orange hue is obvious with the naked eye.

There are several easy doubles here. μ has a 9.8-magnitude companion at separation 72″.7: ε has a companion, separation 110″.3, magnitude 9. Another binary is δ, separation 6″, period 1,200 years. As the companion is of magnitude 8.2, it is not difficult. There are two naked-eye variables: ζ is a typical Cepheid (range 3.7 to 4.1, period 10.15 days) and η is a semiregular (range 3.2–3.9, and a very rough period of 233 days). It has an 8.8-magnitude companion at a separation of 1″.4. It is of the same color as μ, a good comparison star. R Geminorum is a Mira variable (RA 07h 07.4m, dec. +22° 42′, range 6.0–14.0, period 370 days) and lies not far from δ.

Of the several bright open clusters in Gemini, the most prominent is M35 (RA 06h 08.9m, dec. +24° 20′), near μ and η; it is easily visible with the naked eye. An interesting telescopic object is the Eskimo Nebula (NGC 2392) at RA 07h 29.2m, dec. +20° 55′; this planetary has an integrated magnitude of about 10. It also lies not far from δ.

Leading stars	*RA* h m s	*Declination* ° ′ ″	*Magnitude*	*Spectrum*	*Distance* ly
β Pollux	07 45 18.9	+28 01 34	1.14	K0	36
α Castor	07 34 35.9	+31 53 18	1.58	A0	46
γ Alhena	06 37 42.7	+16 23 57	1.93	A0	85
μ Tejat	06 22 57.6	+22 30 49	2.88v	M3	231
ε Mebsuta	06 43 55.9	+25 07 52	2.98	G8	685
η Propus	06 14 52.6	+22 30 24	3.1 (max)	M3	186

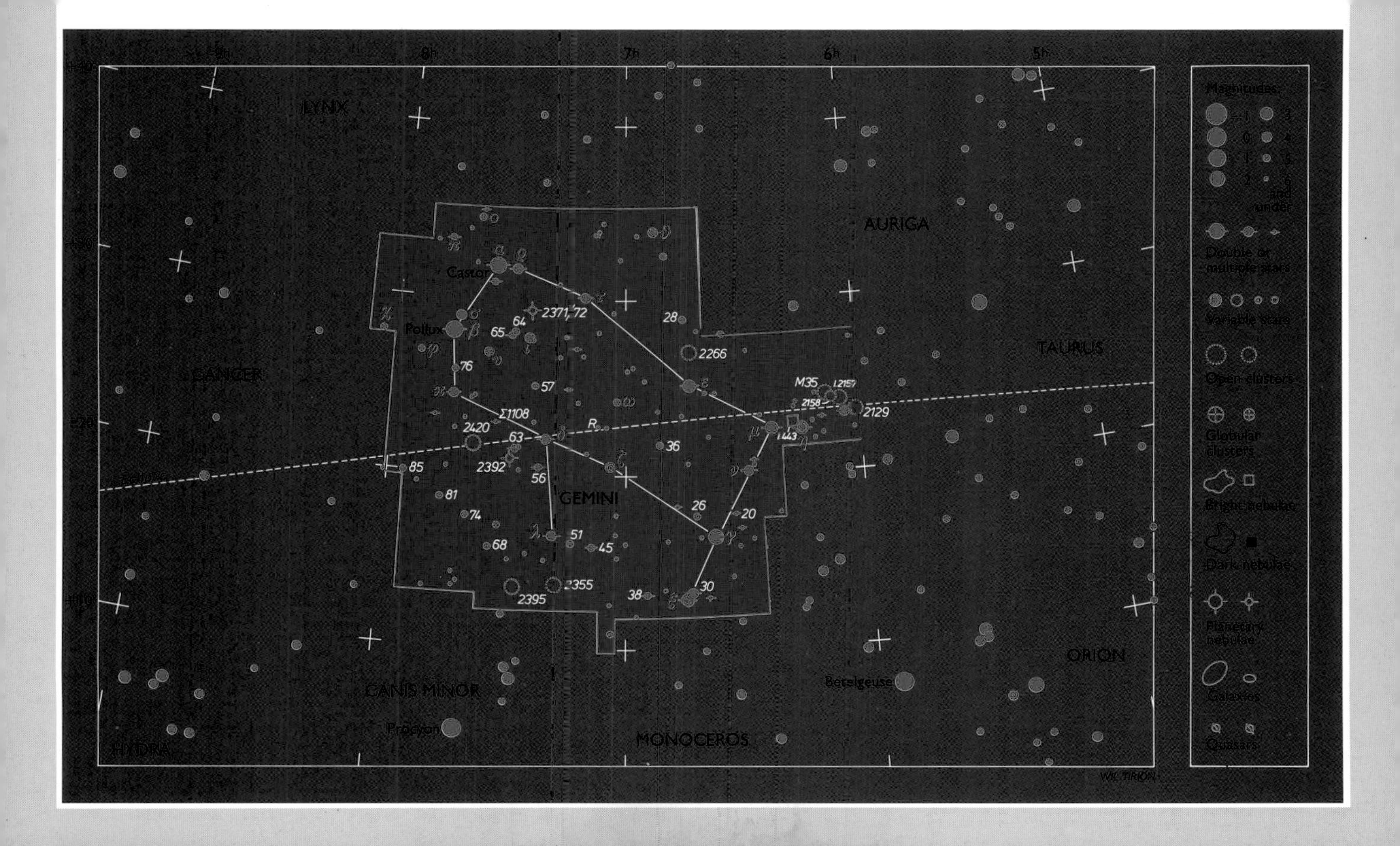
8h
7h
6h
5h
LYNX
AURIGA
TAURUS
CANCER
GEMINI
ORION
CANIS MINOR
MONOCEROS
HYDRA
Castor
Pollux
Procyon
Betelgeuse
M35
2129
2266
2371, 72
2420
2392
2395
2355
Σ1108
R
Magnitudes:
and under
Double or multiple stars
Variable stars
Open clusters
Globular clusters
Bright nebulae
Dark nebulae
Planetary nebulae
Galaxies
Quasars
WIL TIRION

Grus/Piscis Austrinus

Both of these lie well south of the celestial equator. Fomalhaut, the leading star of Piscis Austrinus, can be seen from Britain, the United States and most of Europe, but Grus is to all intents and purposes invisible from the British Isles and the northern United States.

Grus, the Crane, is one of the Southern Birds, and the only one of the four which is really distinctive. A line of faint stars runs from γ to β; two pairs, labeled μ and δ, look like wide doubles, though in fact the components are not genuinely connected with each other. α is bluish-white, β warm orange – a contrast excellently shown with binoculars.

Two Mira variables in Grus, R (RA 21h 48.5m, dec. −46° 55′) and S (RA 22h 26.1m, dec. −48° 26′) are easy binocular objects when at maximum; R ranges in magnitude from 7.4 to 14.9 (period 332 days) and S from 6.0 to 15.0 (401 days). Grus contains a number of galaxies, but none has an integrated magnitude as bright as 10.

Piscis Austrinus, the Southern Fish, is also known as Piscis Australis; no mythological legends seem to be attached to it. It has only one bright star, α or Fomalhaut (RA 22h 38.9s, dec. −29° 37′ 20″), which is easily found by using two of the stars in the Square of Pegasus, β and α, as pointers. It is always very low as seen from Britain, and from northern Scotland it barely rises at all. It is 22 light-years away, and 13 times as luminous as the Sun; it is of type A3 and apparent magnitude 1.16.

Fomalhaut is of special interest because in 1983 a survey by IRAS, the Infra-Red Astronomical Satellite, showed that it is one of those stars associated with cool, possibly planet-forming material. This is not to claim that Fomalhaut is the center of a planetary system, but there is at least a distinct possibility.

There is one easy double in Piscis Austrinus. β (Fum al Samakah), magnitude 4.3, has a 7.9-magnitude companion at a separation of 30″.3. This is an optical pair, not a binary. δ (RA 22h 55.9m, dec. −32° 32′) is double, with magnitudes 4.2 and 9.2, separation 5″.0. There are several galaxies in the constellation, but all are faint.

GRUS					
Leading stars	***RA***	***Declination***	***Magnitude***	***Spectrum***	***Distance***
	h m s	° ′ ″			ly
α Alnair	22 08 13.8	−46 57 40	1.74	B5	68
β Al Dhanab	22 42 39.9	−46 53 05	2.11	M3	173
γ	21 53 55.6	−37 21 54	3.01	B8	228

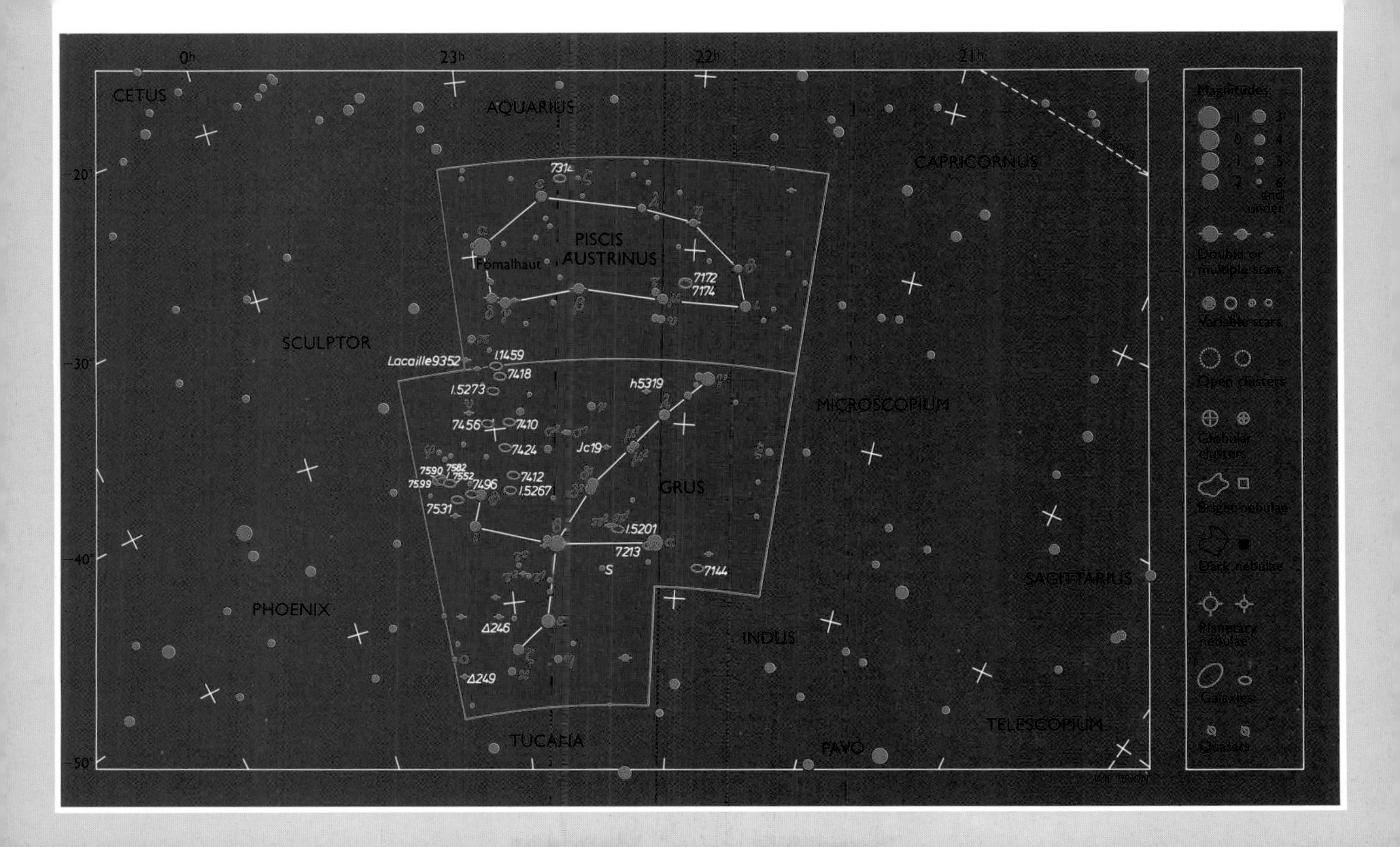
0h
23h
22h
21h
CETUS
AQUARIUS
CAPRICORNUS
Ecliptic
PISCIS
AUSTRINUS
Fomalhaut
731ε
7172
7174
SCULPTOR
Lacaille9352
I.1459
7418
I.5273
h5319
MICROSCOPIUM
7456
7410
7424
Jc19
7590
7582
7552
7599
7496
7412
I.5267
GRUS
7531
I.5201
7213
S
7144
PHOENIX
Δ246
INDUS
Δ249
SAGITTARIUS
TUCANA
PAVO
TELESCOPIUM
-20°
-30°
-40°
-50°
Magnitudes
-1
0
1
2
3
4
5
6
and
under
Double or
multiple stars
Variable stars
Open clusters
Globular
clusters
Bright nebulae
Dark nebulae
Planetary
nebulae
Galaxies
Quasars
Wil Tirion

Hercules

In mythology Hercules was a great hero – most of us have heard of the Labors of Hercules; among his victims were the multiheaded monster Hydra and the Nemaean Lion (Leo). Yet in the sky Hercules is not very bright, even though the constellation covers a very large area; it occupies the large triangle bounded by Vega (α Lyrae), Rasalhague (α Ophiuchi) and Alphekka (α Coronae Borealis). α Herculis (Rasalgethi) lies close to Rasalhague, and is rather divorced from the rest of the constellation.

Rasalgethi is a semiregular variable; the extreme range is said to be from 3.0 to 4.0, but for most of the time the magnitude is between 3.3 and 3.7. There is a very rough period of about 100 days. There is a 5.4-magnitude companion at a separation of 4″.7; by contrast the companion appears decidedly green – a fine object for a small telescope. Other easy doubles are κ (magnitudes 5.3 and 6.5, separation 28″.4); γ (3.8 and 9.8, 41″.6); and δ (3.7 and 8.2, 8″.9 – an optical pair). ζ is a fine binary. The components are of magnitudes 2.9 and 5.5; the separation averages around 1″.6, but as the period is only 34.5 years, the position angle and separation alter quickly.

There are two magnificent globular clusters in Hercules: M13 (RA 16h 41.7m, dec. +36° 28′) and M92 (RA 17h 17.1m, dec. +3° 08′). M13, between η and ζ, is just visible with the naked eye, and is the brightest of all globulars apart from the far-southern ω Centauri and 47 Tucanae. It was discovered by Edmond Halley in 1714; it is around 22,000 light-years away, with a real diameter of at least 160 light-years. The outer parts are not hard to resolve with a small telescope. M92 is scarcely inferior, though its greater distance (35,000 light-years) makes it appear fainter; it too is easy to resolve at its outer edges. Look also for the planetary nebula NGC 6210 (RA 16h 44.5m, dec. +23° 49′), which has an integrated magnitude above 10 and lies some way off a line joining β to δ.

Leading stars	***RA*** h m s	***Declination*** ° ′ ″	***Magnitude***	***Spectrum***	***Distance*** ly
β Kornephoros	16 30 13.1	+21 29 22	2.77	G8	101
ζ Rutilicus	16 41 17.1	+31 36 10	2.81	G0	31
α Rasalgethi	17 14 38.8	+14 23 25	3–4	M5	218
δ Sarin	17 15 01.8	+24 50 21	3.14	A3	91
π	17 15 02.6	+36 48 33	3.16	K3	391

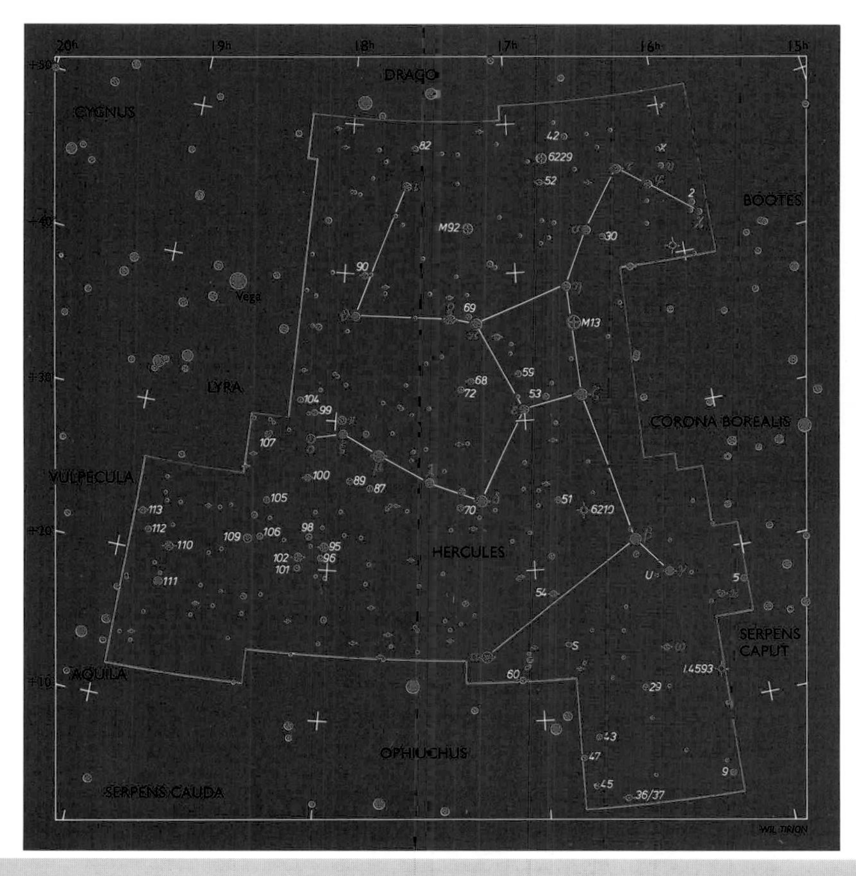
20h
19h
18h
17h
16h
15h
+50
+40
+30
+20
+10
DRACO
CYGNUS
BOOTES
LYRA
Vega
VULPECULA
HERCULES
CORONA BOREALIS
SERPENS CAPUT
AQUILA
OPHIUCHUS
SERPENS CAUDA
M92
M13
6229
6210
I.4593
WIL TIRION

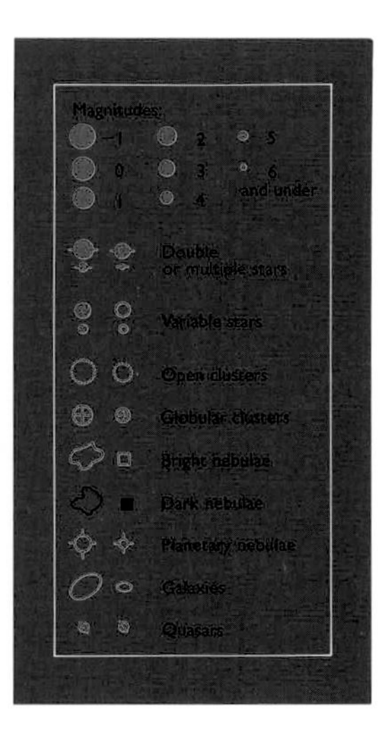
Magnitudes:
−1
0
1
2
3
4
5
6 and under
Double or multiple stars
Variable stars
Open clusters
Globular clusters
Bright nebulae
Dark nebulae
Planetary nebulae
Galaxies
Quasars

Horologium/Reticulum

Two more small and undistinguished far-southern constellations. Horologium (the Clock) has only one star above the fourth magnitude; this is α (RA 04h 14m 00.0s, dec. −42° 17′ 40″, magnitude 3.86). It is of type K1, and distinctly orange. The Mira variable R Horologii (RA 02h 53.9m, dec. −49° 53′) has a period of 404 days; at maximum it may rise to magnitude 4.7, but at minimum sinks to below 14.

There are various faint galaxies in Horologium, but the most interesting object is probably the globular cluster NGC 1261 (RA 03h 12.3m, dec. −55° 13′), which has an apparent diameter of almost 7 minutes of arc. It is about 70,000 light-years away, and is not a difficult object in a small telescope, though it is not too easy to locate.

Reticulum, the Net (originally Reticulum Rhomboidalis, the Rhomboidal Net), was added to the sky by the French astronomer Lacaille in 1752, and may have referred to one of the measuring instruments he used with his telescope. Its brightest star is μ (magnitude 3.35), which has a G6-type spectrum and is slightly yellowish. Next come β (3.85), ε (4.44), γ (4.51) and δ (4.56); all these are orange, which gives Reticulum a fairly distinctive appearance when viewed with binoculars. β has an 8.1-magnitude companion at a separation of 14″.

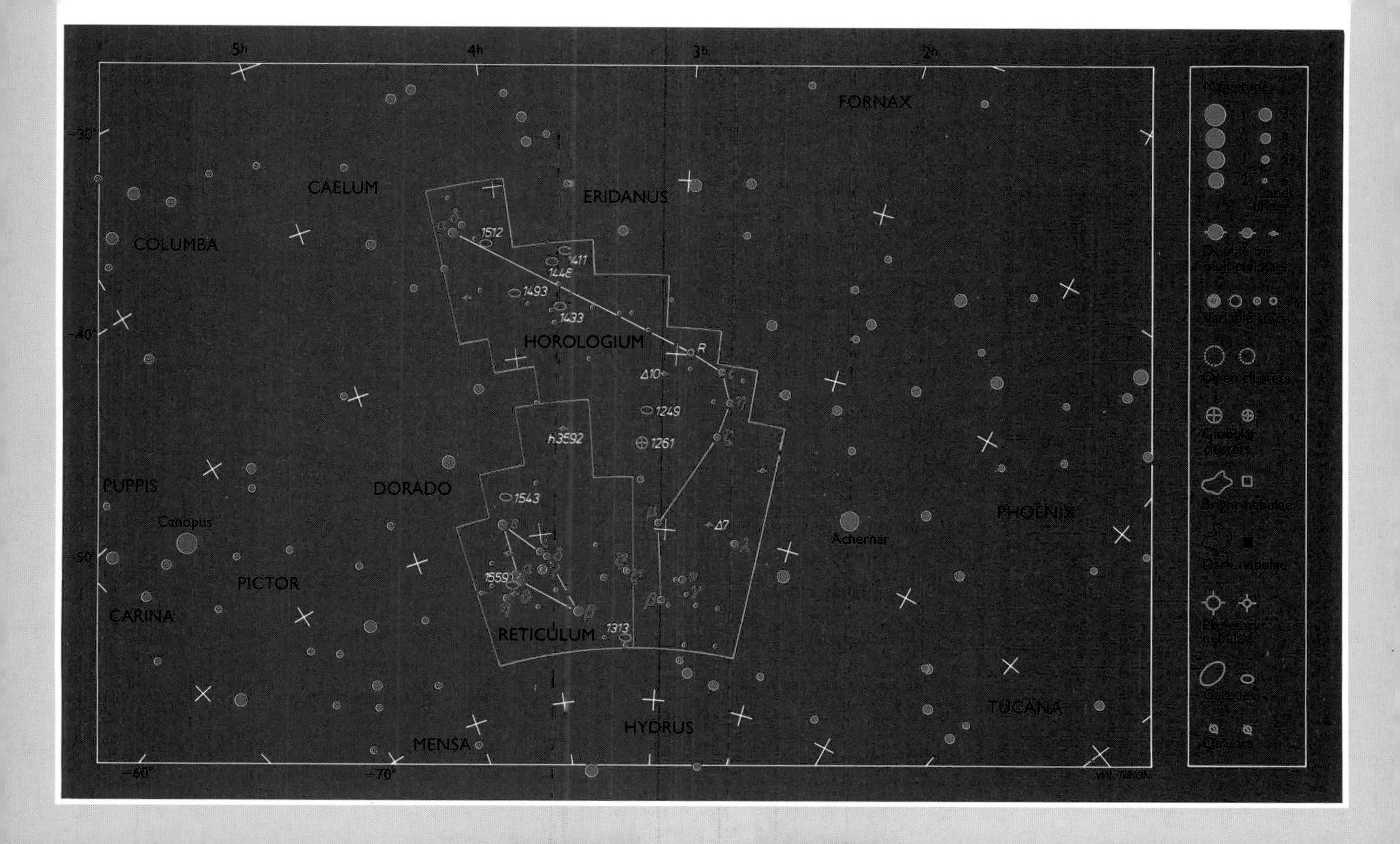

5h
4h
3h
2h
−30°
−40°
−50°
−60°
−70°
FORNAX
CAELUM
ERIDANUS
COLUMBA
HOROLOGIUM
PUPPIS
DORADO
Canopus
PICTOR
CARINA
RETICULUM
PHOENIX
Achernar
TUCANA
HYDRUS
MENSA
1512
1411
1448
1493
1433
R
Δ10
1249
h3592
1261
1543
Δ7
1559
1313
WIL TIRION
Magnitudes
−1
0
1
2
3
4
5
6
and under
Double or multiple stars
Variable stars
Open clusters
Globular clusters
Bright nebulae
Dark nebulae
Planetary nebulae
Galaxies
Quasars

Hydra

Now that Argo Navis has been chopped up, Hydra is the largest constellation in the sky. The name means "watersnake," though the constellation is usually identified with the monster killed by Hercules for one of his labors. Hydra extends over more than six hours of right ascension, from Cancer to Centaurus, but has only one bright star: this is α Hydrae (Alphard), known as the Solitary One because it is so isolated in the sky. An easy way to identify it is to use the Twins in Gemini as pointers: a line from Castor through Pollux, and extended, will show the way to Alphard.

Alphard is of magnitude 1.98, and is decidedly reddish, with a K3-type spectrum; it is 85 light-years away, and 115 times as luminous as the Sun. It has been suspected of being variable over a small range, but its brightness is awkward to estimate, because there are no suitable comparison stars anywhere near it. The next-brightest stars in Hydra are γ (magnitude 3.00) and ζ and ν (each 3.11). The "head," made up of ζ, ε (3.38), η (4.30) and δ (4.16), is fairly distinctive.

R Hydrae (RA 13h 29.7m, dec. −23° 17′) is a very red Mira variable with a period of 390 days: at maximum it may reach magnitude 4, but drops to 10 at minimum. It lies in the southern part of the constellation, near γ. Not far from ν lies U Hydrae (RA 10h 37m.6, dec. −13° 23′), which is another very red star, of type N and range 4.8 to 5.8. This is a semiregular variable, not a Mira star, and is always within binocular range.

There are several nebular objects in Hydra which are worth locating (see table). M48 is a bright open cluster; it is detectable with the naked eye under good conditions, but is rather isolated. M68 is a binocular object, though a telescope is needed to show it well. M83 is a fine face-on spiral, around 20 million light-years away; a 37-centimeter telescope shows the arms well, and they are detectable even with a 30-centimeter instrument. As is to be expected in such a large constellation, Hydra contains many other galaxies, but few are above the 11th magnitude.

Clusters and galaxies	**RA** h m	**Declination** ° ′	**Magnitude**	**Type**
M48	08 13.8	−05 48	5.8	Open cluster
M68	12 39.5	−26 45	8.2	Globular
M83	13 37.0	−29 52	8.2	Spiral galaxy
NGC 3621	11 18.3	−32 49	9.9	Spiral galaxy

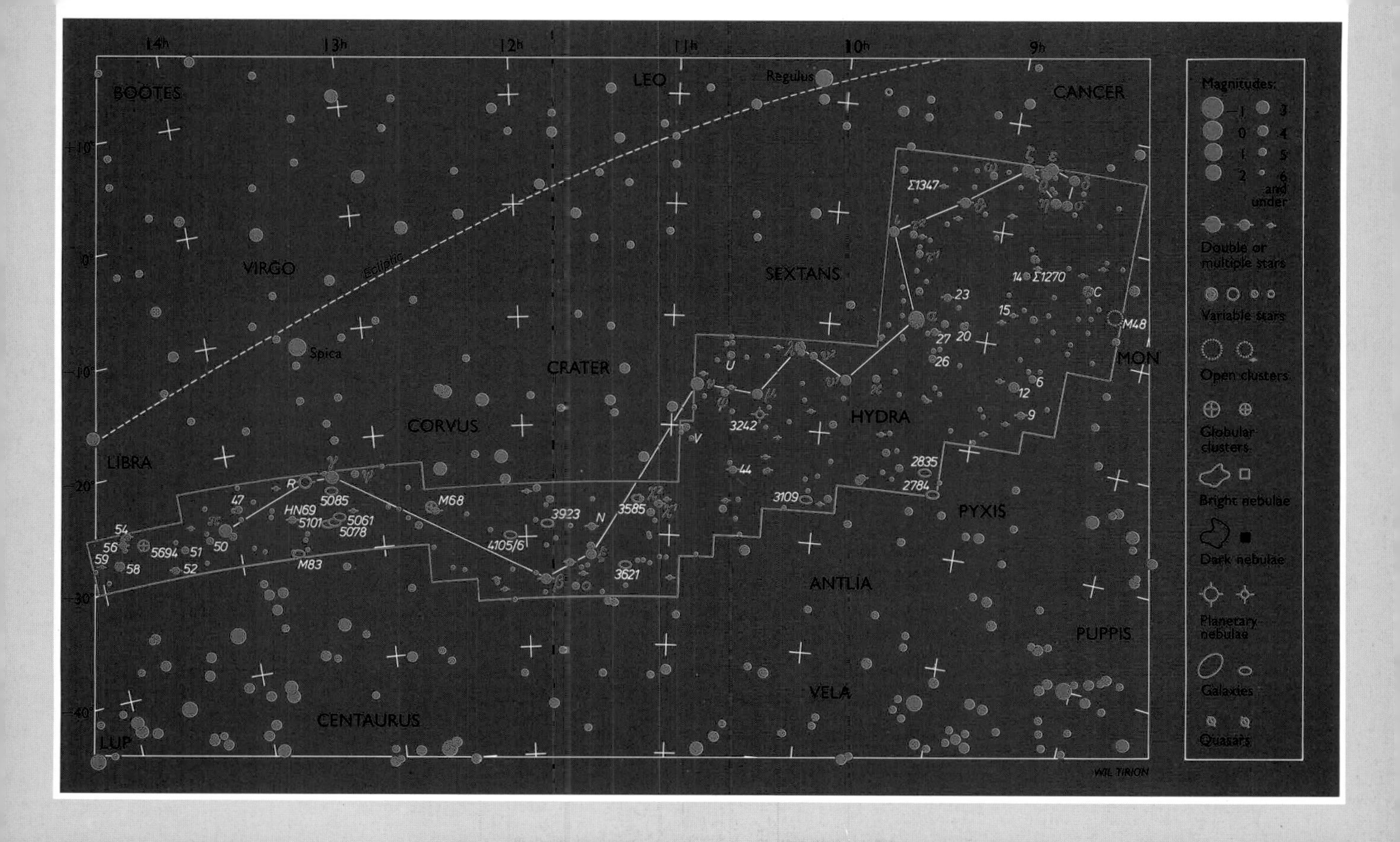
14h
13h
12h
11h
10h
9h
+10°
0°
−10°
−20°
−30°
−40°
BOOTES
VIRGO
LIBRA
LEO
CRATER
CORVUS
SEXTANS
HYDRA
CANCER
MON
PYXIS
ANTLIA
PUPPIS
VELA
CENTAURUS
LUP
Regulus
Spica
Ecliptic
Σ1347
14 Σ1270
23
15
27
20
26
6
12
9
C
M48
U
3242
V
44
2835
2784
3109
3585
3923
N
3621
4105/6
M68
R
47
HN69
5085
5101
5061
5078
M83
54
56
59
58
5694
51
50
52
WIL TIRION
Magnitudes:
−1
0
1
2
3
4
5
6 and under
Double or multiple stars
Variable stars
Open clusters
Globular clusters
Bright nebulae
Dark nebulae
Planetary nebulae
Galaxies
Quasars

Hydrus/Tucana

Hydrus (the Little Snake) is in the far south. Indeed, its brightest star, β – magnitude 2.80 – is the nearest prominent star to the polar point, even though it is over 12 degrees away. It is of type G, and is only just over 20 light-years away, so that it is one of our nearer neighbors. The only other fairly bright stars in Hydrus are α (magnitude 2.86) and γ (3.24).

Tucana, the Toucan, is the dimmest of the four Southern Birds, but contains most of the Small Magellanic Cloud, together with the globular cluster 47 Tucanae. The brightest star is the rather orange α (magnitude 2.86). β is a very wide double, of magnitudes 4.4 and 4.5. Each component is again double, and the group lies in a rich field.

The Small Cloud of Magellan cannot rival the Large Cloud, but is a very easy naked-eye object, and parts of it can be resolved without difficulty. Almost silhouetted against it, though belonging to our Galaxy, is 47 Tucanae (NGC 104, at RA 00h 24.1m, dec. −72° 05′), the brightest of all globular clusters apart from ω Centauri. It is interesting to note that the surface brightness of the cluster is much greater than that of the Cloud. It is fair to say that in a telescope 47 Tucanae is even more striking than ω Centauri, because it is not so large, and all of it can be fitted into the field at the same time. Not far away is another bright globular cluster, NGC 362, at RA 01h 03.2m, dec. −70° 51′; it has an integrated magnitude of 6.6.

The Small Magellanic Cloud, a satellite galaxy to our own, lies in Tucana.

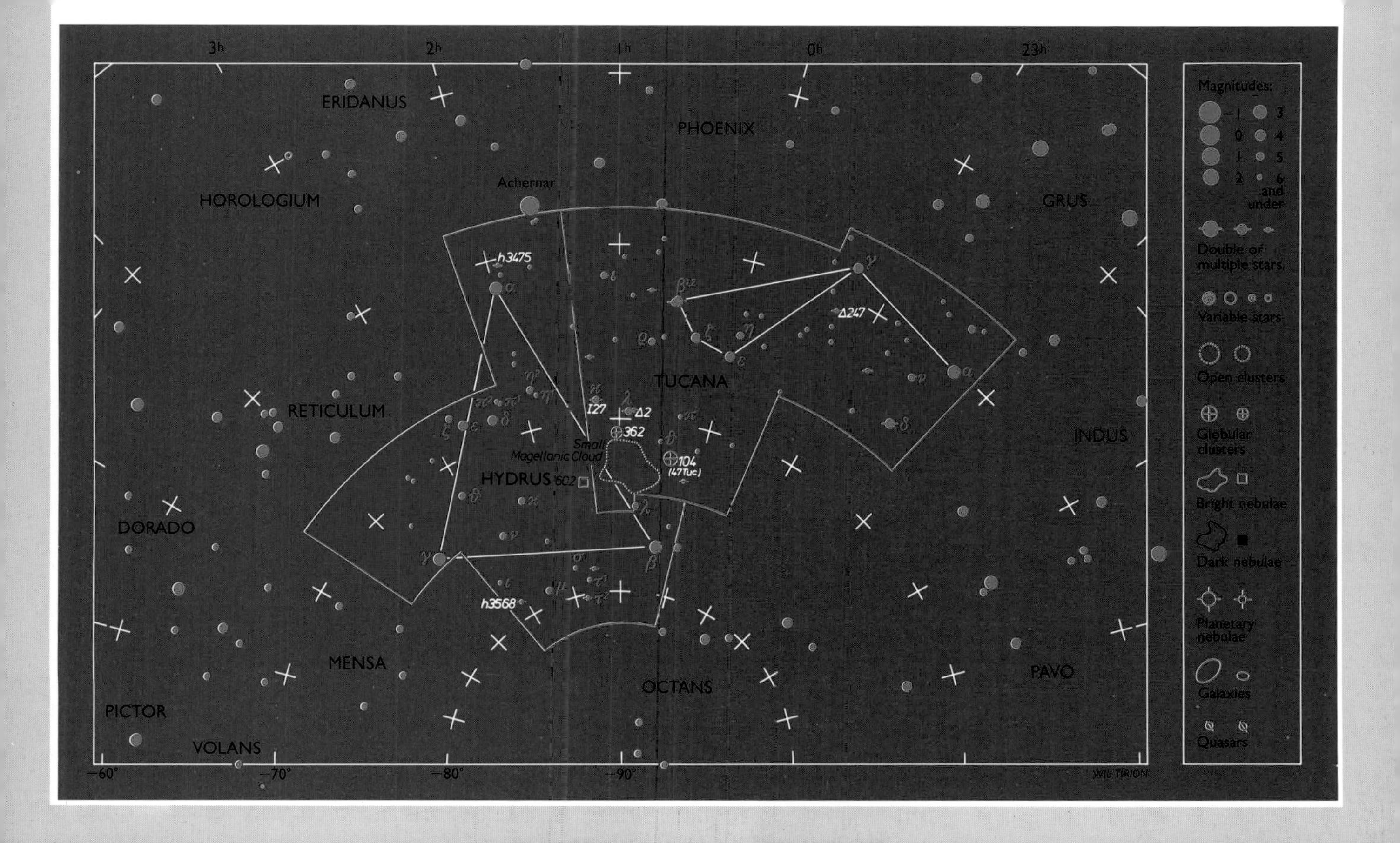
3h
2h
1h
0h
23h
ERIDANUS
PHOENIX
HOROLOGIUM
Achernar
GRUS
h3475
Δ247
RETICULUM
TUCANA
I27
Δ2
362
Small
Magellanic Cloud
104
(47Tuc)
INDUS
HYDRUS
602
DORADO
h3568
MENSA
OCTANS
PAVO
PICTOR
VOLANS
−60°
−70°
−80°
−90°
WIL TIRION
Magnitudes:
−1
0
1
2
3
4
5
6
and
under
Double or
multiple stars
Variable stars
Open clusters
Globular
clusters
Bright nebulae
Dark nebulae
Planetary
nebulae
Galaxies
Quasars

Indus/Pavo/Telescopium

These are three more far-southern constellations, of which only Pavo has any claim to distinction. Indus (the Indian) has two stars above the fourth magnitude: α or Persian (3.11), and β (3.65), but very little of interest. ε, of magnitude 4.7, is one of our near neighbors: it lies at only 11 light-years from us, and is actually the least luminous star visible with the naked eye, as it has only 13 per cent of the luminosity of the Sun. It is a red dwarf, of type K5.

Telescopium (the Telescope) also lacks any bright star: its leader is α (magnitude 3.51). The only object of interest is the erratic variable RR Telescopii (RA 20h 04.2m, dec. −55° 43′), which has an extreme range of magnitude, from 6.5 to 16.5. It is almost certainly a close binary: one component is a red giant, which may even be in the process of producing a planetary nebula.

Pavo, the Peacock, has one bright star, α (magnitude 1.94). It is of type B3, and is 700 times as luminous as the Sun. ϰ is variable, with a magnitude range of 3.9 to 4.7, and period 9.09 days. This is a short-period star, of the type known as Type II Cepheids or W Virginis stars.

The globular cluster NGC 6752 (RA 19h 10.9m, dec. −59° 59′) is a splendid object: the integrated magnitude is 5.4. There is also a barred spiral galaxy, NGC 6744 (RA 19h 09.8m, dec. −63° 51′), which is large but not very bright. The structure is too dim to be made out except with very powerful equipment.

Spiral and elliptical galaxies in a cluster in Pavo, 300 million light-years away.

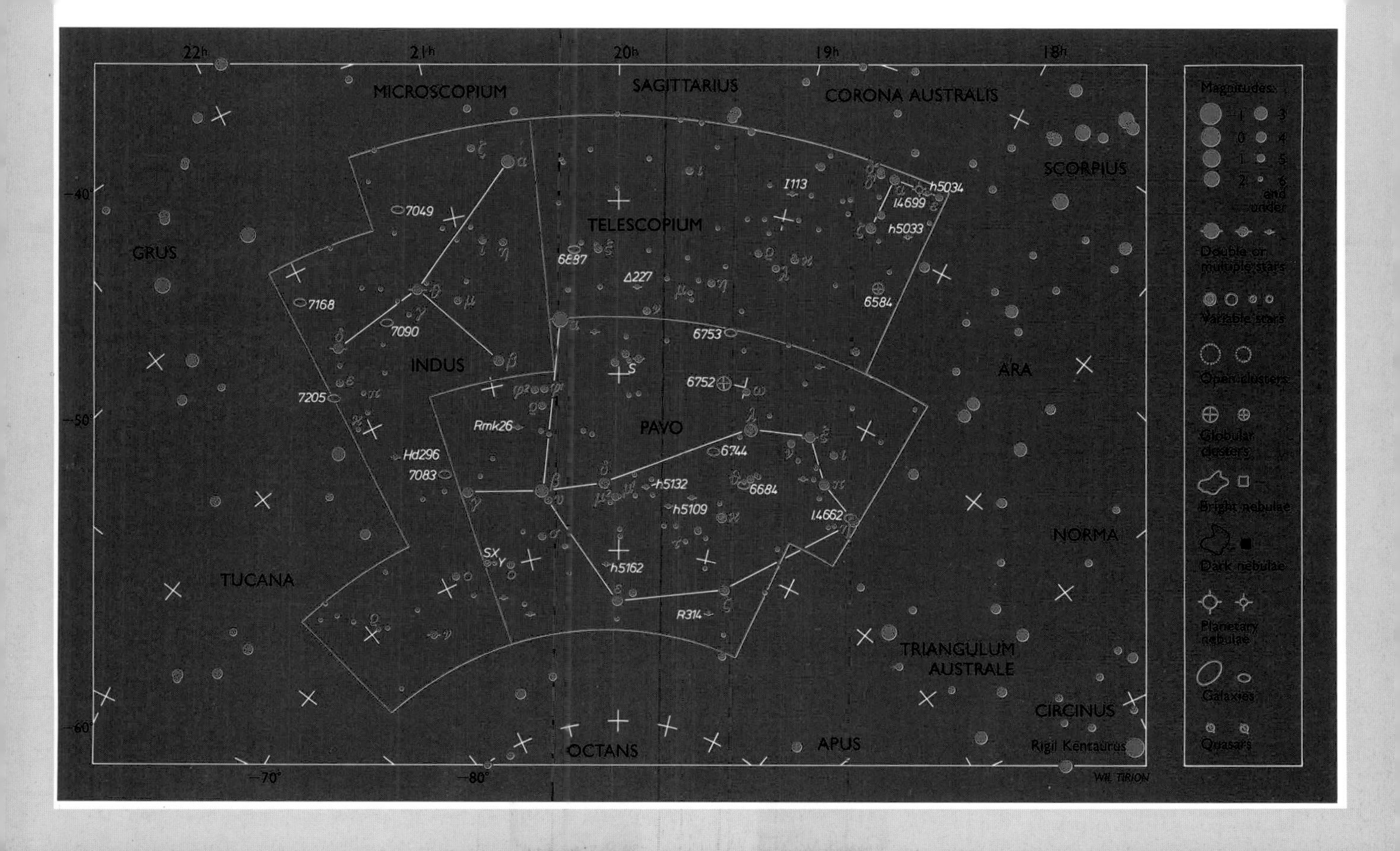
22h
21h
20h
19h
18h
MICROSCOPIUM
SAGITTARIUS
CORONA AUSTRALIS
SCORPIUS
GRUS
TELESCOPIUM
INDUS
PAVO
ARA
NORMA
TUCANA
TRIANGULUM AUSTRALE
CIRCINUS
OCTANS
APUS
Rigil Kentaurus
−40°
−50°
−60°
−70°
−80°
7049
7168
7090
7205
6887
Δ227
I113
14699
h5034
h5033
6584
6753
6752
6744
6684
h5132
h5109
I4662
h5162
R314
Rmk26
Hd296
7083
SX
Y
S
WIL TIRION
Magnitudes:
−1
0
1
2
3
4
5
6 and under
Double or multiple stars
Variable stars
Open clusters
Globular clusters
Bright nebulae
Dark nebulae
Planetary nebulae
Galaxies
Quasars

Lacerta

Lacerta, the Lizard, is one of the ancient constellations but is small and obscure; it adjoins Cepheus. The brightest star is α (magnitude 3.77). The open cluster NGC 7243 contains about 40 stars, and can just be seen with binoculars.

The main object of interest in Lacerta is the strange object BL Lacertae. It was once classed as a variable star, but is now known to be something much more dramatic: it is far beyond our Galaxy, and has given its name to a whole class of objects (usually known, for short, as "BL Lacs"). Apparently they are essentially similar to quasars, and may be powered by massive black holes. However, their great distance means that they are faint; BL Lacertae itself can never become brighter than the 14th magnitude, and is not easy to locate. If you decide to try, the position is: RA 22h 02.7m, dec. +42° 16′ 40″. Visually it looks just like a normal faint star.

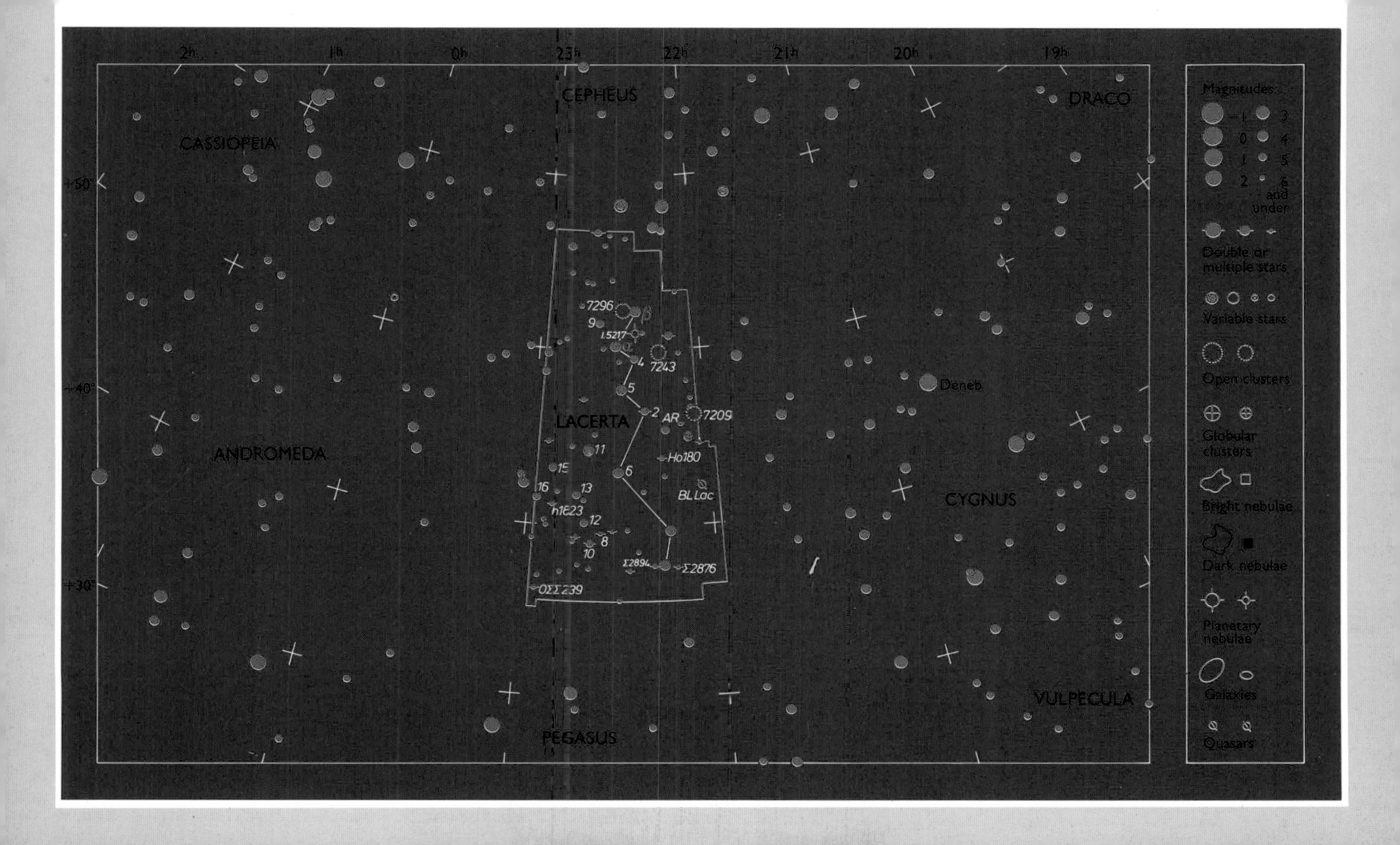
2h
1h
0h
23h
22h
21h
20h
19h
+50°
+40°
+30°
CASSIOPEIA
CEPHEUS
DRACO
ANDROMEDA
LACERTA
CYGNUS
PEGASUS
VULPECULA
Deneb
7296
β
9
I.5217
α
4
7243
5
2
AR
7209
11
Ho180
15
6
BL Lac
16
13
h1823
12
8
10
Σ2894
Σ2876
OΣΣ239
Magnitudes:
–1
0
1
2
3
4
5
6
and under
Double or multiple stars
Variable stars
Open clusters
Globular clusters
Bright nebulae
Dark nebulae
Planetary nebulae
Galaxies
Quasars

Leo/Leo Minor/Sextans

Leo, one of the most imposing of the Zodiacal constellations, represents the Nemaean Lion killed by Hercules. Leo Minor and Sextans have no mythological associations.

The main pattern of Leo is the "Sickle," made up of Regulus, η (magnitude 3.52), γ, ζ, μ (3.88), ε and λ (4.31); it resembles a reversed question mark. The rest of Leo is made up mainly of a triangle (β, δ and θ). β, or Denebola, was ranked of the first magnitude in ancient times; whether any real change has occurred seems doubtful.

γ Leonis is a superb double: period 619 years, magnitudes 2.2 and 3.5, separation 4″.3; a small telescope will show it well. The brighter component is orange. R Leonis is a Mira variable (RA 09h 47.6m, dec. +11° 25′; range 4.4–11.3, period 312 days), lying not far from Regulus. Regulus itself is white, and 130 times as luminous as the Sun.

The physically associated galaxies M65 and M66, 29 million light-years from us, are just within binocular range, and are in the same field with θ Leonis. NGC 3628 is another member of the same group.

LEO

Leading stars	***RA*** h m s	***Declination*** ° ′ ″	***Magnitude***	***Spectrum***	***Distance*** ly
α Regulus	10 08 22.2	+11 58 02	1.35	B7	85
γ Algeiba	10 19 58.3	+19 50 30	1.99	K0+G7, K0+G7	90
β Denebola	11 49 03.5	+14 34 19	2.14	A3	39
δ Zosma	11 14 06.4	+20 31 25	2.56	A4	52
ε Asad Australis	09 45 51.0	+23 46 27	2.98	G0	310

LEO/LEO MINOR/SEXTANS

Galaxies	***RA*** h m	***Declination*** ° ′	***Magnitude***	***Type***
M65	11 18.9	+13 05	9.3	Spiral
M66	11 20.2	+12 59	9.0	Spiral
M95	10 44.0	+11 42	9.7	Barred spiral
M96	10 46.8	+11 49	9.2	Spiral
M105	10 47.8	+12 35	9.3	Elliptical
NGC 3628	11 20.3	+13 36	9.5	Spiral
NGC 3344	10 43.5	+24 55	9.9	Spiral
NGC 3115	10 05.2	−07 43	9.9	Edge-on spiral

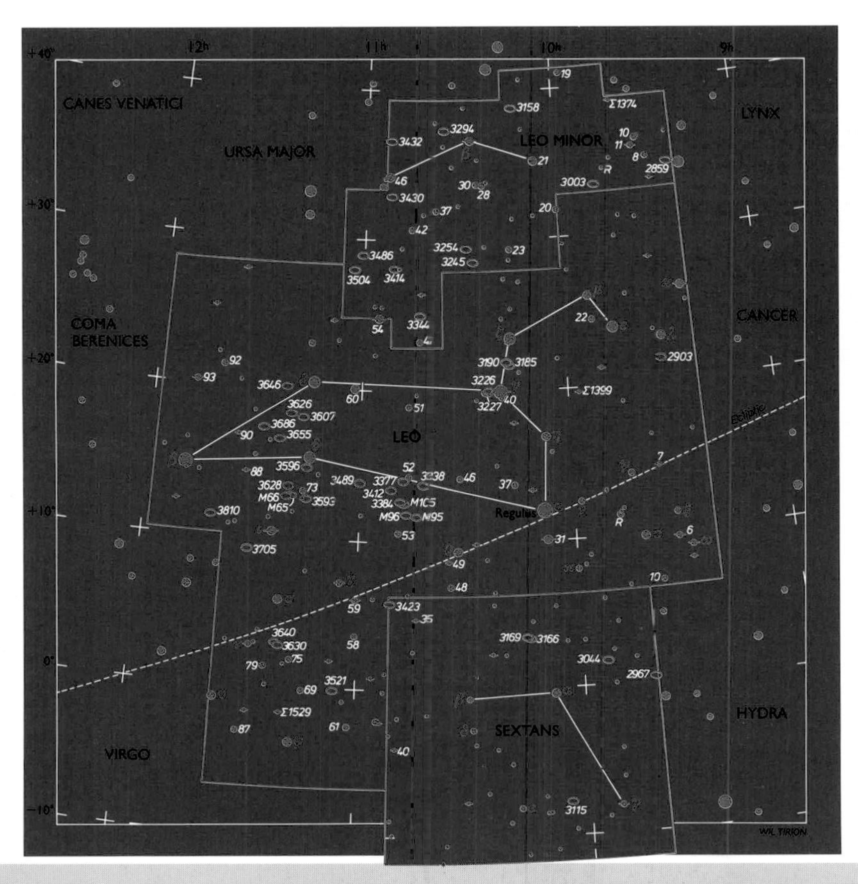
CANES VENATICI
URSA MAJOR
LEO MINOR
LYNX
CANCER
COMA BERENICES
LEO
Regulus
Ecliptic
SEXTANS
HYDRA
VIRGO
WIL TIRION

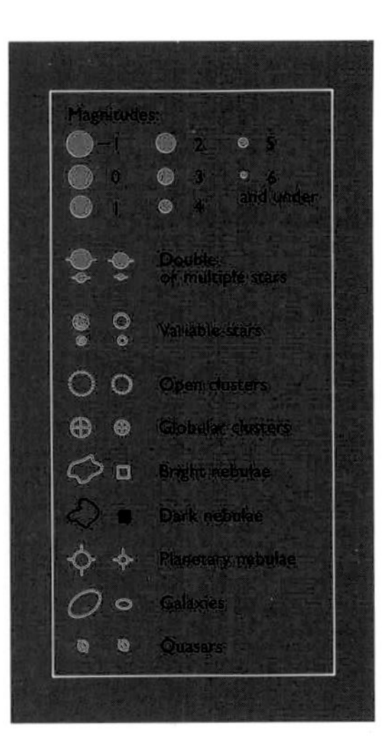
Magnitudes:
−1 0 1 2 3 4 5 6 and under
Double or multiple stars
Variable stars
Open clusters
Globular clusters
Bright nebulae
Dark nebulae
Planetary nebulae
Galaxies
Quasars

Libra

Libra, the Scales or Balance, is one of the least conspicuous of the Zodiacal constellations. No mythological legends are definitely attached to it, though it has been identified rather loosely with Mochis, the inventor of weights and measures. It was formerly known as Chelae Scorpii, the Scorpion's Claws, and the star now known as σ Librae was formerly included in Scorpius, as γ Scorpii.

β is said to be the only single naked-eye star to show a definite green tint. Whether or not this is true must depend upon the observer's judgement, but most people will certainly call it white! It is about 100 times as luminous as the Sun.

α consists of two components of magnitudes 2.8 and 5.2, separated by 231 seconds of arc and therefore easily split with binoculars. There is not much of real interest in the constellation, but δ Librae (RA 15h 01.1m, dec. −08° 31′) is an Algol-type variable with a range from magnitude 4.9 to 5.9 and a period of 2.33 days. There are also several Mira-type stars (Y, S, RS, RU and RR) which exceed the eighth magnitude at maximum.

Leading stars	*RA* h m s	*Declination* ° ′ ″	*Magnitude*	*Spectrum*	*Distance* ly
β Zubenelchemale	15 17 00.3	−09 22 58	2.61	B8	121
α Zubenelgenubi	14 50 52.6	−16 02 30	2.75	A3	72
σ Zubenalgubi	15 04 04.1	−25 16 55	3.29	M4	166

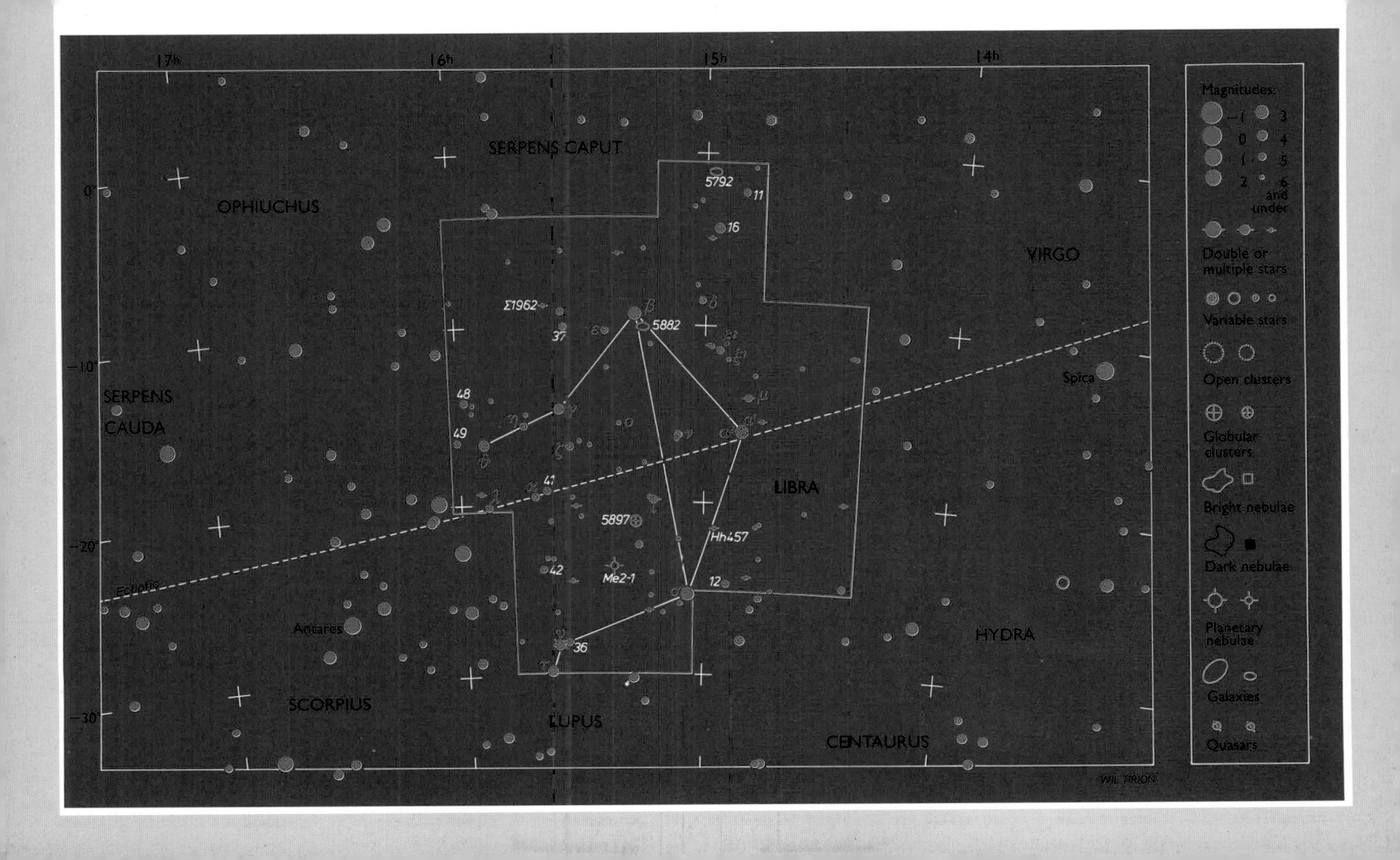
17h
16h
15h
14h
0°
−10°
−20°
−30°
SERPENS CAPUT
OPHIUCHUS
SERPENS
CAUDA
VIRGO
LIBRA
HYDRA
SCORPIUS
LUPUS
CENTAURUS
Spica
Antares
Ecliptic
5792
11
16
Σ1962
37
5882
48
49
41
5897
Hh457
42
Me2-1
12
36
Magnitudes:
−1 0 1 2 3 4 5 6 and under
Double or multiple stars
Variable stars
Open clusters
Globular clusters
Bright nebulae
Dark nebulae
Planetary nebulae
Galaxies
Quasars
WIL TIRION

Lupus/Norma

Lupus, the Wolf, is an ancient constellation, though no legends seem to be attached to it. It lies between Scorpius and Centaurus, and contains several fairly bright stars, of which three are above the third magnitude (see table).

Several of the stars in Lupus are double: π (magnitudes 4.6 and 4.7, separation 1″.4); κ (3.9 and 5.8, 26″.8); ξ (5.3 and 5.8, 10″.4); η (3.6 and 7.6, 15″.0). μ is triple; the main components are of magnitudes 5.1 and 5.2 (separation 1″.2) and there is a 7.2-magnitude star at 23″.7. It is worth looking at the two components of φ: their magnitudes are 3.6 and 4.5. The brighter star is orange, the companion white.

There are also several nebular objects (see table). NGC 5822 is a binocular object lying close to ζ (magnitude 3.41).

Norma was once known as Quadra Euclidis (Euclid's Quadrant); its brightest star, γ, is only of magnitude 4.0. ε is double: magnitudes 4.8 and 7.5, separation 22″.8. There are several fairly bright open clusters in Norma: NGC 4925 (magnitude 8.4), NGC 6031 (8.5), NGC 6067 (5.6), NGC 6134 (7.2), NGC 6152 (8.1), and NGC 6087. The last of these lies round the Cepheid variable S Normae, and has 40 stars and an integrated magnitude of 5.4. Its position is: RA 16h 18.9m, dec. −57° 54′. S Normae has a magnitude range of 6.1 to 6.8 and a period of 9.75 days.

LUPUS					
Leading stars	***RA*** h m s	***Declination*** ° ′ ″	***Magnitude***	***Spectrum***	***Distance*** ly
α Men	14 41 55.7	−47 23 17	2.30	B1	685
β Kekouan	14 58 31.8	−43 08 02	2.68	B2	359
γ	15 35 08.4	−41 10 00	2.78	B3	258

LUPUS				
Non-stellar objects	***RA*** h m	***Declination*** ° ′	***Magnitude***	***Type***
NGC 5822	15 05.2	−54 21	6.5	Open cluster
NGC 5824	15 04.0	−33 04	6.2	Globular cluster
NGC 3986	15 46.1	−37 47	9.8	Globular cluster
NGC 5882	15 16.8	−45 39	10.5	Planetary nebula
NGC 5643	14 32.7	−44 10	10.7	Barred spiral galaxy

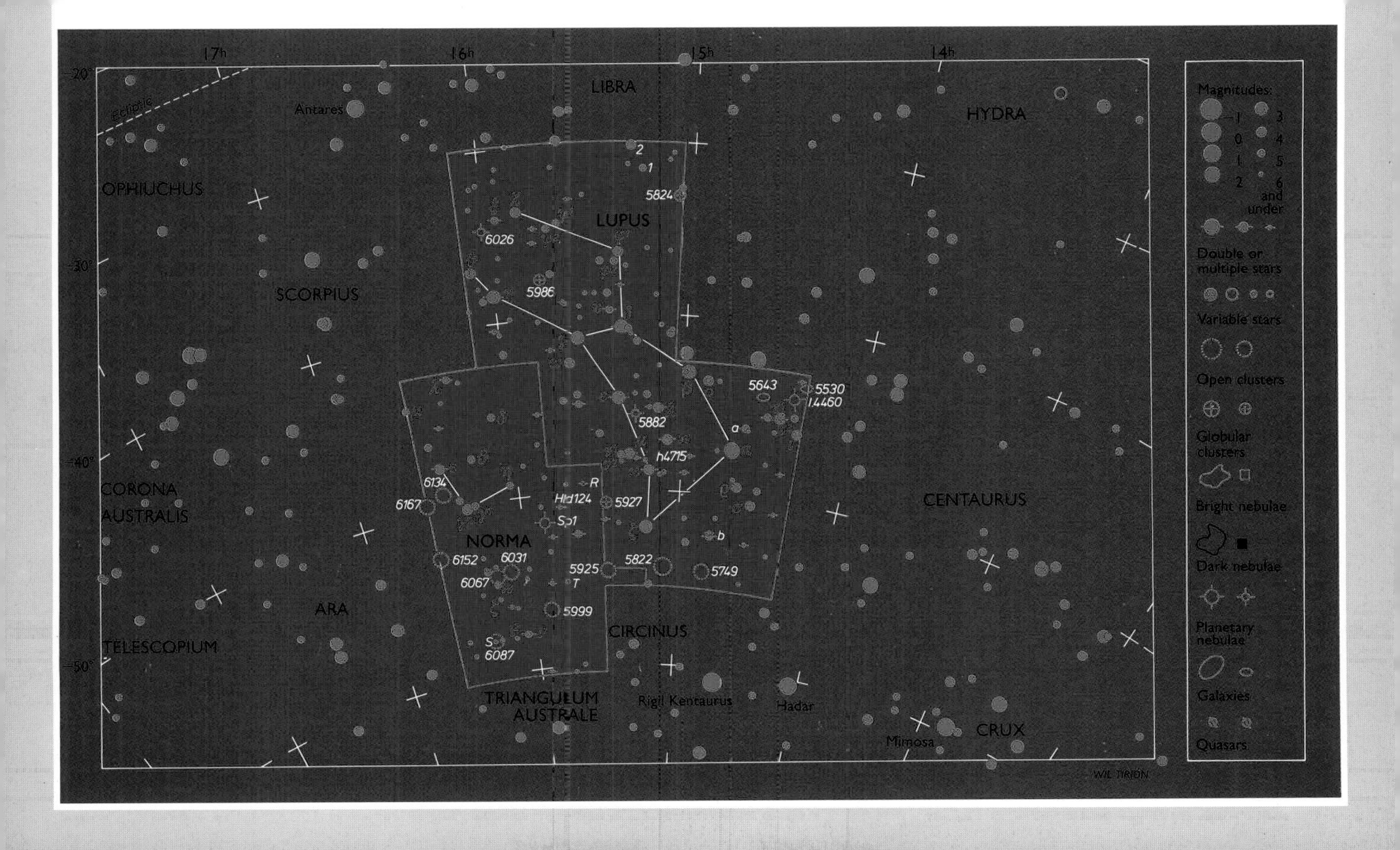

17h
16h
15h
14h
-20°
-30°
-40°
-50°
Ecliptic
OPHIUCHUS
Antares
LIBRA
HYDRA
LUPUS
SCORPIUS
CORONA AUSTRALIS
NORMA
ARA
TELESCOPIUM
CIRCINUS
CENTAURUS
TRIANGULUM AUSTRALE
Rigil Kentaurus
Hadar
Mimosa
CRUX
2
1
5824
6026
5986
5643
5530
I.4460
5882
a
h4715
R
Hd124
5927
Sp1
b
6134
6167
6152
6031
5925
5822
5749
6067
T
5999
S
6087
WIL TIRION
Magnitudes:
−1
0
1
2
3
4
5
6
and under
Double or multiple stars
Variable stars
Open clusters
Globular clusters
Bright nebulae
Dark nebulae
Planetary nebulae
Galaxies
Quasars

Lynx

The Lynx covers a fairly wide area, mainly between Ursa Major and Gemini, but it has only one reasonably bright star, α (magnitude 3.13), at RA 08h 21m 03.2s, dec. +34° 23′ 33″. It is easy to find, because it is decidedly red; it is of type M0, and forms an equilateral triangle with Regulus and Pollux. Lynx was added to the sky by Hevelius, in 1690; it has been said that the name is appropriate, because you need to be lynx-eyed to see anything at all in the area!

12, 19 and 38 are triple stars, but with no special distinction. The Mira variable R Lyncis (RA 07h 01.3m, dec. +55° 20′) can reach magnitude 7.2 at maximum, but sinks to 14.5 at minimum; the period is 379 days. Y Lyncis (RA 07h 28.2m, dec. +45° 59′) is a red M-type semiregular variable, with a magnitude range of 7.8 to 10.3 and a rough period of around 110 days.

There are several galaxies: the brightest of these is NGC 2683 (RA 08h 52.7m, dec. 33° 25′), which has an integrated magnitude of 9.7. It is spiral, but we see it almost edge-on.

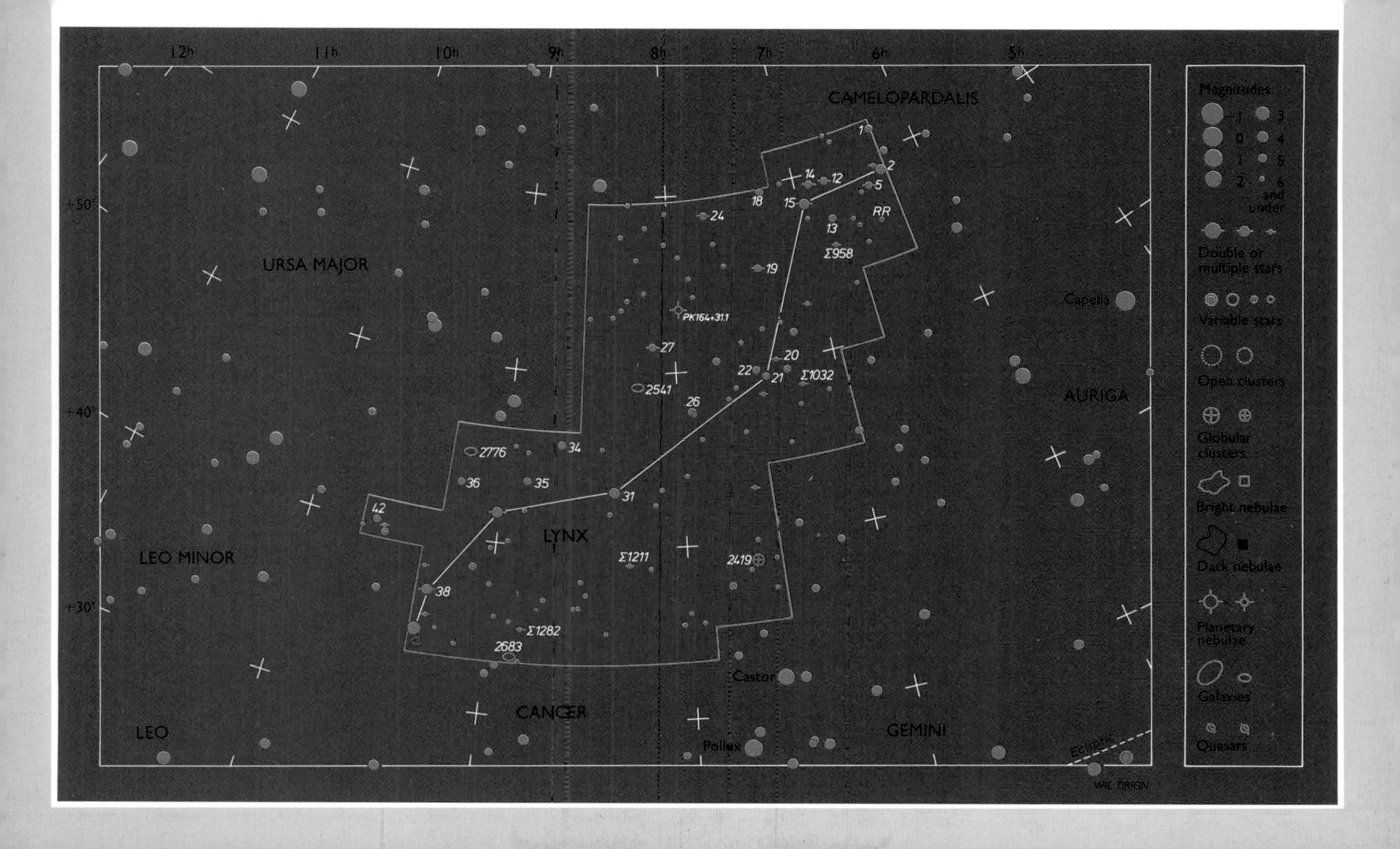
12h
11h
10h
9h
8h
7h
6h
5h
+50°
+40°
+30°
CAMELOPARDALIS
URSA MAJOR
LYNX
LEO MINOR
LEO
CANCER
GEMINI
AURIGA
Capella
Castor
Pollux
Ecliptic
1
2
5
12
13
14
15
18
19
20
21
22
24
26
27
31
34
35
36
38
42
RR
Σ958
Σ1032
Σ1211
Σ1282
PK164+31.1
2419
2541
2683
2776
Magnitudes:
−1
0
1
2
3
4
5
6
and under
Double or multiple stars
Variable stars
Open clusters
Globular clusters
Bright nebulae
Dark nebulae
Planetary nebulae
Galaxies
Quasars
WIL TIRION

Lyra

Lyra represents the harp which Apollo gave to the great musician Orpheus. It is a very small constellation, but contains many interesting objects. Its leader α, or Vega, is of magnitude 0.03 and is the fourth-brightest star in the sky. Vega is of type A and is 26 light-years away, with a luminosity 52 times that of the Sun. It is obviously blue, and is a beautiful sight in binoculars. From countries such as Britain it is almost overhead during summer evenings.

Vega was one of the stars found in 1983 by IRAS to be associated with cool material which may be planet-forming. Whether this indicates a true planetary system is, of course, very uncertain.

There are several interesting variable stars in Lyra (see table). β is the prototype of its class. It is an eclipsing binary but the components, unlike those of Algol, are not very unequal; so alternate deep and shallow minima do not occur – variations are always going on. The fluctuations are easy to follow with the naked eye: suitable comparison stars are γ (magnitude 3.24) and κ (4.33).

Another naked-eye variable is R Lyrae, which is of type M and decidedly red. Its period, like those of all semiregular stars, is very rough. Suitable comparison stars are η and θ (each 4.4).

ε Lyrae, close to Vega, is the celebrated "double-double" star. The components are of magnitude 4.7 and 5.1, 208 seconds of arc apart, so keen-sighted people can split them with the naked eye. Each is again double: ε^1 has magnitudes of 5.0 and 5.1, and separation 2″.6; ε^2 has magnitudes of 5.2 and 5.5, and separation 2″.3.

δ is another easy double: one very red component (magnitude 4.5) is of type M, while the companion (5.5) is white. Another easy double is ζ (4.3), with a 5.9-magnitude companion at 43″.7.

There is a bright globular cluster in Lyra, M56 (RA 19h 16.6m, dec. +30° 11′), with an integrated magnitude of 8.2; the outer edges are not hard to resolve telescopically. Between β and γ is the Ring Nebula, M57 (RA 18h 53.6m, dec. +33° 02′), most famous of the planetaries. It is very hard to see with binoculars, but is a beautiful telescopic sight; the central star is in range of a 30-centimeter reflector.

Variable stars	*RA* h m	*Declination* ° ′	*Magnitude* range	*Period* days	*Type*
β	18 50.1	+33 22	3.3–4.3	12.94	β Lyrae
R	18 55.3	+43 57	3.9–5.0	46	Semiregular
T	18 32.3	+37 00	7.9–9.6	–	Irregular
W	18 14.9	+36 40	7.3–13.0	197	Mira

22h
21h
20h
19h
18h
17h
16h
+40°
+30°
+20°
CEPHEUS
DRACO
Deneb
CYGNUS
16
R
RR
XY
Vega
6791
Σ2470
Σ2474
17
M57
M56
LYRA
HERCULES
VULPECULA
DELPHINUS
SAGITTA
WIL TIRION
Magnitudes:
-1
0
1
2
3
4
5
6
and under
Double or multiple stars
Variable stars
Open clusters
Globular clusters
Bright nebulae
Dark nebulae
Planetary nebulae
Galaxies
Quasars

Microscopium

The celestial Microscope is a very obscure group, whose claim to a separate identity is at best dubious. Its brightest star, γ, is only of magnitude 4.67, and was formerly included in Piscis Austrinus as 1 Piscis Austrini.

θ is a triple, with a close pair (magnitudes 6.4 and 7.0) separated by 0″.5, and a third star of magnitude 10.5 at a distance of 78″.4. There is also a 10th-magnitude companion to α with a separation of 20″.5; the brighter component is of magnitude 5.0 and is of type G6, so that it is somewhat yellowish.

The brightest galaxy in Microscopium is NGC 6925 (RA 20h 34m.3, dec. −31° 59′), which is an almost edge-on spiral. The integrated magnitude is only 11.3, so that it is not a particularly easy object. Its neighbor NGC 6923 is also a spiral, but almost a magnitude fainter.

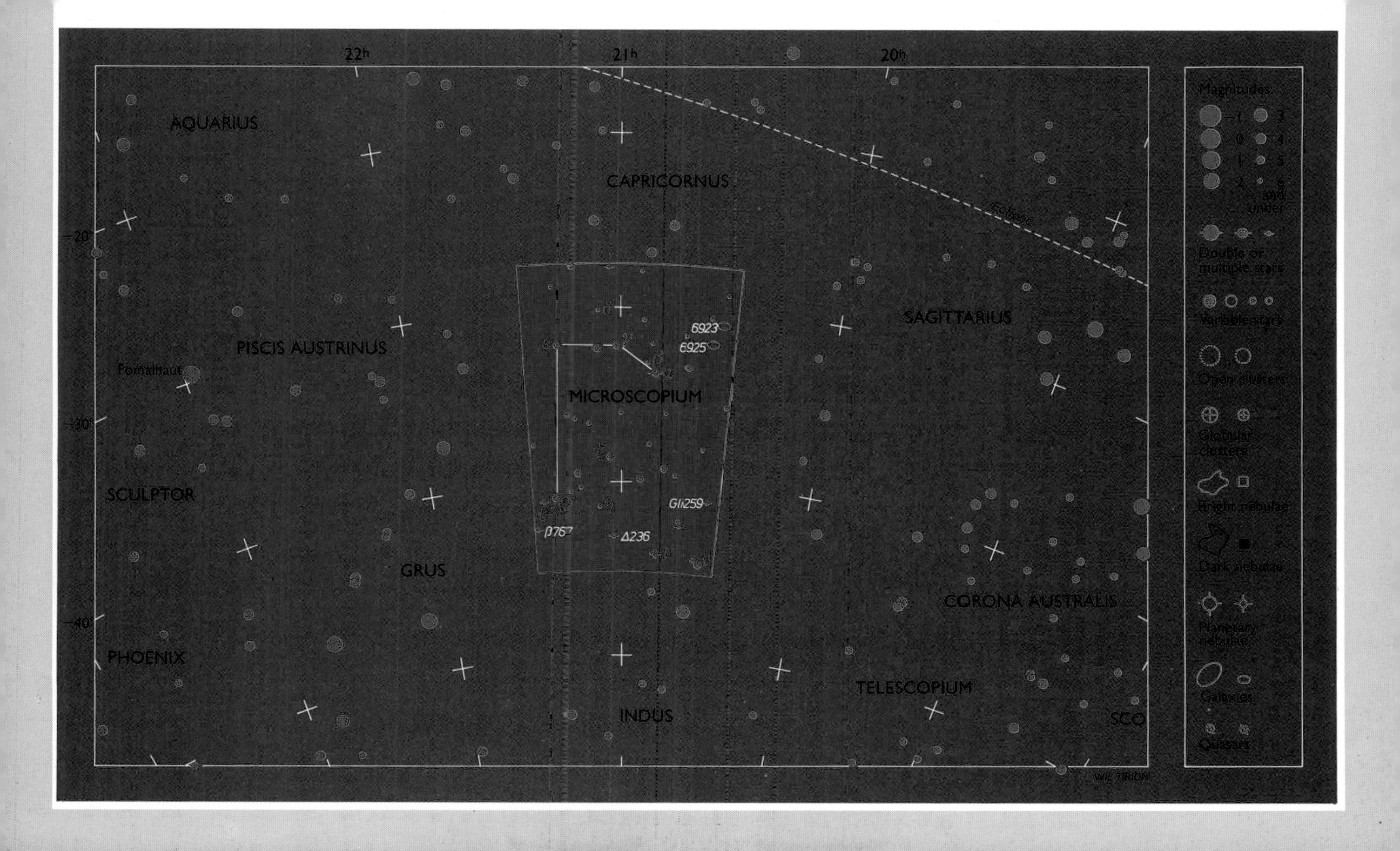
22h
21h
20h
−20°
−30°
−40°
AQUARIUS
CAPRICORNUS
Ecliptic
SAGITTARIUS
PISCIS AUSTRINUS
Fomalhaut
MICROSCOPIUM
6923
6925
Gli259
β767
Δ236
SCULPTOR
GRUS
CORONA AUSTRALIS
PHOENIX
TELESCOPIUM
INDUS
SCO
WIL TIRION
Magnitudes:
−1
0
1
2
3
4
5
6 and under
Double or multiple stars
Variable stars
Open clusters
Globular clusters
Bright nebulae
Dark nebulae
Planetary nebulae
Galaxies
Quasars

Ophiuchus

Ophiuchus, the Serpent Bearer, was formerly known as Serpentarius. In mythology it is associated with the healer Aesculapius, who became so skilled that he was able to bring the dead back to life. To avoid depopulation of the underworld, Jupiter disposed of Aesculapius with a thunderbolt, but relented sufficiently to place him in the sky!

Ophiuchus intrudes into the Zodiac between Scorpius and Sagittarius – to the fury of astrologers, who devoutly wish that the Serpent Bearer did not exist! There are five stars above the third magnitude (see table). α is rather isolated from the rest of the constellation, and lies near α Herculis (Rasalhague). δ is very red, and makes a good pair with the white ε or Yed Post (magnitude 3.24).

There are various doubles in Ophiuchus: η is an interesting close binary, orbital period 84.3 years, magnitudes 3.0 and 3.5, but the separation is on average no more than 0″.5.

The region round ρ Ophiuchi (magnitude 4.59) is a superb complex of dust and gas, IC 4604 – a favorite subject for astrophotographers. Planetary nebulae include NGC 6309 (RA 17h 14.1m, dec. −12° 55′), which resembles a fainter version of the Ring Nebula in Lyra. But it is for globular clusters that Ophiuchus is renowned. There are over 20 of them, including seven Messier objects (see table). Of these, M10 and M12 are the easiest to find: M10 lies near the reddish fifth-magnitude star 30 Ophiuchi. However, the whole area is rich, and it is easy to be misled by the adjacent star fields.

Leading stars	***RA*** h m s	***Declination*** ° ′ ″	***Magnitude***	***Spectrum***	***Distance*** ly
α Rasalhague	17 34 55.9	+12 33 36	2.08	A5	62
η Sabik	17 10 22.5	−15 43 30	2.43	A2	59
ζ Han	16 37 09.4	−10 34 02	2.56	09.5	554
δ Yed Prior	16 14 20.6	−03 41 39	2.74	M1	140
β Cheleb	17 43 28.2	+04 34 02	2.77	K2	121

Globular clusters	***RA*** h m	***Declination*** ° ′	***Magnitude***
M9	17 19.2	−18 31	7.9
M10	16 57.1	−04 06	6.6
M12	16 47.2	−01 57	6.6
M14	17 37.6	−03 15	7.6
M19	17 02.6	−26 16	7.1
M62	17 01.2	−30 07	6.6
M107	16 32.5	−13 03	8.1

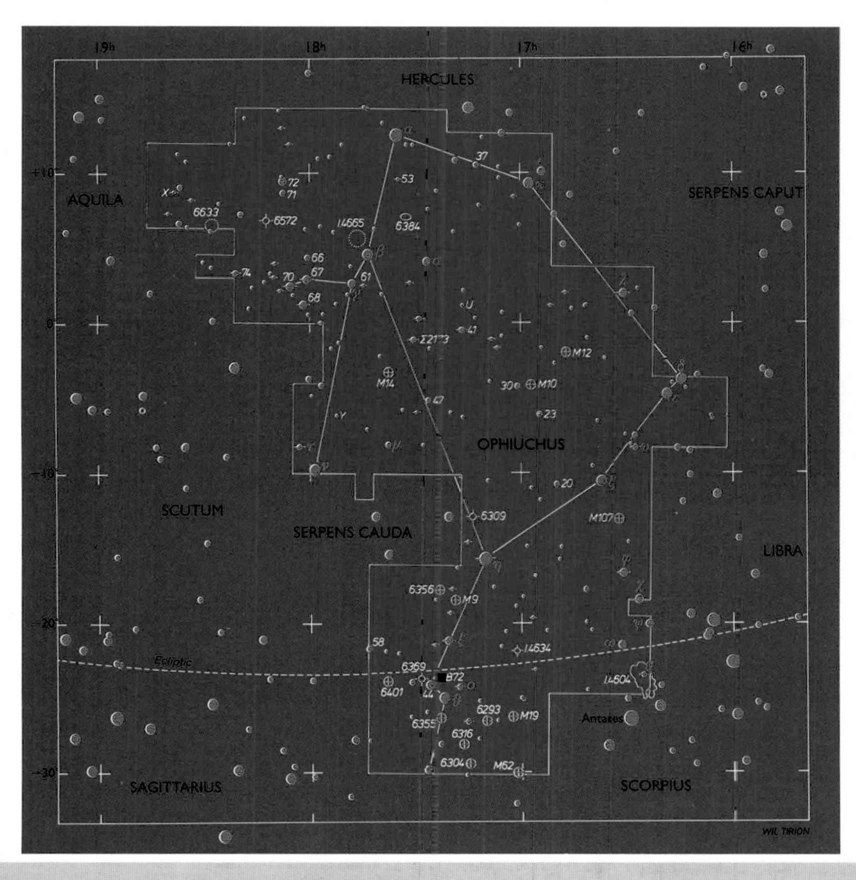
19h
18h
17h
16h
+10°
0°
−10°
−20°
−30°
HERCULES
AQUILA
SERPENS CAPUT
OPHIUCHUS
SCUTUM
SERPENS CAUDA
LIBRA
SAGITTARIUS
SCORPIUS
Antares
Ecliptic
6633
6572
I4665
6384
M12
M10
M14
M107
6309
6356
M9
I4634
6369
B72
6401
I4604
6293
M19
6355
6316
6304
M62
WIL TIRION

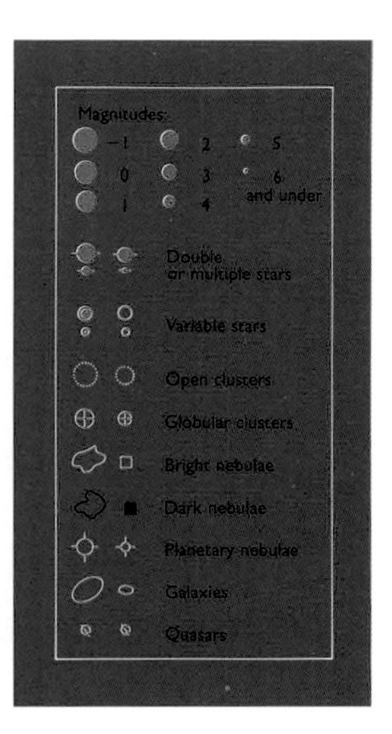
Magnitudes:
−1
0
1
2
3
4
5
6
and under
Double or multiple stars
Variable stars
Open clusters
Globular clusters
Bright nebulae
Dark nebulae
Planetary nebulae
Galaxies
Quasars

Orion

Orion must be the most splendid constellation in the entire sky. It represents the great hunter who boasted that he could kill any creature on earth – but he met his match in the scorpion, which stung him in the heel and killed him.

The main pattern of Orion is unmistakable. Rigel is a true cosmic searchlight, at least 60,000 times as luminous as the Sun; Betelgeux is a vast red supergiant, large enough to contain the whole path of the Earth round the Sun. It is variable, though the magnitude is usually around 0.5 – decidedly brighter than Aldebaran in Taurus. Mintaka is very slightly variable and lies very close to the celestial equator.

There are various double stars. Rigel has a 6.8-magnitude companion at a separation of 9″.5; this is not a difficult object, despite the glare from the brilliant primary. δ also has an easy companion, of magnitude 6.3 and at 52″.6 separation. ι has a 6.9-magnitude companion at 11″.3.

σ and θ are multiple stars. θ, in front of the Great Nebula, M42, is known as the Trapezium from the arrangement of its four main components. In a small telescope it is unlike anything else in the sky. The hot stars of the Trapezium stimulate the nebula to self-luminosity.

M42 itself is the most celebrated nebula of its type. It is in fact merely the brightest part of a huge molecular cloud which covers most of Orion; it is over 1,000 light-years away, and inside it fresh stars are being formed. It is an ideal photographic target. Much more elusive is the dark Horse's-Head Nebula, Barnard 33, silhouetted against the nebula IC 434, not far from ζ (RA 05h 41m.0, dec. −02° 24′).

There is much else of interest in Orion; do not forget two red variables, the Mira star U Orionis in the far north of the constellation (magnitudes 4.8 to 12.6, period 372 days) and the semiregular W (magnitudes 5.9 to 7.7; semiregular, with period around 212 days).

Leading stars	*RA* h m s	*Declination* ° ′ ″	*Magnitude*	*Spectrum*	*Distance* ly
β Rigel	05 14 32.2	−08 12 16	0.12	B8	900
α Betelgeux	05 55 10.2	+07 24 26	0.2–0.8	M2	310
γ Bellatrix	05 25 07.8	+06 20 59	1.64	B2	360
ε Alnilam	05 36 12.7	−01 12 07	1.70	B0	1,200
ζ Alnitak	05 40 45.5	−01 56 34	1.77	O9.5	1,100
κ Saiph	05 47 45.3	−09 40 11	2.06	B0.5	2,200
δ Mintaka	05 32 00.3	−00 17 57	2.23	O9.5	2,350
ι Hatysa	05 35 25.9	−05 54 36	2.76	O9	1,860

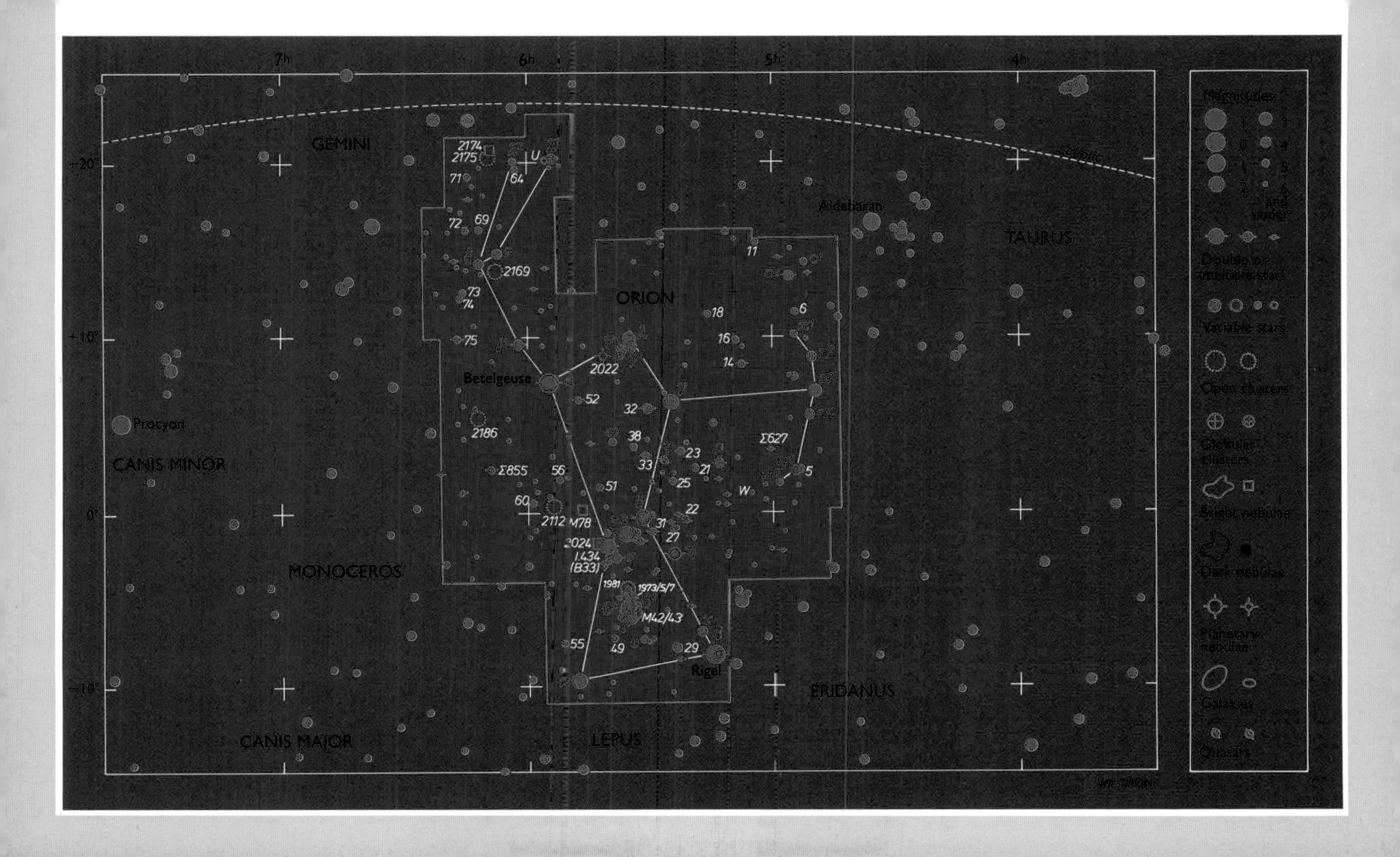
7h
6h
5h
4h
+20°
+10°
0°
–10°
Ecliptic
GEMINI
TAURUS
ORION
Aldebaran
Betelgeuse
Rigel
Procyon
CANIS MINOR
MONOCEROS
CANIS MAJOR
LEPUS
ERIDANUS
2174
2175
U
71
64
72
69
2169
73
74
75
11
18
6
16
14
2022
52
32
2186
38
Σ627
23
Σ855
56
33
21
51
25
W
5
60
22
2112
M78
31
27
2024
1434
(B33)
1981
1973/5/7
M42/43
55
49
29
Magnitudes
1
0
1
2
3
4
5
6
and under
Double or multiple stars
Variable stars
Open clusters
Globular clusters
Bright nebulae
Dark nebulae
Planetary nebulae
Galaxies
Quasars
Wil Tirion

Pegasus

Pegasus, the Flying Horse, commemorates the steed upon which the hero Bellerophon rode to do battle with a particularly nasty fire-breathing monster, the Chimaera. The four main stars make up a square, but for some curious reason one of the four stars, Alpheratz, has been transferred to the neighboring constellation of Andromeda – it used to be δ Pegasi, but is now α Andromedae. Its magnitude is 2.06, so it is the brightest of the stars in the Square.

β is a semiregular variable with a rough period of between 35 and 40 days; α and γ make good comparison stars. ε is some way from the Square; it has been suspected of variability, but there is no definite proof. η has a 9.9-magnitude companion at a separation of 90″.4, so that this is an easy pair; the faint component is itself a close binary.

There are several faint galaxies in Pegasus, but the main object of interest is the globular cluster M15 (RA 21h 30m.0, dec. +12° 10′). It is not far below naked-eye visibility, and is easy in binoculars; a telescope shows it to be one of the finest globulars in the whole of the sky. It lies in a direct line with θ (magnitude 3.53) and ε, and there is a sixth-magnitude star close beside it.

Leading stars	*RA* h m s	*Declination* ° ′ ″	*Magnitude*	*Spectrum*	*Distance* ly
ε Enif	21 44 11.0	+09 52 30	2.38	K2	522
β Scheat	23 03 46.3	+28 04 58	2.4–2.9	M2	178
α Markab	28 04 45.5	+15 12 19	2.49	B9	101
γ Algenib	00 13 14.1	+15 11 01	2.83	B2	520
η Matar	22 43 00.0	+30 13 17	2.94	G2	173

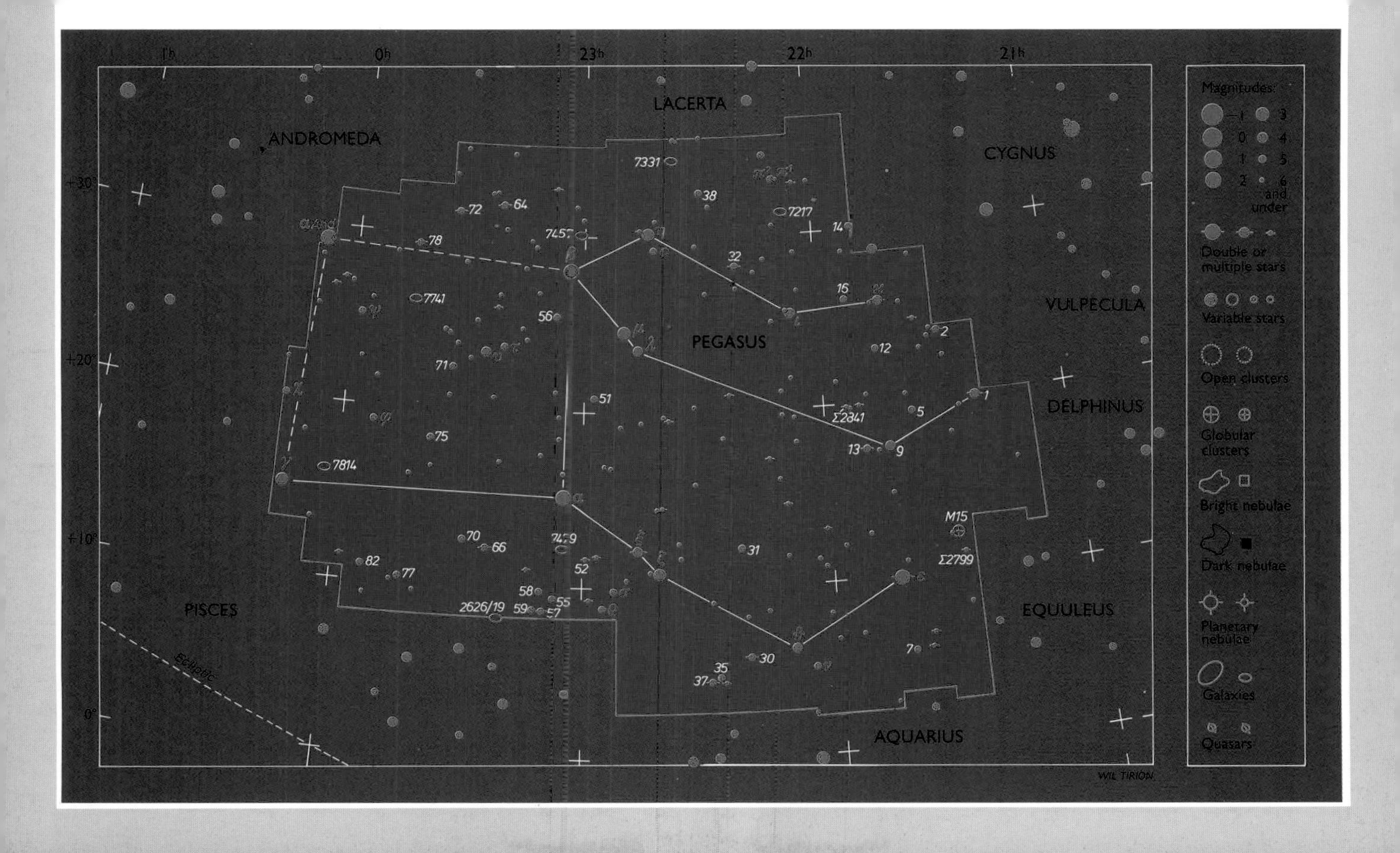

1h
0h
23h
22h
21h
+30°
+20°
+10°
0°
ANDROMEDA
LACERTA
CYGNUS
VULPECULA
PEGASUS
DELPHINUS
EQUULEUS
PISCES
AQUARIUS
Ecliptic
7331
38
7217
14
72
64
78
32
16
7741
56
2
12
71
51
5
Σ2841
75
13
9
1
7814
M15
70
66
31
Σ2799
82
77
52
58
55
2626/19
59
57
30
7
35
37
Magnitudes:
Double or multiple stars
Variable stars
Open clusters
Globular clusters
Bright nebulae
Dark nebulae
Planetary nebulae
Galaxies
Quasars
WIL TIRION

Perseus

This is a large and important northern constellation, commemorating the hero who rescued Princess Andromeda from the unwelcome attentions of a sea monster. It has four stars above the third magnitude.

Algol is the prototype eclipsing binary, with a range of magnitude from 2.1 to 3.3 and a period of 2.7 days; the secondary minimum is too slight to be detected with the naked eye. Suitable comparison stars are ζ, ε, ϰ (3.80) and β Trianguli (3.00). Avoid ϱ Persei, a red semiregular variable (magnitude range 3–4 and a rough period of 35–55 days).

γ has a 10.6-magnitude companion at a separation of 57″ and ε has an 8.1-magnitude companion at 8″.8 separation. η is a good and easy pair, with magnitudes 3.3 and 8.5, and separation 28″.3, while ζ is in an interesting group – there is a 9.5-magnitude companion at 12″.9, and two other stars in the same field.

Perseus is crossed by the Milky Way, and there are many rich star fields and several bright open clusters (see table). The Sword Handle is a superb sight in binoculars or a low-power telescope. M34, over 1,000 light-years away, is a fairly easy binocular object.

There is one interesting planetary nebula, M76 (RA 01h 42.4m, dec. +51° 34′), known as the Little Dumbbell. It is faint, but not hard to identify; telescopically it appears as a hazy, slightly elliptical patch.

The nebula NGC 1499, known because of its shape as the California Nebula, is at RA 04h 00.7m, dec. +36° 37′. It has low surface brightness but is quite easy to photograph.

Leading stars	*RA* h m s	*Declination* ° ′ ″	*Magnitude*	*Spectrum*	*Distance* ly
α Mirphak	03 24 19.3	+49 51 40	1.80	F5	620
β Algol	03 08 10.1	+40 57 21	2.12	B8	95
ζ Atik	03 54 07.8	+31 53 01	2.85	B1	1,108
γ	03 57 51.1	+40 00 37	2.89	B0.5	110
ε	03 04 47.7	+53 30 23	2.93	G8	678

Open clusters	*RA* h m	*Declination* ° ′	*Magnitude*	*Notes*
NGC 869	02 19.0	+57 09	4.3	Double cluster:
NGC 884	02 22.4	+57 07	4.4	the Sword Handle
M34	02 40.0	+42 47	5.2	Contains 60 stars
NGC 1245	03 14.7	+47 15	8.4	Contains 200 stars
NGC 1528	04 15.4	+51 14	6.4	Contains 40 stars

7h
6h
5h
4h
3h
2h
1h
0h
+50°
+40°
+30°
CAMELOPARDALIS
CASSIOPEIA
ANDROMEDA
PERSEUS
AURIGA
TRIANGULUM
TAURUS
ARIES
PISCES
Capella
Ecliptic
957
869
884
9
744
1
4
M76
6
14-4
1624
1528
1545
43
1513
Σ331
34
29
31
53
OΣΣ44
36
1245
59
57
14
30
32
M34
1275
Σ552
52
1003
12
1023
OΣ531
16
20
1342
1579
1499
54
24
17
40
42
21
Magnitudes:
−1
0
1
2
3
4
5
6 and under
Double or multiple stars
Variable stars
Open clusters
Globular clusters
Bright nebulae
Dark nebulae
Planetary nebulae
Galaxies
Quasars
W.L. TIRION

Pictor

This faint southern constellation – originally Equuleus Pictoris, the Painter's Easel – lies close to Canopus. The brightest star is α, magnitude 3.27. An easy double is ι, with magnitudes 5.6 and 6.4, separation 12″.3.

There is little of obvious interest in the Painter, but attention has been focused upon the second star of the constellation, β Pictoris. It is 78 light-years away and of type A5, with a luminosity 60 times that of the Sun. It was one of the stars found by the Infra-Red Astronomical Satellite, IRAS, to be associated with cool material – and subsequently this material was actually photographed, from the Las Campañas Observatory in Chile, with the 2.5-meter Irénée du Pont reflector. The disk may consist of materials similar to those involved in planetary formation, and it is fair to say that at the moment β Pictoris seems to be the most promising candidate as the center of a planetary system.

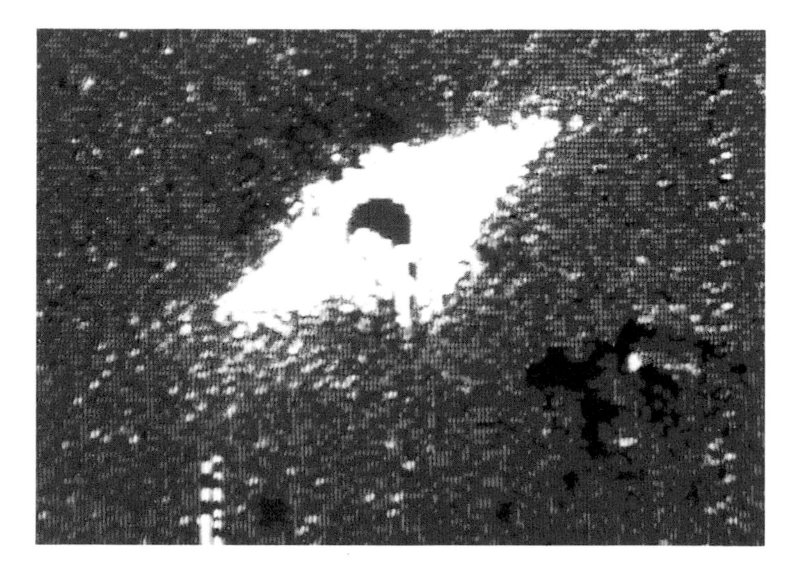

A possible solar system – the disk of dust and ice (white) around β Pictoris.

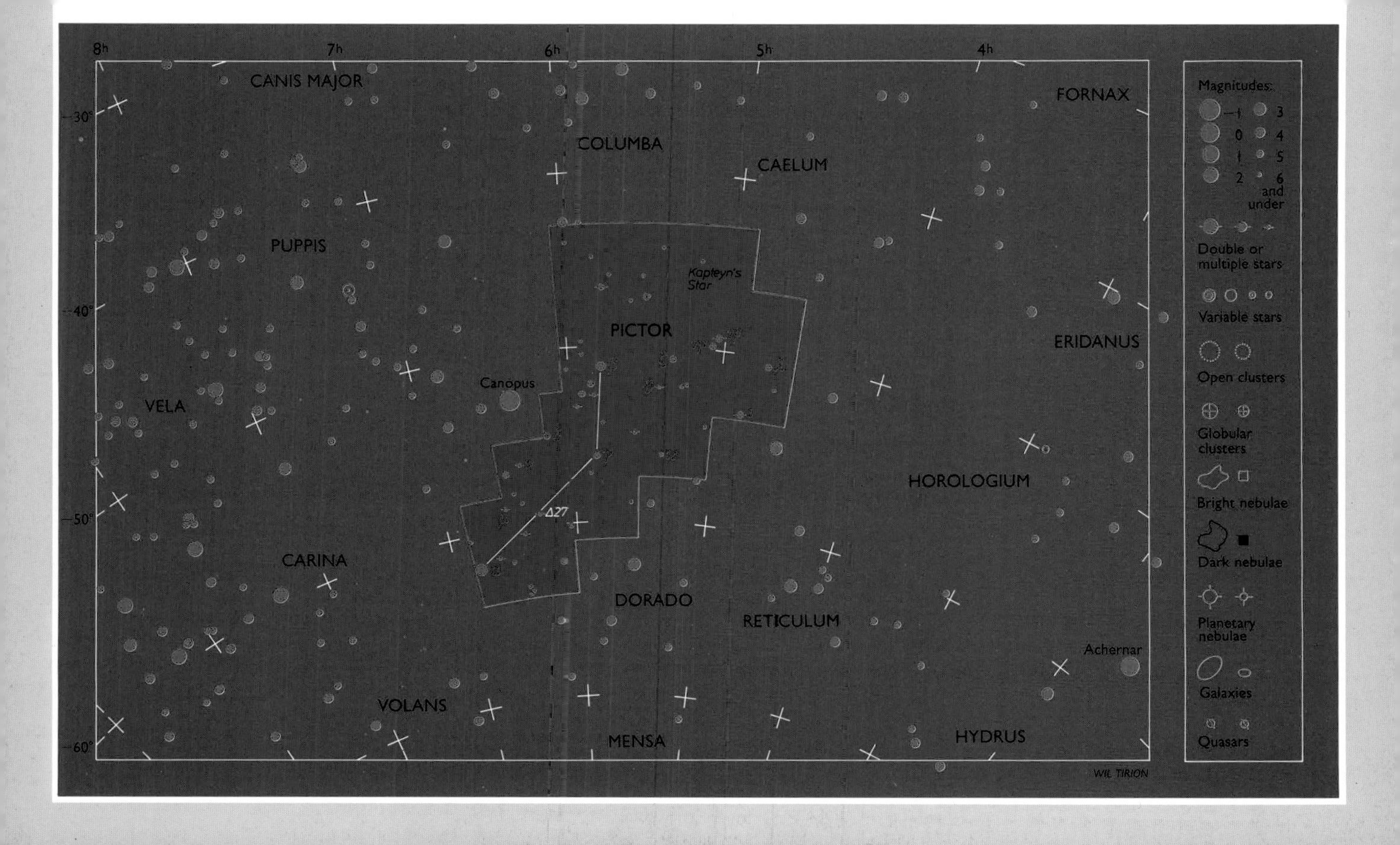
8h
7h
6h
5h
4h
-30°
-40°
-50°
-60°
CANIS MAJOR
COLUMBA
CAELUM
FORNAX
PUPPIS
Kapteyn's Star
PICTOR
ERIDANUS
Canopus
VELA
HOROLOGIUM
Δ27
CARINA
DORADO
RETICULUM
Achernar
VOLANS
MENSA
HYDRUS
WIL TIRION
Magnitudes:
-1
0
1
2
3
4
5
6
and under
Double or multiple stars
Variable stars
Open clusters
Globular clusters
Bright nebulae
Dark nebulae
Planetary nebulae
Galaxies
Quasars

Pisces

The Fishes may represent the forms into which Venus and Cupid once changed themselves to avoid the unwelcome attentions of the monster Typhon. Pisces now contains the First Point of Aries, the vernal equinox, which is not marked by any bright star. Pisces is in fact decidedly barren, and is marked by a straggly line of faint stars adjoining the Square of Pegasus; the brightest is η, magnitude 3.62.

α (Al Rischa) is double, with magnitudes 4.2 and 5.1 and present separation 1″.9 – but this is becoming less, as the pair makes up a binary system with a period of 933 years, and we see them at a less and less favorable angle each year.

Perhaps the most interesting object in Pisces is the irregular variable TX (RA 23h 46.4m, dec. +03° 29′). Its magnitude ranges from 6.9 to 7.7, and there seems to be no semblance of a period. The spectrum is of type N, and TX Piscium is one of the very reddest stars within binocular range. It is in the same low-power binocular field with ι (4.1) and λ (4.5).

M74, an open spiral galaxy, lies near η, at RA 01h 36.7m, dec. 15° 47′. The integrated magnitude is 9.2, but it can be surprisingly difficult to identify and locate.

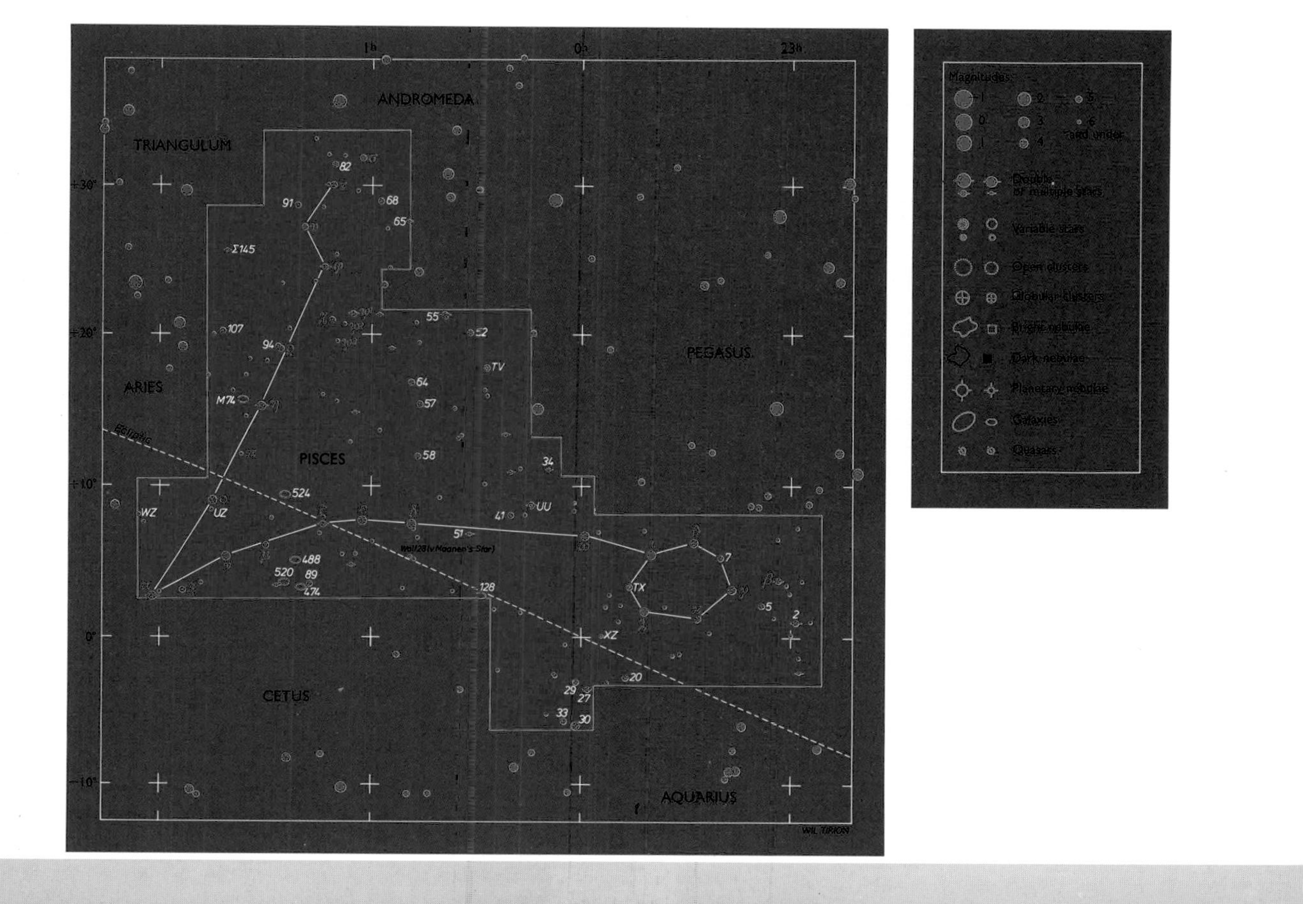
1h
0h
23h
ANDROMEDA
TRIANGULUM
ARIES
PISCES
PEGASUS
CETUS
AQUARIUS
Ecliptic
+30°
+20°
+10°
0°
−10°
82
91
68
65
Σ145
107
94
55
52
TV
64
57
M74
58
34
524
UZ
WZ
UU
41
51
Wolf28 (v.Maanen's Star)
488
520
89
474
128
TX
7
5
2
XZ
20
29
27
33
30
WIL TIRION
Magnitudes:
−1
0
1
2
3
4
5
6
and under
Double or multiple stars
Variable stars
Open clusters
Globular clusters
Bright nebulae
Dark nebulae
Planetary nebulae
Galaxies
Quasars

Puppis

This is the poop of the old Ship Argo; parts of it can attain a reasonable altitude from Britain and the northern United States, but the southern part never rises.

L^2 is a bright semiregular variable: RA 07h 13.5m, dec. −44° 39′. It is of type M, and can reach magnitude 2.6 at maximum; it never descends below 6.2. The period is of the order of 140 days. Other variables include Z Puppis (RA 07h 32.6m, dec. −20° 40′; magnitude range 7.2–14.6, period 500 days, Mira type) and V Puppis (RA 07h 58.2m, dec. −49° 15′; magnitude range 4.7–5.2, period 1.45 days, β Lyrae eclipsing binary).

There are several planetary nebulae but Puppis is most celebrated for its open clusters (see table); there are many of these, of which three are far enough north to be included in Messier's list. M46 and M47 lie close together; M47 is much the more conspicuous of the two, and is visible with the naked eye. M93 is fairly bright and condensed: the brighter arms are spread out in straggling arms recalling the shape of a starfish. NGC 2477 is so compact that it was once classed as a globular cluster; it looks not unlike the Wild Duck, M11 in Scutum, though it is much fainter.

Leading stars	***RA*** h m s	***Declination*** ° ′ ″	***Magnitude***	***Spectrum***	***Distance*** ly
ζ Suhail Hadar	08 03 35.0	−40 00 12	2.25	O5.8	2,410
π	07 17 08.5	−37 05 51	2.70	K5	130
ϱ Turais	08 07 32.6	−24 18 15	2.81	F6	300
τ	06 49 56.1	−50 36 53	2.93	K0	82

Clusters	***RA*** h m	***Declination*** ° ′	***Magnitude***	***Notes***
M46	07 41.8	−14 49	6.1	100 stars
M47	07 36.6	−14 30	4.4	At least 30 stars
M93	07 44.6	−23 52	6.2	80 stars
NGC 2477	07 52.3	−38 33	5.8	Over 150 stars
NGC 2539	08 10.7	−12 50	6.5	Over 50 stars

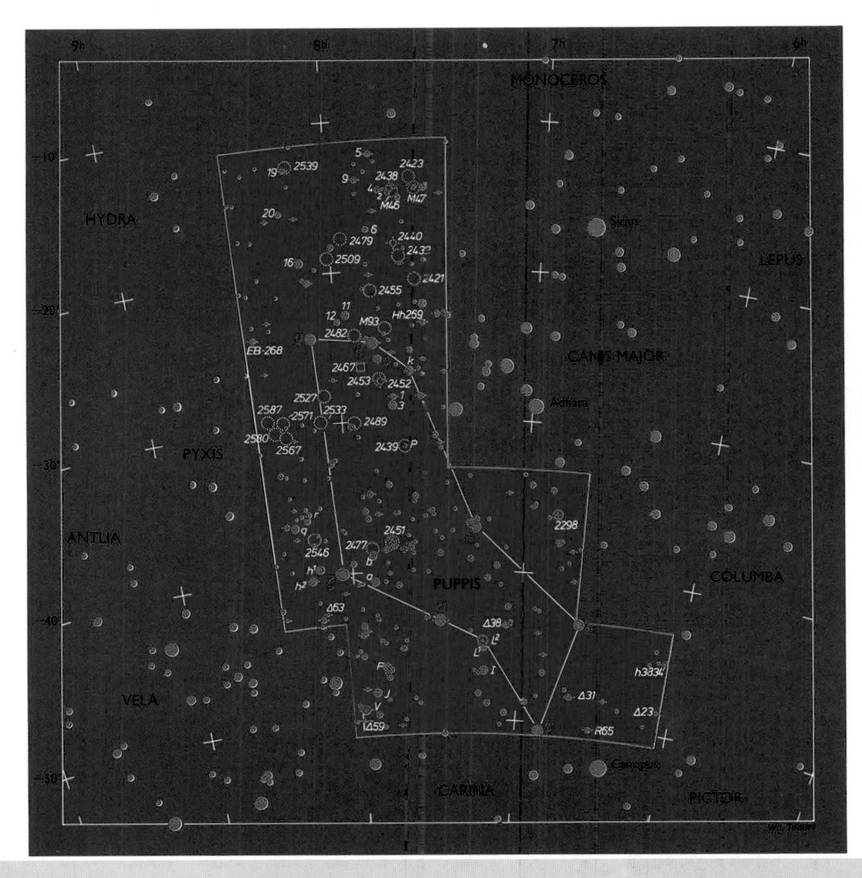
9h
8h
7h
6h
MONOCEROS
HYDRA
LEPUS
CANIS MAJOR
Sirius
Adhara
PYXIS
ANTLIA
PUPPIS
COLUMBA
VELA
CARINA
PICTOR
Canopus
-10°
-20°
-30°
-40°
-50°
2539
2423
2438
M46
M47
2479
2440
2509
2421
2455
M93
Hh259
2482
EB-268
2467
2453
2452
2527
2571
2533
2489
2587
2580
2567
2439
2298
2451
2477
2546
Δ63
Δ38
h3834
Δ31
Δ23
R65
Δ59
WIL TIRION

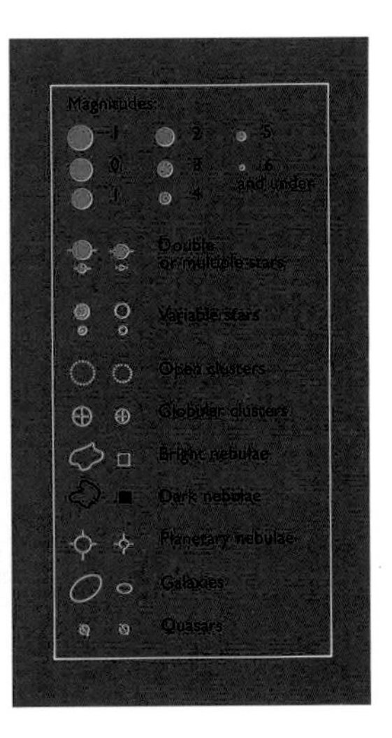
Magnitudes:
-1
0
1
2
3
4
5
6
and under
Double or multiple stars
Variable stars
Open clusters
Globular clusters
Bright nebulae
Dark nebulae
Planetary nebulae
Galaxies
Quasars

Sagitta/Vulpecula

Sagitta, the Arrow, is an ancient constellation, possibly linked either with Cupid's bow or with an arrow used by Apollo against the one-eyed Cyclops. The brightest star, γ, is only of magnitude 3.47. The shape, made up by γ together with α (magnitude 4.37), β (also 4.37) and δ (3.82), really does recall that of an arrow. U Sagittae (RA 19h 18.8m, dec. +19° 37′) is an Algol-type eclipsing binary, of magnitude range 6.6 to 9.2 and period 3.38 days. This is a good target for a small telescope, as the changes near the start and end of minimum are noticeable over periods of a few minutes.

M71 lies between γ and δ. It has been classed as an open cluster, but it seems to be a long way away (around 18,000 light-years) and may really be a globular. It is very hard to see with binoculars, but easy enough in a telescope, though it shows little detail.

Vulpecula, the Fox (originally Vulpecula et Ansa, the Fox and Goose) adjoins Sagitta. Its brightest star is α (magnitude 4.44). The main object here is M27, the Dumbbell Nebula, which is a fine planetary; with most binoculars it is in the field with γ Sagittae (the position is RA 19h 59.6m, dec. +22° 43′). It is a favorite subject for astrophotographers: it was discovered by Messier himself in 1764. The central star is of the 12th magnitude.

The open cluster NGC 6940 (RA 20h 34.6m, dec. +28° 18′) is near the Cygnus boundary; it has low surface brightness, but is an easy binocular or telescopic object, showing up as an oval blur which a higher power will resolve into dim stars.

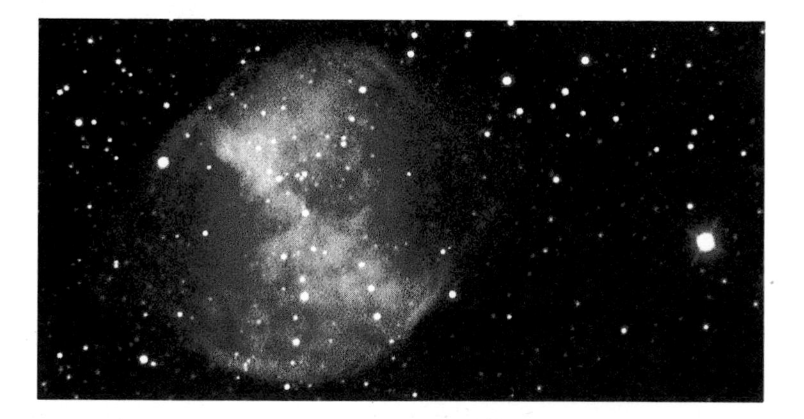

The Dumbbell Nebula was thrown out by a dying star 50,000 years ago.

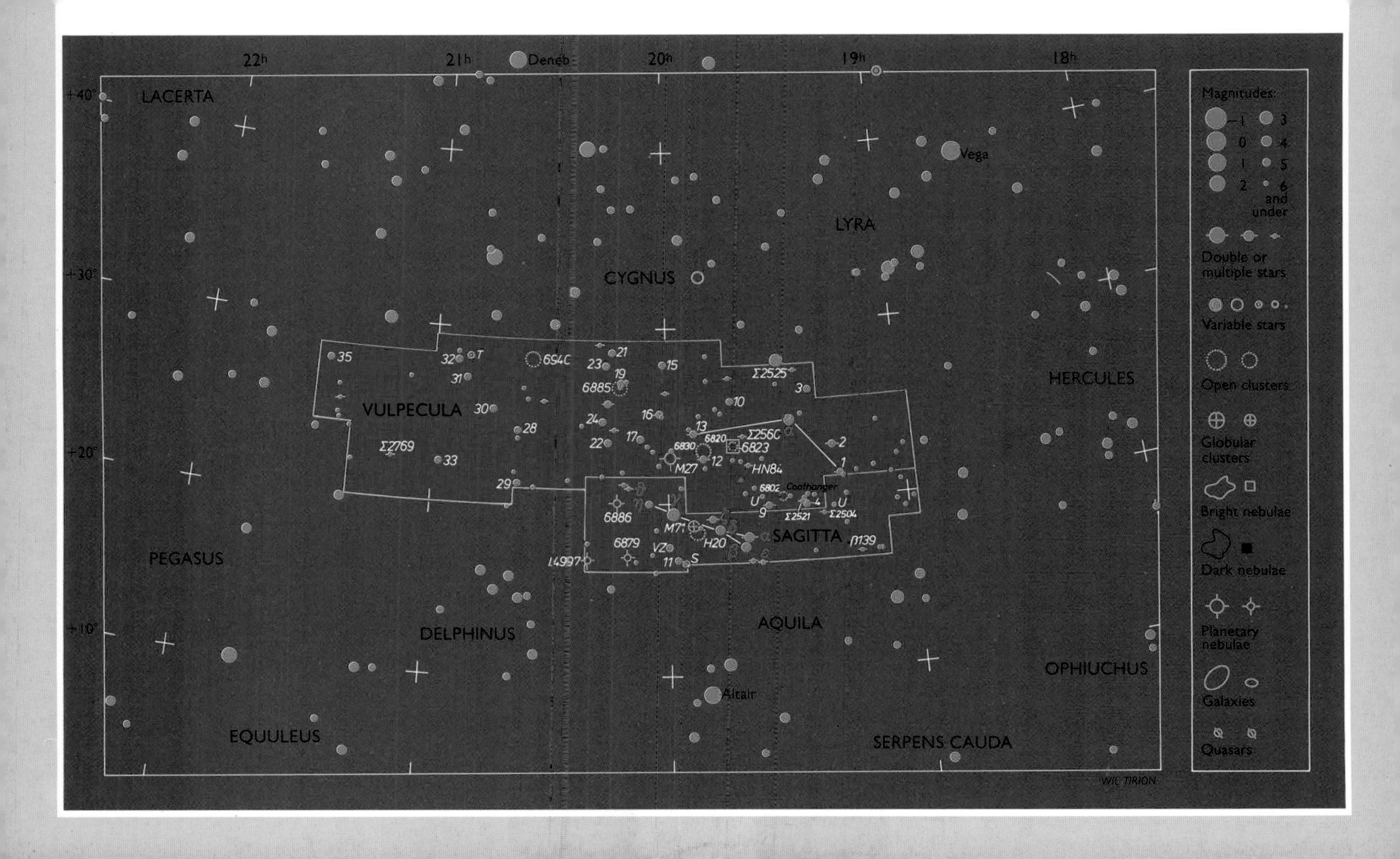
22h
21h
Deneb
20h
19h
18h
+40°
+30°
+20°
+10°
LACERTA
Vega
LYRA
CYGNUS
HERCULES
VULPECULA
SAGITTA
PEGASUS
DELPHINUS
AQUILA
OPHIUCHUS
EQUULEUS
SERPENS CAUDA
Altair
Coathanger
M27
M71
H20
HN84
6820
6823
6830
6802
6885
6940
6886
6879
Σ2525
Σ2560
Σ2521
Σ2504
Σ2769
β139
I4997
WIL TIRION
Magnitudes:
Double or multiple stars
Variable stars
Open clusters
Globular clusters
Bright nebulae
Dark nebulae
Planetary nebulae
Galaxies
Quasars

Sagittarius

Sagittarius, the Archer, is a brilliant Zodiacal constellation – even though it has no first-magnitude stars. It has sometimes been identified with the wise centaur Chiron (as has Centaurus). Sagittarius contains the richest part of the Milky Way, and its glorious star clouds hide the center of the Galaxy, 30,000 light-years away from us.

α (Rukbat) is only of magnitude 4.0, and β (Arkab) of magnitude 3.9. β is made up of two components, separable with the naked eye: the brighter has an 8th-magnitude companion, at a separation of 28″.3. η is an easy double: magnitudes 3.2 and 7.8, separation 3″.6. ξ is a close binary (magnitudes 3.3 and 3.4, period only 21.2 years; however, the separation is never as much as 0″.5, so this is a difficult pair.

The most interesting star is RY Sagittarii (RA 19h 16.5m, dec. −33° 31′, magnitude range 6–15). There is no period, as this is an R Coronae-type star – the brightest example apart from R Coronae itself. It dims unpredictably when soot accumulates in its atmosphere.

Among the non-stellar objects there is an embarrassment of riches (see table), and the whole area will repay sweeping time and again. The nebulae M8, M17 and M20 are photographic favorites.

Leading stars	RA h m s	Declination ° ′ ″	Magnitude	Spectrum	Distance ly
ε Kaus Australis	18 24 10.2	−34 23 05	1.85	B9	85
σ Nunki	18 55 15.7	−26 17 48	2.02	B3	209
ζ Ascella	19 02 36.5	−29 52 49	2.59	A2	78
δ Kaus Meridionalis	18 20 59.5	−29 49 42	2.70	K2	85
λ Kaus Borealis	18 27 58.1	−25 25 18	2.81	K2	98
π Albaldah	19 09 45.6	−21 01 25	2.89	F2	310
γ Alnasr	18 05 48.3	−30 25 26	2.99	K0	117

Non-stellar objects	RA h m	Declination ° ′	Magnitude	Notes
M18	18 19.9	−17 08	6.9	Open cluster, 20 stars
M21	18 04.6	−22 30	5.9	Open cluster, at least 70 stars
M23	17 56.8	−19 01	5.5	Loose cluster, around 150 stars
M22	18 36.4	−23 54	5.1	Very fine globular cluster
M55	19 40.0	−30 58	6.9	Globular cluster
M8	18 03.8	−24 23	6.0	Lagoon Nebula
M17	18 20.8	−16 11	7.0	Omega Nebula
M20	18 02.6	−23 02	7.6	Trifid Nebula

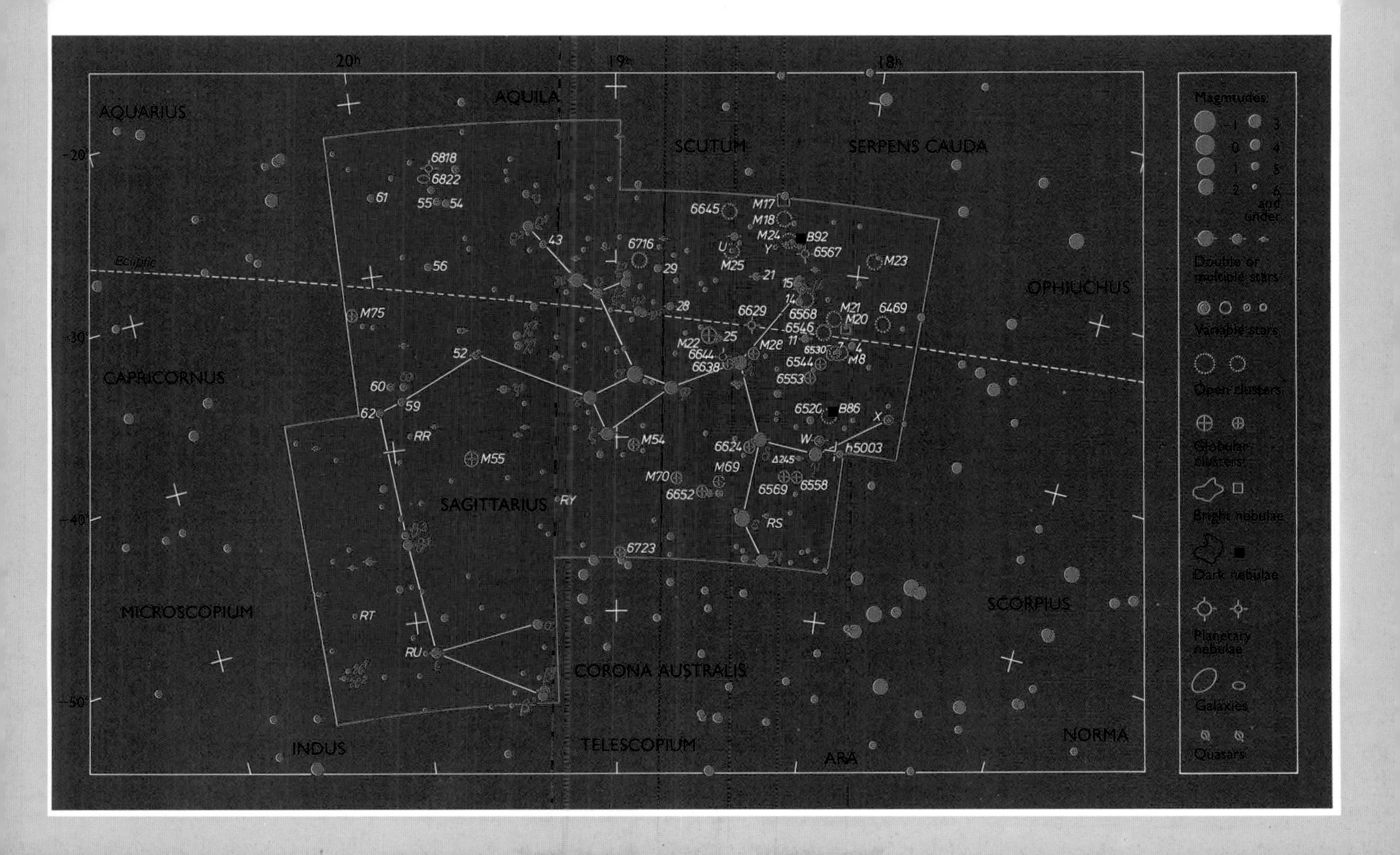

20h
19h
18h
-20°
-30°
-40°
-50°
AQUARIUS
AQUILA
SCUTUM
SERPENS CAUDA
OPHIUCHUS
CAPRICORNUS
SAGITTARIUS
MICROSCOPIUM
CORONA AUSTRALIS
SCORPIUS
INDUS
TELESCOPIUM
ARA
NORMA
Ecliptic
6818
6822
61
55
54
56
43
6716
29
28
M75
52
60
59
62
RR
M55
RT
RU
RY
M54
6723
M70
6652
M69
6624
6569
6558
RS
6645
M17
M18
M24
B92
Y
U
6567
M25
21
M23
15
14
6629
6568
6546
M21
M20
6469
M22
25
6644
6638
M28
11
6530
7
4
M8
6544
6553
6520
B86
X
W
h5003
Δ245
Magnitudes
-1
0
1
2
3
4
5
6 and under
Double or multiple stars
Variable stars
Open clusters
Globular clusters
Bright nebulae
Dark nebulae
Planetary nebulae
Galaxies
Quasars

Scorpius

Scorpius is perhaps the only constellation that rivals Orion in magnificence. Its long line of bright stars really suggests a scorpion – the creature which bit Orion and caused his untimely death.

The brilliant red supergiant Antares (see table) is called the "rival to Mars (Ares)"; it is flanked on either side by fainter stars, τ and σ. The head of Scorpius is made up of β, ν (magnitude 4.0) and the double ω (magnitudes 4.0 and 4.3). The "sting" is marked by λ and υ, which look like a wide double even though they are quite unconnected; υ is much the more remote and luminous. ζ is a naked-eye double (magnitudes 3.6 and 4.7) made up of two unassociated stars; β is a beautiful telescopic double (magnitudes 2.6 and 4.9, separation 13″.6 – the brighter star is itself a close binary).

There are many clusters in Scorpius: M6 and M7 are two superb naked-eye clusters, and splendid sights in a telescope. M4 is one of the most prominent of all globular clusters; it is not far below naked-eye visibility (and, incidentally, very rich in short-period variable stars). It is much more prominent than the adjacent M80, which has a condensed nucleus and is difficult to resolve into stars. Well-equipped observers will enjoy looking for the faint and elusive little Bug Nebula!

Leading stars	***RA*** h m s	***Declination*** ° ′ ″	***Magnitude***	***Spectrum***	***Distance*** ly
α Antares	16 29 24.3	−26 25 55	0.96	M1	330
λ Shaula	17 33 36.4	−37 06 14	1.63	B2	274
θ Sargas	17 37 19.0	−42 59 52	1.87	F0	900
ε Wei	16 50 09.7	−34 17 36	2.29	K2	65
δ Dschubba	16 00 19.9	−22 37 18	2.32	B0	554
κ Girtab	17 42 29.0	−39 01 48	2.41	B2	390
β Graffias	16 05 26.1	−19 48 19	2.64	B+B	815

Non stellar objects	***RA*** h m	***Declination*** ° ′	***Magnitude***	***Notes***
M6	17 40.1	−32 13	4.2	Open cluster (Butterfly Cluster)
M7	17 53.9	−34 49	3.3	Open cluster; over 80 stars
M4	16.23.6	−26 32	5.9	Globular cluster, near Antares
NGC 6124	16 25.6	−40 40	5.8	Open cluster; about 100 stars
NGC 6242	16 55.6	−39 30	6.4	Bright loose cluster
NGC 6302	17 13.7	−37 06	12.8	Planetary nebula (Bug Nebula)

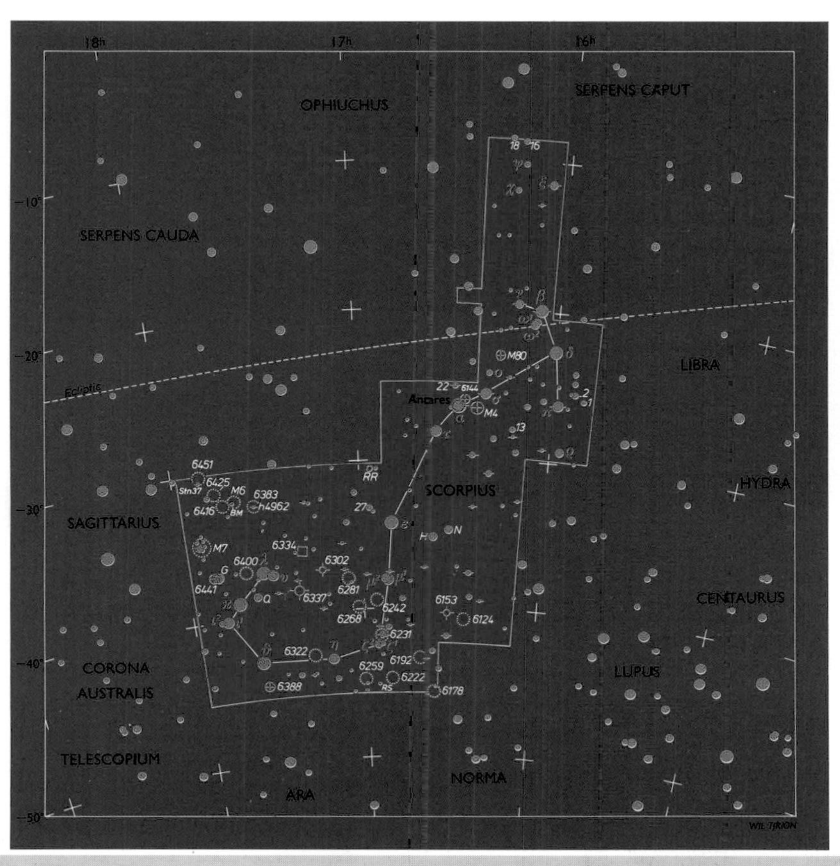

18h
17h
16h
OPHIUCHUS
SERPENS CAPUT
SERPENS CAUDA
LIBRA
SCORPIUS
Antares
SAGITTARIUS
HYDRA
CENTAURUS
LUPUS
CORONA AUSTRALIS
TELESCOPIUM
ARA
NORMA
Ecliptic
−10°
−20°
−30°
−40°
−50°
WIL TIRION

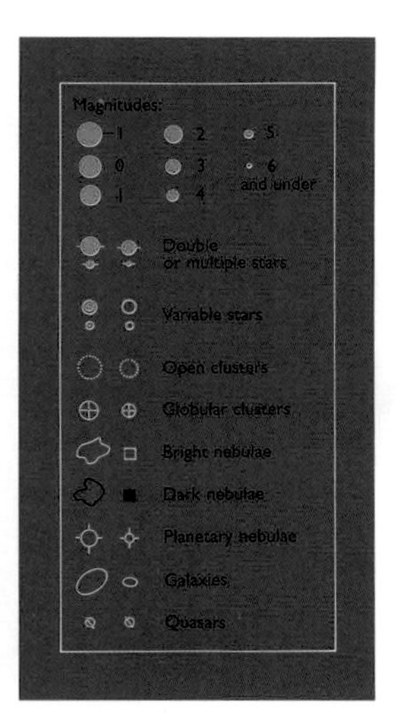

Magnitudes:
−1
0
1
2
3
4
5
6
and under
Double or multiple stars
Variable stars
Open clusters
Globular clusters
Bright nebulae
Dark nebulae
Planetary nebulae
Galaxies
Quasars

Scutum/Serpens Cauda

Scutum (originally Scutum Sobieskii, Sobieski's Shield) adjoins Aquila, and was formerly included in it. It has only one star above the fourth magnitude – α (3.85) – but it is very rich, as the Milky Way flows right through it.

R Scuti (RA 18h 47.5m, dec. −05° 42′) is an interesting variable within binocular range; it fluctuates between magnitudes 4.4 and 8.2, with a main period around 140 days. It is the brightest member of the RV Tauri class: it seems to be oscillating in at least two superimposed periods, and at its peak is at least 8,000 times as powerful as the Sun.

Also in the constellation lies M11, the Wild Duck cluster (RA 18h 51.1m, dec. −06° 16′). This is a splendid object in a small telescope: it is somewhat fan-shaped, and is visible with the naked eye. The distance is about 5,500 light-years. A moderate power will resolve it. M26 is a globular cluster close to δ (RA 18h 53.1m, dec. −08° 42′).

Serpens Cauda, the body of the serpent, is distinct from the head. In Cauda the brightest star is η (magnitude 3.26), but the main points of interest are the double star θ, or Alya, and the Eagle Nebula, M16.

θ is a good example of a double in which the components are identical twins; both are of magnitude 4.5, and of type A5. They are certainly associated, though they are a long way apart. Good binoculars will show them separately, and in any telescope they are superb.

M16 is at RA 18h 18.8m, dec. −13° 47′. It is a large, scattered cluster immersed in a vast diffuse nebula; there are areas of dark nebulosity, and the object is also a source of radio waves. A telescope of at least 20 centimeters aperture is needed to do it real justice.

Star fields in Scutum and Sagittarius, toward the Galactic center.

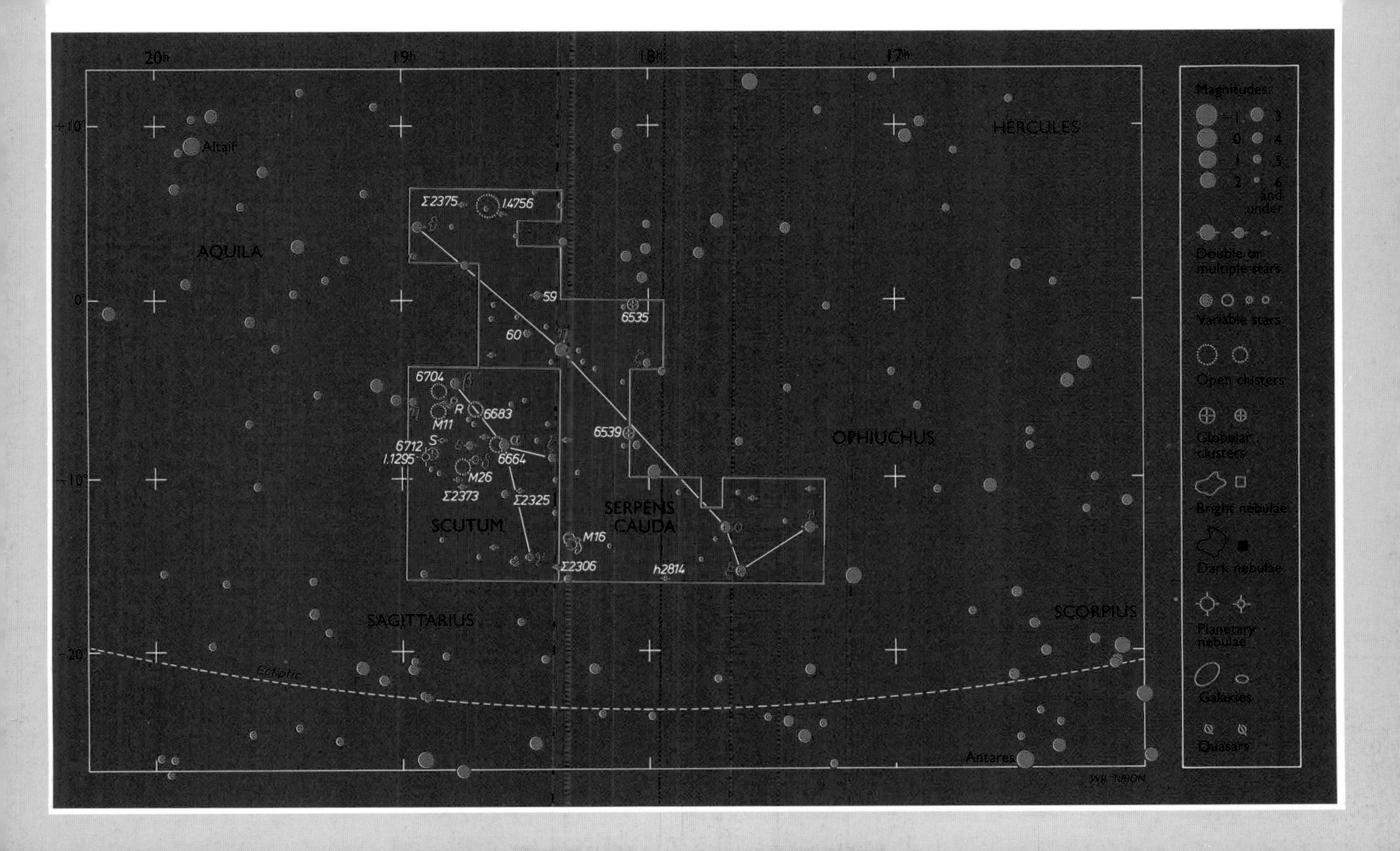
20h
19h
18h
17h
+10°
0°
−10°
−20°
Altair
AQUILA
HERCULES
OPHIUCHUS
SERPENS
CAUDA
SCUTUM
SAGITTARIUS
SCORPIUS
Antares
Ecliptic
Σ2375
I.4756
59
60
6535
6539
6704
6683
R
M11
S
6712
I.1295
6664
M26
Σ2373
Σ2325
M16
Σ2306
h2814
Magnitudes:
−1
0
1
2
3
4
5
6
and
under
Double or
multiple stars
Variable stars
Open clusters
Globular
clusters
Bright nebulae
Dark nebulae
Planetary
nebulae
Galaxies
Quasars
WIL TIRION

Taurus

Taurus represents the bull into which Jupiter transformed himself when he carried off Europa, daughter of the King of Crete. The red leader of the constellation, Aldebaran, is known as "the Eye of the Bull." The second brightest star, Al Nath or β Tauri, was removed from Auriga; it had previously been known as γ Aurigae.

λ Tauri is an Algol-type eclipsing binary, with a range of magnitude from 3.3 to 3.8 and a period of 3.95 days; useful comparison stars are γ (magnitude 3.6), μ (4.3), ο (3.6) and ξ (3.7). λ Tauri is really much more luminous than Algol but is much further away (330 light-years).

M1, the Crab Nebula, is the famous supernova remnant (RA 05h 34.5m, dec. +22° 01′); good binoculars show it as a dim patch, close to ζ, but photography is needed to bring out its immensely complicated form. It is the remnant of a supernova seen in 1054; deep inside it is a pulsar – its "powerhouse." It is 6,000 light-years away.

The Hyades cluster, round Aldebaran, is bright; the stars making up the little V-formation are σ (magnitude 4.7), θ (3.4 and 3.8), γ (3.6), δ (3.8 and 3.4) and ε (3.5). θ is a naked-eye double; the brighter component is white, the companion orange. The two are 15 light-years apart, though they may well have had a common origin. δ makes up a wide pair with 68 Tauri (magnitude 4.8). The Hyades are somewhat overpowered by Aldebaran, which lies about midway between the cluster and ourselves.

The Pleiades or Seven Sisters make up the most famous of star clusters. They are hot and bluish-white, and by stellar standards they are very young; the leaders are Alcyone (magnitude 2.9), Electra (3.7), Atlas (3.6), Merope (4.2), Maia (3.9), Taygete (4.3), Pleione (5.1, but variable), Celaeno (5.4) and Asterope (5.6). Atlas and Pleione are close together, but binoculars will split them. If you can see a dozen Pleiads with the naked eye on a clear night, you are doing very well indeed. The total population of the cluster is at least 400. The considerable nebulosity in the Pleiades is not hard to record photographically.

Leading stars	*RA*	*Declination*	*Magnitude*	*Spectrum*	*Distance*
	h m s	° ′ ″			ly
α Aldebaran	04 35 55.2	+16 30 33	0.85	K5	68
β Al Nath	05 26 17.5	+28 36 27	1.65	B7	130
η Alcyone	03 47 29.0	+24 06 18	2.87	B7	240
ζ Alheka	05 37 38.6	+21 08 33	3.00	B2	490

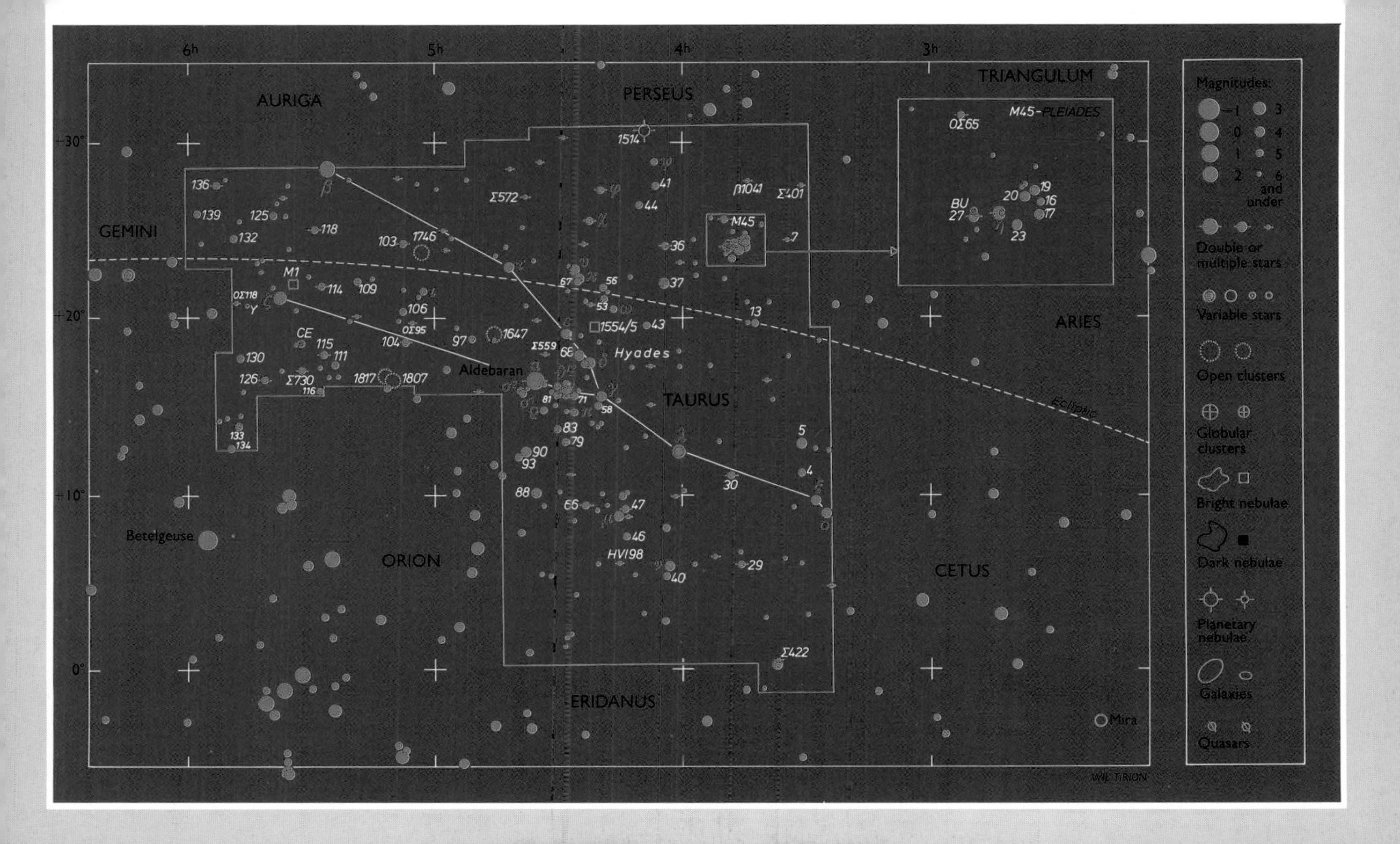
6h
5h
4h
3h
+30°
+20°
+10°
0°
AURIGA
PERSEUS
TRIANGULUM
GEMINI
ARIES
TAURUS
Hyades
ORION
CETUS
ERIDANUS
Aldebaran
Betelgeuse
Mira
Ecliptic
M45-PLEIADES
M45
M1
CE
Y
OΣ118
OΣ95
OΣ65
BU 27
Σ572
Σ559
Σ730
Σ401
Σ422
β1041
HV198
1514
1554/5
1647
1746
1807
1817
Magnitudes:
−1
0
1
2
3
4
5
6 and under
Double or multiple stars
Variable stars
Open clusters
Globular clusters
Bright nebulae
Dark nebulae
Planetary nebulae
Galaxies
Quasars
WIL TIRION

Triangulum

This is an ancient constellation, adjoining Andromeda and Aries. No legends are attached to it, but at least its three main stars, β (magnitude 3.00), α (3.41) and γ (4.01), really do form a triangle. There is one fairly bright Mira variable: R Trianguli (RA 02h 37.0m, dec. +34° 16′), with a range of magnitude of 5.4 to 12.6 and a period of 267 days.

The most important object is the spiral galaxy M33, which is a member of the Local Group. It is a rather loose spiral, lying at a distance of 2,300,000 light-years – slightly further away than the Andromeda Spiral. It is also a smaller system, so that it is much fainter, and its surface brightness is rather low. It is not hard to see with binoculars, between α and β Andromedae, but it can be surprisingly elusive when it is being sought with a telescope. However, a 37-centimeter reflector will show the form quite well. Predictably, it contains objects of all kinds: clusters, variable stars and nebulae, for example. It is sometimes nicknamed the Pinwheel Galaxy.

A fellow member of our Local Group – M33, the Pinwheel Galaxy.

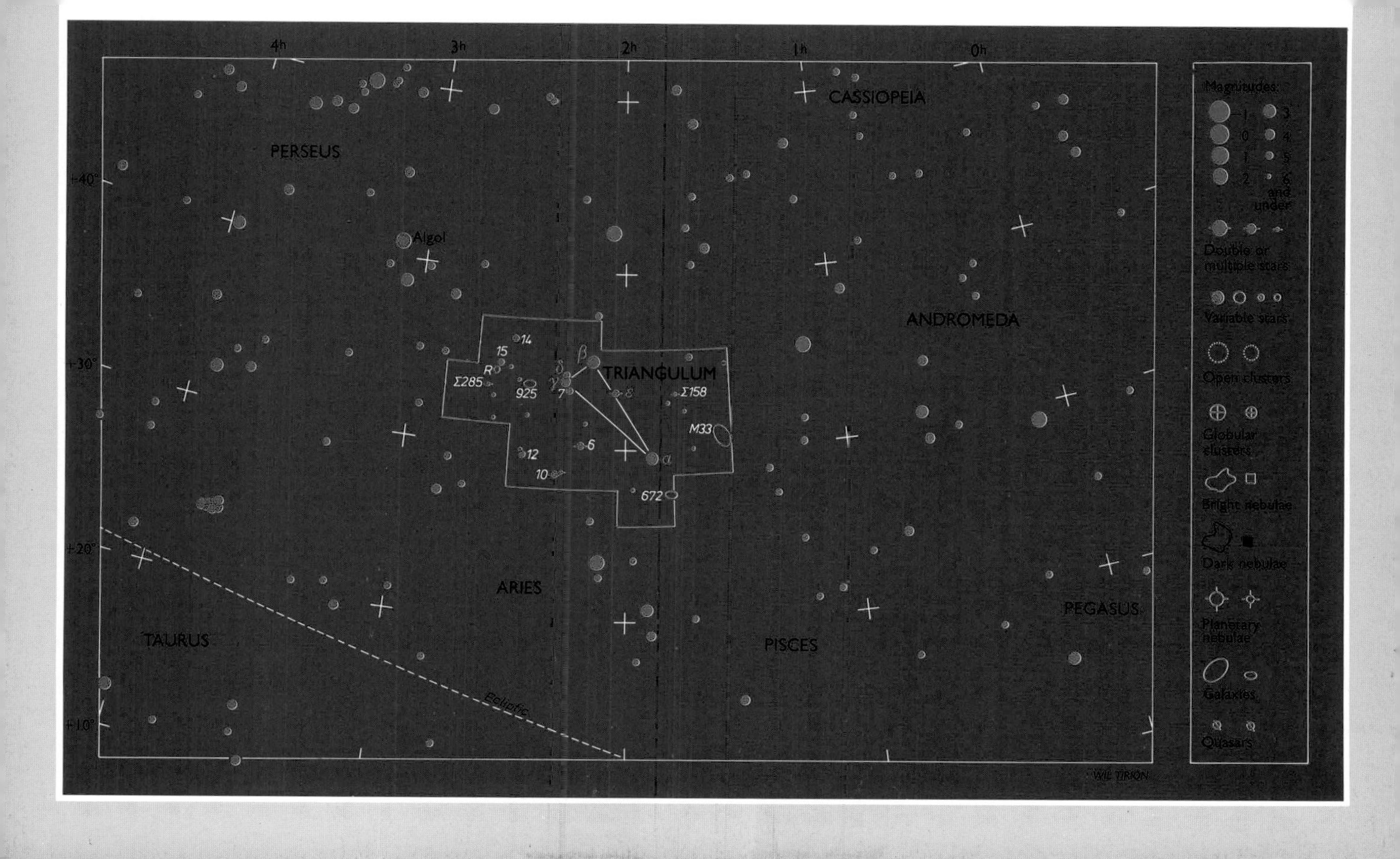
4h
3h
2h
1h
0h
+40°
+30°
+20°
+10°
PERSEUS
CASSIOPEIA
ANDROMEDA
TRIANGULUM
ARIES
PISCES
PEGASUS
TAURUS
Algol
Ecliptic
M33
672
Σ158
Σ285
925
14
15
R
12
10
6
7
α
β
γ
δ
WIL TIRION
Magnitudes:
-1
0
1
2
3
4
5
6 and under
Double or multiple stars
Variable stars
Open clusters
Globular clusters
Bright nebulae
Dark nebulae
Planetary nebulae
Galaxies
Quasars

Ursa Major

This most famous of constellations is named in honor of Callisto, daughter of King Lycaon of Arcadia, who was turned into a bear by Juno, the jealous wife of Jupiter. Years later Callisto's son, Arcas, was out hunting and met the bear; he was about to shoot it when Jupiter intervened, changed Arcas into a bear also, and swung both creatures into the sky by their tails as Ursa Major and Ursa Minor. Presumably this is why both celestial bears have tails of somewhat un-ursine length!

Five of the seven stars nicknamed the Big Dipper, King Charles' Wain, or, in Britain, the Plough, make up a "moving cluster." The exceptions are Dubhe and Alkaid, which are moving in the opposite direction. Dubhe and Merak point to the Pole Star; they are clearly different in color, Merak pure white and Dubhe a warm orange.

Mizar is the celebrated double star. It makes up a naked-eye pair with Alcor or 80 Ursae Majoris (magnitude 4.0); telescopically Mizar itself is seen to be double (magnitudes 2.3 and 4.0, separation 14″.4). Each component is a spectroscopic binary – and so is Alcor, so that altogether we are dealing with six suns! Between Mizar and Alcor is an 8th-magnitude star, named Sidus Ludovicianum in 1723 by courtiers of the Emperor Ludwig V.

There are many galaxies in Ursa Major, notably: M81 (RA 09h 55.6m, dec. +69° 04′), M82 (09h 55.8m, +69° 41′) and M101 (14h 03.2m, +54° 21′). M81 is a spiral; M82 is irregular, and is a radio source – it seems to be perturbed by the pull of its larger neighbor. M101 is a large face-on spiral, one of the easiest to see well with a fairly small telescope.

The planetary nebula M97, between Merak and Phad (RA 11h 14.8m, dec. +55° 01′) is known as the Owl Nebula, because when properly seen it does look rather like an owl's face. However, it is faint – the integrated magnitude is about 12 – and can be hard to locate. The central star is well below the 15th magnitude.

Leading stars	*RA* h m s	*Declination* ° ′ ″	*Magnitude*	*Spectrum*	*Distance* ly
ε Alioth	12 54 01.7	+55 57 35	1.77	A0	62
α Dubhe	11 03 43.6	+61 45 03	1.79	K0	75
η Alkaid	13 47 32.3	+49 18 48	1.86	B3	108
ζ Mizar	13 23 05.5	+54 55 31	2.09	A2 + A6	59
β Merak	11 01 50.4	+56 22 56	2.37	A1	62
γ Phad	11 53 49.7	+53 41 41	2.44	A0	75
δ Megrez	12 15 25.5	+57 01 57	3.31	A3	65

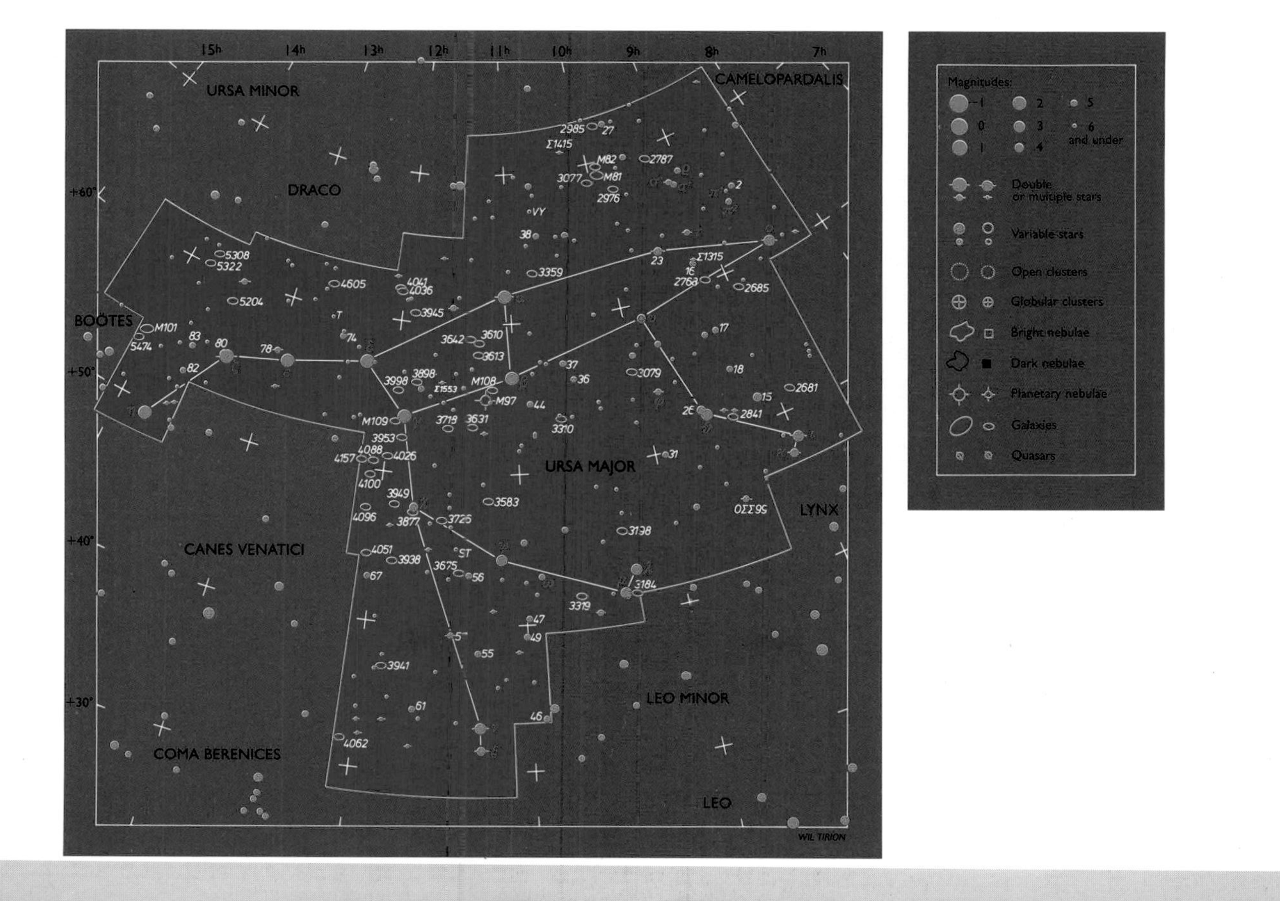
URSA MINOR
DRACO
CAMELOPARDALIS
BOÖTES
URSA MAJOR
LYNX
CANES VENATICI
LEO MINOR
COMA BERENICES
LEO
WIL TIRION
Magnitudes:
−1
0
1
2
3
4
5
6 and under
Double or multiple stars
Variable stars
Open clusters
Globular clusters
Bright nebulae
Dark nebulae
Planetary nebulae
Galaxies
Quasars

Vela

Vela represents the sails of the old Argo Navis, and adjoins Carina, the Keel. The "False Cross," so often confused with the Southern Cross, is made up of ϰ and δ Velorum together with ι and ε Carinae.

γ is an unusual star of the Wolf-Rayet type; it is very hot and unstable, and is ejecting material at an amazing rate. It is about 3,800 times as luminous as the Sun, but will not endure for nearly so long; it may even explode as a supernova, though perhaps not for many millions of years yet. It has a 4.2-magnitude companion at a separation of 41″.2, so that this is a most imposing pair; also in the field are stars of magnitudes 8.2, 9.1 and 12.5.

δ has a companion of magnitude 5.1 at a distance of 2″.6, and here also there are two much fainter stars in the field (magnitudes 11.0 and 13.5). μ is a binary, with magnitudes 2.7 and 6.4, period 116 years and separation 2″.3.

There are several open clusters in the constellation. The most prominent of these is IC 2391, surrounding the star o Velorum (RA 08h 40.2m, dec. −53° 04′); it contains 30 stars and is easily visible with the naked eye. Also conspicuous is NGC 2547 (RA 08h 10.7m, dec. −49° 16′), which contains at least 80 stars and is also a naked-eye object, close to γ.

NGC 3132 is a bright planetary nebula, lying at RA 10h 07.7m, dec. −40° 26′), close to the border with Antlia. The central star is unusually bright – around magnitude 10 – and the integrated magnitude of the nebula itself is about 8.2.

Vela also contains the Gum Nebula (named in honor of the Australian astronomer Colin Gum, who died tragically early in a skiing accident). This is a supernova remnant, and contains a pulsar which has been detected also at optical wavelengths. It is a pity that the supernova responsible for it flared up in prehistoric times; it must have been really imposing – far brighter than any supernova seen since records began. The estimated date of its appearance is 9000 BC.

Leading stars	RA h m s	Declination ° ′ ″	Magnitude	Spectrum	Distance ly
γ Regor	08 09 31.9	−47 20 12	1.78	WC7	520
δ Koo She	08 44 42.2	−54 42 30	1.96	AO	669
λ Al Suhail al Wazn	09 07 59.7	−43 25 57	2.21	K5	490
ϰ Markeb	09 22 06.8	−55 00 38	2.50	B2	390
μ	10 46 46.1	−49 25 12	2.69	G5	98

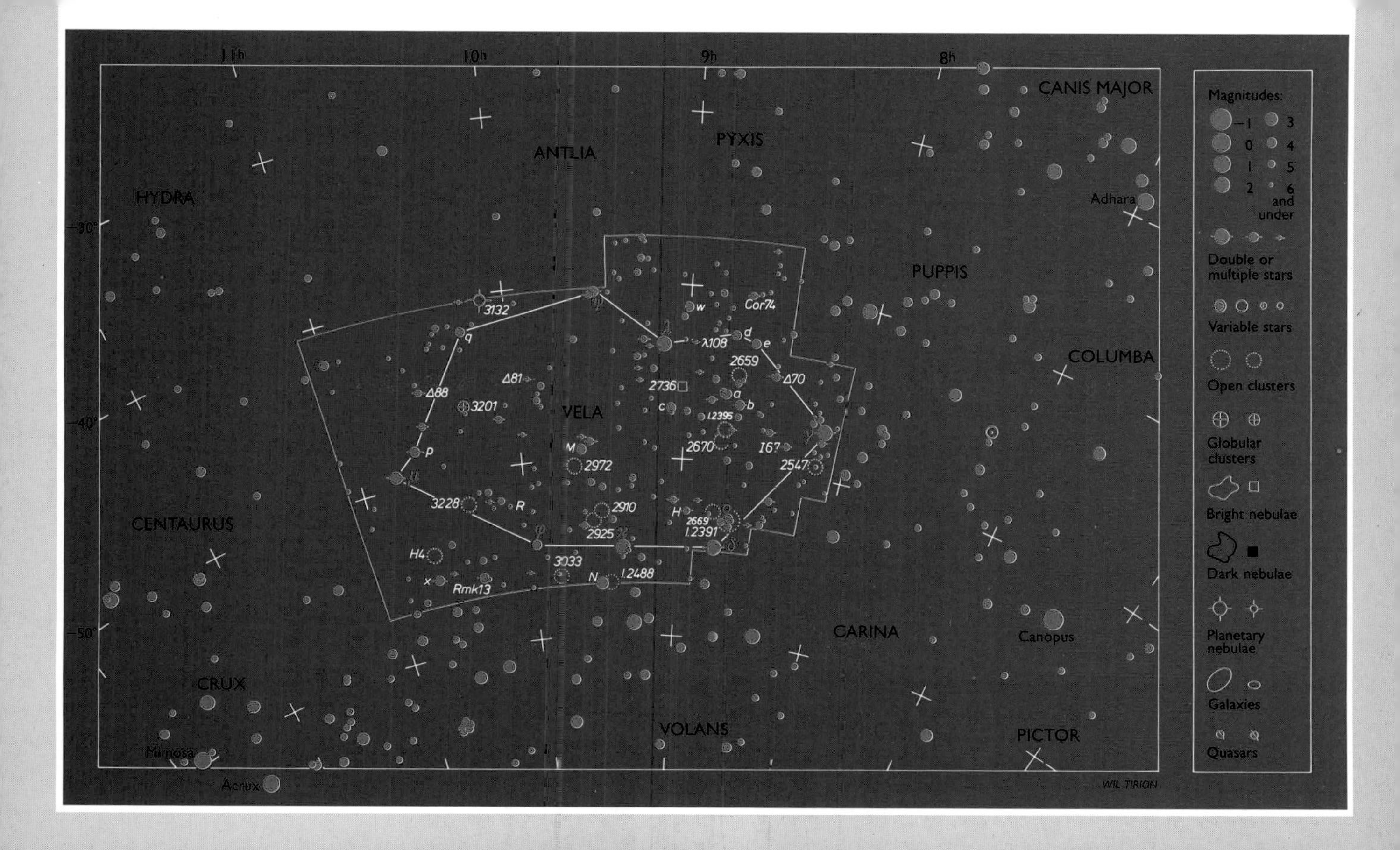
11h
10h
9h
8h
-30°
-40°
-50°
HYDRA
ANTLIA
PYXIS
PUPPIS
CANIS MAJOR
COLUMBA
CENTAURUS
VELA
CARINA
CRUX
VOLANS
PICTOR
Adhara
Canopus
Mimosa
Acrux
3132
q
Δ88
3201
p
3228
R
H4
x
Rmk13
Δ81
M
2972
2910
2925
3033
N
I.2488
2736
c
w
Cor74
λ108
d
e
2659
a
b
I.2395
2670
I67
Δ70
2547
H
2669
I.2391
WIL TIRION
Magnitudes:
−1
0
1
2
3
4
5
6 and under
Double or multiple stars
Variable stars
Open clusters
Globular clusters
Bright nebulae
Dark nebulae
Planetary nebulae
Galaxies
Quasars

Virgo

Virgo, the Virgin, is a very large constellation said to represent Astraea, goddess of justice, daughter of Jupiter and Themis. The leading star is first-magnitude Spica – yet only three stars are above magnitude 3.

Virgo is Y-shaped, with Spica at the foot and γ Virginis at the bottom of the "bowl." The "bowl," formed by γ, δ (magnitude 3.38), ε, η (3.89), β (3.61), and β Leonis, is rich in galaxies: the great Virgo Cluster lies here, about 50 million light-years away.

γ is a splendid double with identical components, each of magnitude 3.5 and of type F0. The orbital period is 171.4 years. The pair is now closing, and by the year 2016 will be apparently single except with a very powerful instrument. R Virginis, in the "bowl," is a Mira variable, at RA 12h 38.5m, dec. +06° 59′, magnitude range 6.0 to 12.1, period 146 days. Predictably, it is a red star, of type M.

Of the 11 Messier galaxies in Virgo, the most important is M87, a giant elliptical system that is also a powerful radio source, Virgo A. From it issue striking jets of material; but even the main jet can be seen only with large telescopes (say, above 37 centimeters aperture).

M104, the Sombrero Hat Galaxy, is near the border with Corvus; a dark dust-lane gives it its distinctive appearance, best seen on photographs. Photographers will enjoy themselves in taking pictures of this whole area, which is crowded with galaxies of all kinds.

Leading stars	*RA* h m s	*Declination* ° ′ ″	*Magnitude*	*Spectrum*	*Distance* ly
α Spica	13 25 11.5	−11 09 41	0.98	B1	260
γ Arich	12 41 39.5	−01 26 57	2.75	F+F	36
ε Vindemiatrix	13 02 10.5	+10 57 33	2.83	G9	104

Galaxies	*RA* h m	*Declination* ° ′	*Magnitude*	*Notes*
M49	12 29.8	+08 00	8.4	Elliptical
M58	12 37.7	+11 49	9.8	Spiral
M59	12 42.0	+11 39	9.8	Elliptical
M60	12 43.7	+11 33	8.8	Elliptical
M61	12 21.9	+04 28	9.7	Loose spiral
M86	12 26.2	+12 57	9.2	Elliptical
M87	12 30.8	+12 44	8.6	Giant elliptical (Virgo A)
M90	12 36.8	+13 10	9.5	Spiral
M104	12 40.0	+11 37	8.3	Spiral (Sombrero Hat Galaxy)

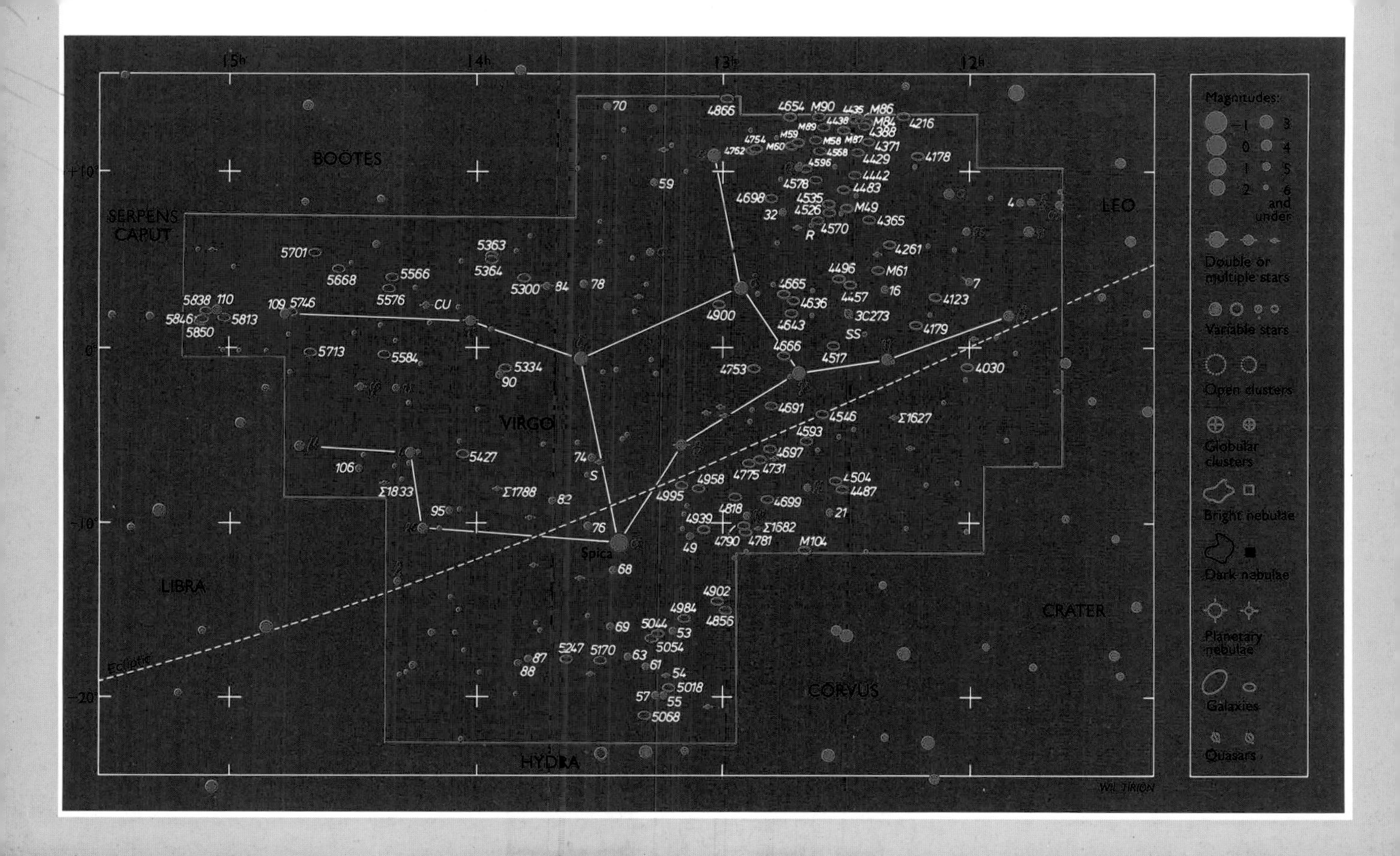
15h
14h
13h
12h
+10°
0°
−10°
−20
BOÖTES
SERPENS CAPUT
VIRGO
LEO
LIBRA
CRATER
CORVUS
HYDRA
Spica
Ecliptic
M104
M49
M61
3C273
Magnitudes:
−1
0
1
2
3
4
5
6 and under
Double or multiple stars
Variable stars
Open clusters
Globular clusters
Bright nebulae
Dark nebulae
Planetary nebulae
Galaxies
Quasars
WIL TIRION

TIME CHART OF ASTRONOMY

1543 Nicolas Copernicus' book published, in which he claimed that the Earth moves round the Sun.

1609 First serious telescopic observations by Galileo Galilei.

1618 Kepler's laws of planetary motion published.

1668 First reflector made, by Sir Isaac Newton (this is the probable date; it was certainly in existence by 1671).

1675 Royal Greenwich Observatory founded.

1687 Sir Isaac Newton's *Principia* published.

1781 Uranus discovered, by Sir William Herschel. Charles Messier's catalog of clusters and nebulae published.

1786 First approximately accurate idea of the shape of the Galaxy discovered, by Sir William Herschel.

1801 First asteroid discovered: Ceres, by G. Piazzi at Palermo.

1802 Existence of binary star systems established, by Sir William Herschel.

1814 Dark lines in the solar spectrum studied, by J. Fraunhofer.

1834/8 First extensive observations of the stars in the far Southern Hemisphere made, by Sir John Herschel.

1838 First measurement of distance of a star, by F. W. Bessel.

1845 Spiral nature of the galaxies discovered, by Lord Rosse with his 182 cm reflector in Ireland.

1846 Neptune discovered.

1859 Dark lines in the solar spectrum explained, by G. Kirchhoff.

1863/4 First classifications of the spectra of stars, by A. Secchi in Italy and Sir W. Huggins in England.

1912 Period-luminosity law of the Cepheid variables discovered, by Miss H. Leavitt at Harvard.

1915 Albert Einstein's general theory of relativity published.

1917 The 254 cm reflector at Mount Wilson completed.

1923 Proof given, by E. Hubble, that the galaxies are independent systems rather than parts of our Galaxy.

1929 Velocity/distance relationship of galaxies announced, by E. Hubble.

1930 Pluto discovered, by Clyde Tombaugh.

1931 First experiments on receiving radio waves from the sky, by K. Jansky.

1938 New (and correct) theory of stellar energy proposed independently, by H. Bethe and G. Gamow.

1942 Radio waves from the Sun detected, by J. S. Hey and his colleagues.

1946 Radio source Cygnus A identified.

1948 The 508 cm reflector at Palomar in California completed.

1952 Error in the Cepheid scale discovered, by W. Baade.

1955 The 76-meter radio "dish" at Jodrell Bank completed.

1963 Quasars identified, by M. Schmidt at Palomar.

1965/6 The 3K microwave radiation detected, by A. Penzias and R. Wilson.

1967 First pulsar identified, by Miss J. Burnell at Cambridge.

1969 First optical identification of a pulsar, in the Crab Nebula.

1976 600 cm reflector, USSR, completed.

1977 Rings of Uranus discovered. Discovery of Chiron. Discovery of Charon (satellite of Pluto).

1983 Full survey of the sky in infra-red, by IRAS. Discovery of "infra-red excesses" with certain stars, possibly indicating planet-forming material.

1985/6 Return of Halley's Comet.

1987 Supernova SN 1987A exploded in the Larger Magellanic Cloud.

1992 "Ripples" in the cosmic microwave radiation discovered, supporting current views on the Big Bang. Completion of the first Keck telescope on Mauna Kea.

1995 Report of a planet orbiting the star 51 Pegasi.

1996 Identification of a Brown Dwarf star (Gliese 229B). Appearance of two bright comets (Hyakutake and Hale–Bopp). Completion of the second Keck telescope. Reports of primitive life forms on an Antarctic meteorite said to come from Mars.

TIME CHART OF SPACE RESEARCH

1895 First important papers about space flight published, by K. E. Tsiolkovskii.
1926 First liquid-propelled rocket launched, by R. H. Goddard.
1949 First step-rocket fired, from White Sands in New Mexico.
1957 First artificial satellite, Sputnik 1 (USSR), launched on 4 October.
1958 First successful American artificial satellite, Explorer 1. Its instruments detected the Earth's Van Allen belts.
1959 First successful lunar probes (USSR).
1961 First manned space flight: Yuri Gagarin in Vostok 1 (USSR).
1962 First American to orbit the Earth: John Glenn. First successful planetary probe: Mariner 2, to Venus (USA).
1964 First good close-range pictures of the Moon, obtained from the American probe, Ranger 7.
1965 First successful Mars probe: Mariner 4 (USA). Close-range pictures obtained, altering all existing ideas about Mars.
1966 First soft landing on the Moon by an automatic probe: Luna 9 (USSR). The first Lunar Orbiter (USA) sent back high-quality, close-range pictures of the Moon's surface.
1967 First soft landing of an unmanned probe on Venus: Venera 7 (USSR).
1968 First manned flight around the Moon: F. Borman, J. Lovell and W. Anders in Apollo 8 (USA).
1969 21 July. First manned landing on the Moon: N. Armstrong and E. Aldrin in Apollo 11 (USA). Improved pictures of Mars sent from Mariners 6 and 7 (USA).
1971 First capsule landed on Mars, from the Soviet probe Mars 2; loss of contact prevented any useful results. Mariner 9 put into orbit around Mars, and, until mid-1972, sent back thousands of pictures; the great volcanoes were seen for the first time.
1972 Apollo program ended with the Apollo 17 mission.
1973/4 Operational lifetime of the American space station Skylab, manned by three successive crews. Much astronomical work carried out.
1973 First successful probe to Jupiter: Pioneer 10 (USA) bypassed the planet and sent back information from close range.
1974 First pictures of Venus from close range: Mariner 10 (USA). First successful probe to Mercury: also Mariner 10, which made three active passes of the planet, two in 1974–5. Second successful Jupiter probe: Pioneer 11 (USA).
1975 First pictures received from surface of Venus: automatic probes Veneras 9 and 10 (USSR).
1976 First successful soft landings on Mars: Vikings 1 and 2 (USA). Much information was obtained.
1978 Several probes sent to Venus: Veneras 11 and 12 (USSR) and two Pioneers (USA). Further information about the planet obtained, but no more pictures from the surface.
1979 Voyager 1 bypassed Jupiter; active volcanoes discovered on Io.
1980 Voyager 1 passed Saturn.
1981 First flight of the NASA Space Shuttle.
1985 Launch of probes to Halley's Comet.
1989 Voyager 2 encountered Neptune and its major satellite Triton. The Cosmic Background Explorer (COBE) satellite was launched to study the 3K universal microwave radiation.
1990 Hubble Space Telescope launched. Magellan space probe went into orbit around Venus. Ulysses space probe (USA) launched: it will swing by Jupiter and then fly over the Sun's poles.
1993 First Japanese lunar probe (Hiten). First servicing mission to the Hubble Space Telescope.
1994 Launch of Clementine probe to the Moon. Loss of the US probe Mars Observer.
1995 Galileo probe reaches Jupiter, and the entry section plunges into the Jovian clouds.
1996 Details on Jupiter's satellites obtained from Galileo.

GLOSSARY

Albedo The reflecting power of a non-luminous body. A perfect reflector would have an albedo of 100 per cent.

Ångström unit The hundred-millionth part of a centimeter.

Antoniadi scale A roman numeral indicates the quality of seeing according to the following scale:

I Perfect seeing, without a quiver
II Slight undulations, with moments of calm lasting several seconds
III Moderate seeing, with larger air tremors
IV Poor seeing, with constant troublesome undulations
V Very bad seeing, scarcely allowing the making of a rough sketch

Aperture The diameter of an opening through which light passes in an optical instrument.

Aphelion The position of a planet (or other body) when it is at its furthest from the Sun.

Apparent magnitude The apparent brightness of a celestial object: the lower the magnitude, the brighter the object.

Aurora (polar lights) A diffuse glow in the upper air caused by electrified particles emitted from the Sun.

Axis An imaginary line about which a body rotates. The polar diameter of a planet marks the axis of rotation.

Binary Two stars that move around their common center of gravity.

Black hole The remains of a massive star after its final collapse and contraction into a state in which the gravitational pull is so strong that not even light can escape.

Caldera A volcanic crater.

Celestial sphere An imaginary sphere surrounding the Earth, concentric with the Earth's center.

Cepheid variable A variable star of short period. The fluctuations are regular and are linked with its real luminosity: the longer the period, the more luminous the star.

Chromosphere That part of the Sun's atmosphere that lies just above its visible surface, or photosphere.

Circumpolar star A star that never sets. Ursa Major, for example, is circumpolar over the British Isles and Crux Australis is circumpolar over New Zealand.

Conjunction (1) A celestial body is said to be in conjunction with another body when it is apparently close to it in the sky. (2) The planets within the Earth's orbit, Mercury and Venus, are at inferior conjunction when lined up between the Earth and the Sun, and at superior conjunction when on the far side of the Sun. The planets outside the Earth's orbit can only reach superior conjunction.

Constellation A group of stars within an imaginary outline.

Corona The outermost part of the Sun's atmosphere. Made up of very thin gas, it is invisible to the naked eye except during a total eclipse.

Cosmology The study of the universe.

Culmination The maximum altitude of a celestial body.

Declination The angular distance of a celestial body from the celestial equator.

Density The mass, or quantity of matter, contained within a unit of volume.

Direct motion The movement of a celestial body from west to east – that is, in the same direction as that of the Earth around the Sun.

Doppler effect The apparent change in the wavelength of light according to the motion of the body emitting it, in relation to the observer. With an approaching light source the wavelength is shortened ("too blue"); with a receding source it is lengthened ("too red").

Double star A pair of stars. A double may be a genuine physical association (when they form what is known as a binary star) or an optical trick: two stars appearing to be close together but in fact just happening to lie in almost exactly the same line when seen from Earth.

Eccentricity A measure of how closely a planet's orbit approximates to a perfect circle.

Eclipse, lunar The passage of the Moon through the shadow cast by the Earth. Lunar eclipses may be either total or partial.

Eclipse, solar The covering of the Sun by the Moon, when seen from Earth. Solar eclipses may be either total, partial or annular. An annular eclipse occurs when the Moon is close to its point of maximum recession from the Earth and so is too small to hide the Sun completely.

Eclipsing binary A binary star, one component of which is seen to pass in front of the other, thereby cutting out some or all of its light.

Ecliptic The apparent yearly path of the Sun against the stars.

Elongation The angular distance of a planet from the Sun or of a satellite from its primary planet.

Equator, celestial The projection of the Earth's equator on to the celestial sphere, dividing the sky into equal hemispheres.

Equinox The two points at which the Sun crosses the celestial equator; the spring equinox (First Point of Aries) is reached about 21 March and the autumnal equinox about 22 September.

Escape velocity The minimum velocity that an object must possess to escape from the surface of a planet or other body.

Extinction The apparent reduction in the brightness of a star or planet when low over the horizon because more of its light is absorbed by the Earth's atmosphere.

Eyepiece The lens (or lenses) at the eye-end of a telescope responsible for enlarging the image produced by the object-glass (for a refractor) or mirror (for a reflector).

Faculae The bright patches on the Sun's photosphere.

Finder A small, wide-field telescope attached to a larger one for locating objects in the sky.

Flare, solar Brilliant outbreaks in the solar atmosphere, normally detectable only by spectroscopic methods.

Flare star A faint red star that has short-lived explosions on its surface. These explosions cause the star to appear temporarily brighter.

Fraunhofer lines The dark lines in the spectrum of the Sun.

Galaxy A star system. Most galaxies are so remote that their light takes millions of years to reach Earth.

Inclination Measure of the tilt of a planet's orbital plane, in relation to that of the Earth.

Inferior planet Either of the two planets, Mercury and Venus, that orbit between the Sun and the Earth.

Libration An effect caused by the apparent slight "wobbling" of the Moon from side to side, as seen from Earth. As a result a total of 59 per cent of the Moon's surface can be observed, although no more than 50 per cent at any one time. The remaining 41 per cent remained unknown to observers from Earth until the Space Age explorations.

Light-year An astronomical unit equal to the distance traveled by light in one year: that is, 9.46 million million kilometers.

Limb An edge or border, as of the Sun, Moon or any planet.

Local Group The group of which our Galaxy is a member. It contains more than two dozen systems, including the Andromeda Spiral and the two Clouds of Magellan.

Luminosity The amount of light that is emitted by a star.

Lunation The interval between one new Moon and the next: that is, 29 days 12 hours 44 minutes.

Magnetosphere The area around a planet in which its magnetic field is dominant.

Magnitude Brightness, according to a scale in which fainter stars are represented by lower numbers, and the most brilliant stars are of the first magnitude.

Meridian An imaginary circle through the north and south poles of the celestial sphere.

Meteor A small particle that burns away in the Earth's upper air; it is often known as a shooting star.

Meteorite A natural body, probably associated with asteroids, that is able to reach ground level without being destroyed.

Nebula A cloud of dust and gas in space, from which fresh stars are created.

Neutron A fundamental particle with no electrical charge.

Neutron star A star made up of neutrons. Theoretically, it is the remnant of a massive star that has exploded. Neutron stars that send out rapidly varying radio waves are known as pulsars.

Nova A star that suddenly flares up to many times its normal brightness and remains brilliant for a limited period before fading back to obscurity.

Object-glass (or objective) The main lens in a refracting telescope.

Oblateness The degree of flattening at the poles of a celestial body.

Occultation The concealment of one celestial body by another. Strictly speaking, a solar eclipse is an occultation of the Sun by the Moon.

Opposition The position of a superior planet when exactly opposite the Sun in the sky, as seen from Earth. The planet is then best placed for observation.

Orbit The path of a celestial body.

Parallax The apparent shift in position of an object when viewed from two different positions.

Perihelion The position of a planet (or other body) when it is at its closest to the Sun.

Periodic time *see Sidereal period*

Phase The apparent shape of the Moon and inferior planets, according to the amount of sunlit hemisphere turned toward the Earth. New Moon, for example, occurs when the unlit side of the Moon is presented. Full Moon occurs when its surface is fully exposed as viewed from Earth.

Photosphere The brilliant visible surface of the Sun.

Planetary nebula A shell of gas that has been thrown off from a star, the small, hot core of which remains at the center of the nebula.

Poles, celestial The north and south points of the celestial sphere.

Precession The apparent slow movement of the celestial poles caused by a shift in the direction of the Earth's axis.

Prominence A mass of glowing gas, chiefly hydrogen, above the Sun's photosphere.

Pulsar *see Neutron star*

Quasar A very remote, superluminous object. The nature of quasars is still uncertain.

Radial velocity The movement of a celestial body toward or away from the observer.

Radiant The point in the sky from which a meteor shower appears to emanate.

Red giant The stage in the evolution of an ordinary star when the core contracts, the surface expands to about 50 solar radii and its temperature drops, giving the star its red color.

Retrograde motion The movement of a celestial body from east to west; that is, in the direction opposite to that of the Earth.

Right ascension The time that elapses between the culmination of the First Point of Aries and the culmination of a celestial body.

Seyfert galaxy A galaxy that has a small, bright nucleus and faint spiral arms. It is often a strong radio source.

Sidereal period The revolution period of a planet around the Sun or of a satellite around its primary. Also known as periodic time

Solar wind Charged particles from the Sun that travel into the Solar System at about 1.5 million kph.

Spectroscopic binary A very close double that is recognizable only by the periodic splitting of lines in the combined spectrum of the two stars, owing to the opposite Doppler effects resulting from their motions.

Spectrum The range of color produced when light is split up by a prism or diffraction grating.

Supergiant The stage in the evolution of a massive star when its core contracts, its surface expands to about 500 solar radii and its temperature drops, giving the star its red color.

Superior planet Any planet beyond the orbit of the Earth in the Solar System.

Supernova (Type I) A gigantic stellar explosion involving the destruction of a white dwarf belonging to a binary pair.

Supernova (Type II) A massive star that reaches a peak of luminosity and then explodes in a cataclysmic outburst, leaving a neutron star surrounded by a cloud of expanding gas.

Synodic period The interval between successive oppositions,conjunctions, etc., of a celestial body.

Terminator The boundary between the illuminated and dark portions of a planet or satellite.

Transit (1) The passage of a celestial body across the observer's meridian. (2) The apparent passage of a smaller body across the disk of a larger one.

Van Allen belts Radiation zones of charged particles surrounding the Earth.

Variable star A star that fluctuates in brilliancy. Eclipsing binaries fall under this heading.

White dwarf A small, very dense star that has used up its nuclear energy and collapsed. White dwarfs have been described as "bankrupt stars."

Yellow dwarf An ordinary star such as the Sun at a comparatively stable and long-lived stage in its evolution.

Zenith The observer's overhead point (celestial latitude 90°).

Zodiac A belt stretching 8° to either side of the ecliptic, in which the Sun, Moon and all planets are always to be found.

Zodiacal light A faint cone of light rising from the horizon after sunset or before sunrise. It is caused by sunlight reflected from thinly spread interplanetary material lying in the main plane of the Solar System.

CONSTELLATION NAMES

Abbreviation	Name	Genitive	Popular name
And	Andromeda	Andromedae	The Princess of Ethiopia
Ant	Antlia	Antliae	The Air Pump
Aps	Apus	Apodis	The Bird of Paradise
Aqr	Aquarius	Aquarii	The Water Carrier
Aql	Aquila	Aquilae	The Eagle
Ara	Ara	Arae	The Altar
Ari	Aries	Arietis	The Ram
Aur	Auriga	Aurigae	The Charioteer
Boo	Boötes	Boötis	The Bear Driver
Cae	Caelum	Caeli	The Graving Tool
Cam	Camelopardalis	Camelopardalis	The Giraffe
Cnc	Cancer	Cancri	The Crab
CVn	Canes Venatici	Canum Venaticorum	The Hunting Dogs
CMa	Canis Major	Canis Majoris	The Great Dog
CMi	Canis Minor	Canis Minoris	The Little Dog
Cap	Capricornus	Capricorni	The Sea Goat
Car	Carina	Carinae	The Keel
Cas	Cassiopeia	Cassiopeiae	Cassiopeia
Cen	Centaurus	Centauri	The Centaur
Cep	Cepheus	Cephei	Cepheus
Cet	Cetus	Ceti	The Sea Monster, Whale
Cha	Chamaeleon	Chamaeleontis	The Chamaeleon
Cir	Circinus	Circini	The Compasses
Col	Columba	Columbae	The Dove
Com	Coma Berenices	Comae Berenicis	Berenice's Hair
CrA	Corona Australis	Coronae Australis	The Southern Crown
CrB	Corona Borealis	Coronae Borealis	The Northern Crown
Crv	Corvus	Corvi	The Crow
Crt	Crater	Crateris	The Cup
Cru	Crux	Crucis	The Southern Cross
Cyg	Cygnus	Cygni	The Swan
Del	Delphinus	Delphini	The Dolphin
Dor	Dorado	Doradûs	The Swordfish
Dra	Draco	Draconis	The Dragon
Equ	Equuleus	Equulei	The Foal
Eri	Eridanus	Eridani	The River Eridanus
For	Fornax	Fornacis	The Furnace
Gem	Gemini	Geminorum	The Twins
Gru	Grus	Gruis	The Crane

Her	Hercules	Herculis	Hercules
Hor	Horologium	Horologii	The Pendulum Clock
Hya	Hydra	Hydrae	The Sea Serpent
Hyi	Hydrus	Hydri	The Water Snake
Ind	Indus	Indi	The Indian
Lac	Lacerta	Lacertae	The Lizard
Leo	Leo	Leonis	The Lion
LMi	Leo Minor	Leonis Minoris	The Lion Cub
Lep	Lepus	Leporis	The Hare
Lib	Libra	Librae	The Scales
Lup	Lupus	Lupi	The Wolf
Lyn	Lynx	Lyncis	The Lynx
Lyr	Lyra	Lyrae	The Lyre
Men	Mensa	Mensae	The Table Mountain
Mic	Microscopium	Microscopii	The Microscope
Mon	Monoceros	Monocerotis	The Unicorn
Mus	Musca	Muscae	The Fly
Nor	Norma	Normae	The Level
Oct	Octans	Octantis	The Octant
Oph	Ophiuchus	Ophiuchi	The Serpent Holder
Ori	Orion	Orionis	The Great Hunter
Pav	Pavo	Pavonis	The Peacock
Peg	Pegasus	Pegasi	The Winged Horse
Per	Perseus	Persei	Perseus
Phe	Phoenix	Phoenicis	The Phoenix
Pic	Pictor	Pictoris	The Painter
Psc	Pisces	Piscium	The Fishes
PsA	Piscis Austrinus	Piscis Austrini	The Southern Fish
Pup	Puppis	Puppis	The Stern
Pyx	Pyxis	Pyxidis	The Mariner's Compass
Ret	Reticulum	Reticuli	The Net
Sge	Sagitta	Sagittae	The Arrow
Sgr	Sagittarius	Sagittarii	The Archer
Sco	Scorpius	Scorpii	The Scorpion
Scl	Sculptor	Sculptoris	The Sculptor
Sct	Scutum	Scuti	The Shield
Ser	Serpens	Serpentis	The Serpent
Sex	Sextans	Sextantis	The Sextant
Tau	Taurus	Tauri	The Bull
Tel	Telescopium	Telescopii	The Telescope
Tri	Triangulum	Trianguli	The Triangle
TrA	Triangulum Australe	Trianguli Australis	The Southern Triangle
Tuc	Tucana	Tucanae	The Toucan
UMa	Ursa Major	Ursae Majoris	The Great Bear
UMi	Ursa Minor	Ursae Minoris	The Little Bear
Vel	Vela	Velorum	The Sails
Vir	Virgo	Virginis	The Virgin
Vol	Volans	Volantis	The Flying Fish
Vul	Vulpecula	Vulpeculae	The Fox

INDEX

Page numbers in *italics* refer to illustrations or their captions.